The Princeton Review®

AP® ENVIRONMENTAL SCIENCE

PREP

2021 Edition

The Staff of The Princeton Review

PrincetonReview.com

Penguin
Random
House

The Princeton Review
110 East 42nd St, 7th Floor
New York, NY 10017
Email: editorialsupport@review.com

Published in the United States by Penguin Random House LLC, New York, and in Canada by Random House of Canada, a division of Penguin Random House Ltd., Toronto.

ISBN: 978-0-525-56954-1
eBook ISBN: 978-0-525-56991-6
ISSN: 2690-540X

The material in this book is up-to-date at the time of publication. However, changes may have been instituted by the testing body in the test after this book was published.

If there are any important late-breaking developments, changes, or corrections to the materials in this book, we will post that information online in the Student Tools. Register your book and check your Student Tools to see if there are any updates posted there.

Editor: Orion McBean
Production Editors: Lyssa Mandel and Sarah Litt
Production Artist: Steph Calvert
Content Contributor: Ali Landreau

Printed in the United States of America.

10 9 8 7 6 5 4 3 2 1

2021 Edition

Editorial

Rob Franek, Editor-in-Chief
David Soto, Director of Content Development
Stephen Koch, Student Survey Manager
Deborah Weber, Director of Production
Gabriel Berlin, Production Design Manager
Selena Coppock, Managing Editor
Aaron Riccio, Senior Editor
Meave Shelton, Senior Editor
Chris Chimera, Editor
Eleanor Green, Editor
Orion McBean, Editor
Brian Saladino, Editor
Patricia Murphy, Editorial Assistant

Penguin Random House Publishing Team

Tom Russell, VP, Publisher
Rebecca Holland, Publishing Director
Amanda Yee, Associate Managing Editor
Ellen L. Reed, Production Manager
Suzanne Lee, Designer

Acknowledgments

The Princeton Review would like to give a special thanks to Ali Landreau for her incredible work and thorough review of this year's edition.

We would also like to extend our thanks to Steph Calvert, Lyssa Mandel, and Sarah Litt for their time and attention while producing this title.

Contents

Get More (Free) Content

at PrincetonReview.com/prep

As easy as 1·2·3

1 Go to PrincetonReview.com/prep and enter the following ISBN for your book:

9780525569541

2 Answer a few simple questions to set up an exclusive Princeton Review account. *(If you already have one, you can just log in.)*

3 Enjoy access to your **FREE** content!

Once you've registered, you can...

- Get our take on any recent or pending updates to the AP Environmental Science Exam

- Take a full-length practice SAT and/or ACT

- Get valuable advice about the college application process, including tips for writing a great essay and where to apply for financial aid

- If you're still choosing between colleges, use our searchable rankings of *The Best 386 Colleges* to find out more information about your dream school.

- Access comprehensive study guides and a variety of printable resources, including Key Terms lists

- Check to see if there have been any corrections or updates to this edition

Need to report a potential **content** issue?

Contact **EditorialSupport@review.com** and include:

- full title of the book
- ISBN
- page number

Need to report a **technical** issue?

Contact **TPRStudentTech@review.com** and provide:

- your full name
- email address used to register the book
- full book title and ISBN
- Operating system (Mac/PC) and browser (Firefox, Safari, etc.)

Look For These Icons Throughout The Book

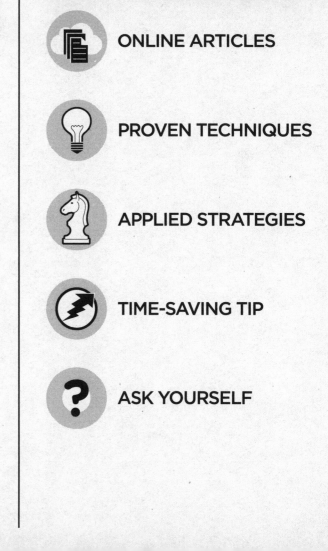

ONLINE ARTICLES

PROVEN TECHNIQUES

APPLIED STRATEGIES

TIME-SAVING TIP

ASK YOURSELF

Part I
Using This Book
to Improve
Your AP Score

- Preview: Your Knowledge, Your Expectations
- Your Guide to Using This Book
- How to Begin

PREVIEW: YOUR KNOWLEDGE, YOUR EXPECTATIONS

Your route to a high score on the AP Environmental Science Exam depends a lot on how you plan to use this book. Respond to the following questions:

1. Rate your level of confidence about your knowledge of the content tested by the AP Environmental Science Exam.
    A.   Very confident—I know it all
    B.   I'm pretty confident, but there are topics for which I could use help
    C.   Not confident—I need quite a bit of support
    D.   I'm not sure

2. Circle your goal score for the AP Environmental Science Exam.
    5   4   3   2   1   I'm not sure yet

3. What do you expect to learn from this book? Circle all that apply to you.
    A.   A general overview of the test and what to expect
    B.   Strategies for how to approach the test
    C.   The content tested by this exam
    D.   I'm not sure yet

YOUR GUIDE TO USING THIS BOOK

This book is organized to provide as much—or as little—support as you need, so you can use this book in whatever way will be most helpful to improving your score on the AP Environmental Science Exam. Please note that this book should be used in conjunction with your AP course textbook.

- The remainder of **Part I** will provide guidance on how to use this book and help you determine your strengths and weaknesses.

- **Part II** of this book contains your first practice test, answers and explanations, and a scoring guide. (Bubble sheets can be found in the very back of the book for easy tear-out.) This is where you should begin your test preparation in order to realistically determine:
    o your starting point right now
    o which question types you're ready for and which you might need to practice
    o which content topics you are familiar with and which you will want to carefully review

Once you have nailed down your strengths and weaknesses with regard to this exam, you can focus your preparation and be efficient with your time.

- **Part III** of this book will:
    o provide information about the structure, scoring, and content of the AP Environmental Science Exam
    o help you make a study plan
    o point you toward additional resources

- **Part IV** of this book will explore the following strategies:
  - o how to attack multiple-choice questions
  - o how to approach free-response questions
  - o how to manage your time to maximize the number of points available to you

- **Part V** of this book covers the content you need to know for the AP Environmental Science Exam.

- **Part VI** of this book contains Practice Test 2 and Practice Test 3, with answers and explanations and a scoring guide. (Bubble sheets can be found in the very back of the book for easy tear-out.) If you skipped Practice Test 1, we recommend that you do all three tests (with at least a day or two between them) so that you can compare your progress between the three. Additionally, this will help to identify any external issues. If you get a certain type of question wrong both times, you probably need to review it. If you only got it wrong once, you may have run out of time or been distracted by something. In either case, this will allow you to focus on the factors that caused the discrepancy in scores and to be as prepared as possible on the day of the test.

> **Want More?**
> Check out your online Student Tools for some great ways to extend your prep, including study guides. Step-by-step instructions for your Student Tools can be found on the Get More (Free) Content page at the beginning of this book.

HOW TO BEGIN

1. **Take a Test**

    Before you can decide how to use this book, you need to take a practice test. Doing so will give you insight into your strengths and weaknesses and help you make an effective study plan. If you're feeling test-phobic, remind yourself that a practice test is a tool for diagnosing yourself—it's not how well you do that matters, but how you use information gleaned from your performance to guide your preparation.

    So, before you read further, take Practice Test 1 starting on page 9 of this book. Be sure to do so in one sitting, following the instructions that appear before the test.

2. **Check Your Answers**

    Using the answer key on page 35, count how many multiple-choice questions you got right and how many you missed. Don't worry about the explanations for now, and don't worry about why you missed questions. We'll get to that soon.

3. **Reflect on the Test**

    After you take your first test, respond to the following questions:
    - How much time did you spend on the multiple-choice questions?
    - How much time did you spend on each free-response question?
    - How many multiple-choice questions did you miss?
    - Do you feel you had the knowledge to address the subject matter of the free-response questions?

- Check the content areas that were most challenging for you and draw a line through the ones in which you felt confident/did well.
  - Geological eras
  - Atmosphere
  - Soil dynamics
  - Population
  - Pollution (specifically solid waste, acid rain, ozone depletion, and thermal pollution)
  - Water conservation
  - Conservation and economic impact
  - Interpreting charts and graphs
  - Interpreting experimental data
  - Analyzing experiments and experimental reasoning
  - Ecosystems and biodiversity
  - Biogeochemical cycles
  - Global climate change
  - Environmental regulation and international treaties
  - Energy resources and consumption
  - Energy science
  - Mining

4. **Read Part III of this Book and Complete the Self-Evaluation**

   As discussed in the Guide section above, Part III will provide information on how the test is structured and scored. It will also outline areas of content that are tested.

   As you read Part III, re-evaluate your answers to the questions you just answered. At the end of Part III, you will revisit and refine your answers to the previous questions. You will then be able to make a study plan, based on your needs and available time, that will allow you to use this book most effectively.

5. **Engage with Parts IV and V as Needed**

   Notice the word *engage*. You'll get more out of this book if you use it intentionally than if you read it passively, hoping for an improved score through osmosis.

   Strategy chapters will help you think about your approach to the question types on this exam. Part IV will open with a reminder to think about how you approach questions now, and then close with a reflection section asking you to think about how/whether you will change your approach in the future.

   Content chapters are designed to provide a review of the content tested on the AP Environmental Science Exam, including the level of detail you need to know and how the content is tested. You will have the opportunity to assess your mastery of the content of each chapter through test-appropriate questions and a reflection section.

6. Take Practice Tests 2 and 3. Assess Your Performance

Once you feel you have developed the strategies you need and gained the knowledge you lacked, you should take Practice Test 2 and then Practice Test 3. Take each test in one sitting, with maybe a day or two in between.

When you are done, check your answers to the multiple-choice sections. See if a teacher will read your responses to the free-response questions and provide feedback.

Once you have taken the test, reflect on what areas you still need to work on, and revisit the chapters in this book that address those deficiencies. Through this type of reflection and engagement, you will continue to improve.

7. Keep Working

As Part III will discuss, there are other resources available to you, including a wealth of information on AP Students. You can continue to explore areas where you can stand to improve and to engage in those areas right up to the day of the test.

> **AP Students**
> Check out the College Board's homepage for more information about the updated exam at https://apstudents.collegeboard.org/courses/ap-environmental-science

Part II
Practice Test 1

Practice Test 1
Practice Test 1: Answers and Explanations

Practice Test 1

AP® Environmental Science Exam

SECTION I: Multiple-Choice Questions

DO NOT OPEN THIS BOOKLET UNTIL YOU ARE TOLD TO DO SO.

At a Glance

Total Time
1 hour and 30 minutes
Number of Questions
80
Percent of Total Grade
60%
Writing Instrument
Pencil required

Instructions

Section I of this examination contains 80 multiple-choice questions. Fill in only the ovals for numbers 1 through 80 on your answer sheet.

Indicate all of your answers to the multiple-choice questions on the answer sheet. No credit will be given for anything written in this exam booklet, but you may use the booklet for notes or scratch work. After you have decided which of the suggested answers is best, completely fill in the corresponding oval on the answer sheet. Give only one answer to each question. If you change an answer, be sure that the previous mark is erased completely. Here is a sample question and answer.

Sample Question   Sample Answer

Chicago is a

   (A) state
   (B) city
   (C) country
   (D) continent

Use your time effectively, working as quickly as you can without losing accuracy. Do not spend too much time on any one question. Go on to other questions and come back to the ones you have not answered if you have time. It is not expected that everyone will know the answers to all the multiple-choice questions.

About Guessing

Many candidates wonder whether or not to guess the answers to questions about which they are not certain. Multiple-choice scores are based on the number of questions answered correctly. Points are not deducted for incorrect answers, and no points are awarded for unanswered questions. Because points are not deducted for incorrect answers, you are encouraged to answer all multiple-choice questions. On any questions you do not know the answer to, you should eliminate as many choices as you can, and then select the best answer among the remaining choices.

GO ON TO THE NEXT PAGE.

ENVIRONMENTAL SCIENCE
Section I
Time—90 minutes
80 Questions

Directions: Each of the questions or incomplete statements below is followed by four suggested answers or completions. Select the one that is best in each case and then fill in the corresponding oval on the answer sheet.

1. Which of the following best explains how clearcutting contributes to climate change on a global scale?

    (A) Clearcutting requires prescribed burning, or setting controlled forest fires to reduce the occurrence of natural fires.

    (B) Clearcutting results in a loss of biodiversity and adds to the problems of endangered species and extinctions.

    (C) Clearcutting involves the burning of cut trees, which releases the greenhouse gas carbon dioxide.

    (D) Clearcutting removes plants and their root systems, leading to soil erosion and flooding.

2. Which of the following does NOT result from the melting of polar icecaps?

    (A) Increase in ocean acidification

    (B) Loss of habitat for ice-dwelling species

    (C) Release of the greenhouse gas methane

    (D) Decrease in albedo causing more warming

3. The release of stored energy when stress overcomes a locked fault results in which of the following?

    (A) Earthquake

    (B) Volcanic eruption

    (C) Mountain creation

    (D) Hot spot formation

4. The table below shows the four categories of ecosystem services.

Type	What is provided
Provisioning services	water, food, medicinal resources, raw materials, energy, ornaments
Regulating services	waste decomposition and detoxification, purification of water and air, pest and disease control and regulation of prey populations through predation, carbon sequestration
Cultural services	use of nature for science and education, therapeutic and recreational uses, spiritual and cultural uses
Supporting services	primary production, nutrient recycling, soil formation, pollination

Based on the information in the table, which of the following gives an example of the possible human consequences of the disruption of a provisioning service?

    (A) Agricultural use of former parklands resulting in fewer recreational areas

    (B) Rainforest destruction resulting in the loss of potential medicinal plant species

    (C) Decline in wild bee populations resulting in less pollination and decreased crop yields

    (D) Destruction of wetland habitats for real estate resulting in decreased carbon sequestration

GO ON TO THE NEXT PAGE.

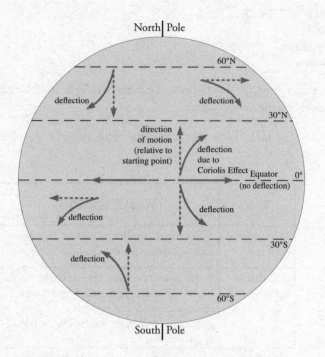

North Pole

South Pole

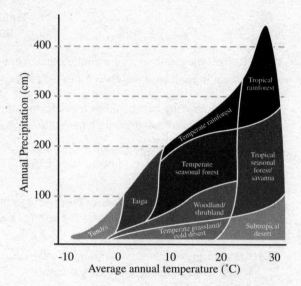

5. Which of the following phenomena are NOT influenced by the Coriolis effect shown in the diagram above?

   (A) Cyclones

   (B) Hadley cells

   (C) Tradewinds

   (D) El Niño–Southern Oscillation

6. Which of the following is the name for the land area that drains into a particular stream?

   (A) Delta

   (B) Estuary

   (C) Watershed

   (D) Headwaters

7. Which of the following factors limiting human population growth is density-independent?

   (A) Access to clean air and water

   (B) Natural disaster frequency

   (C) Disease transmission

   (D) Food availability

8. According to the graph, which biome would likely be found in a location with an average annual temperature above 20°C and average annual precipitation below 50 cm?

   (A) Tundra

   (B) Savanna

   (C) Subtropical desert

   (D) Temperate grassland

9. Which of the following does NOT help explain the relationship between temperature, precipitation, and biome shown in the graph?

   (A) Some vegetation types thrive only in very wet conditions.

   (B) The coldest and hottest temperatures both restrict plant life.

   (C) Drier regions cannot support as great a variety of vegetation as wetter ones can.

   (D) There are temperature extremes in which even the hardiest species cannot grow.

10. Which of the following types of vegetation would likely be found in a location with an average annual temperature below 10°C and average annual precipitation above 100 cm?

    (A) Conifer trees, fungi, mosses, lichens

    (B) Shrubs, grasses, herbs, tubers

    (C) Mosses, heath, lichens, algae

    (D) Cacti, succulents, grasses

GO ON TO THE NEXT PAGE.

Questions 11–13 refer to the following diagrams.

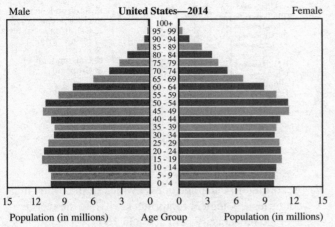

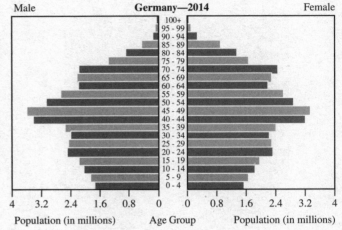

11. Which comparison is valid according to the age-structure diagrams above?

   (A) In 2014 the US had a slower-growing population than Germany did.

   (B) In 2014 Germany had a higher proportion of its population under 45 than the US did.

   (C) In 2014 Germany's population included more people in the 45–49 age range than did the US population.

   (D) In 2014 Germany's population showed negative growth while that of the US showed close to zero growth.

12. Based on the information in the diagram above, where is it likely that Germany falls in the Demographic Transition Model?

   (A) Preindustrial State (Phase I)

   (B) Transitional State (Phase II)

   (C) Industrial State (Phase III)

   (D) Postindustrial State (Phase IV)

13. Based on the information in the diagram above, which of the following was likely true about the US population in 2014?

   (A) The total fertility rate was between 1 and 2.

   (B) The total fertility rate was between 2 and 3.

   (C) The total fertility rate was between 3 and 4.

   (D) The total fertility rate was between 4 and 5.

14. Catalytic converters on automobiles use chemical reactions to convert pollutants from exhaust into less harmful substances. In order for a catalytic converter to function optimally when regular gasoline is used, a ratio of 14.7 air to 1 fuel is needed. However, certain kinds of fuel require different ratios. For example, E85, an ethanol fuel blend, requires 34% more fuel to balance the reaction. What is the approximate optimal ratio of air to E85 fuel?

   (A) 0.09 to 1

   (B) 1.34 to 1

   (C) 10.97 to 1

   (D) 14.7 to 1

15. Which of the following is NOT an aspect of the phosphorous cycle?

   (A) Precipitation returning atmospheric phosphorous to land and water

   (B) Geological uplift pushing phosphorous-containing rock to the ocean surface

   (C) Phosphorous from the soil and farming reaching the ocean via leaching and runoff

   (D) Decomposers breaking down phosphorous-containing plant and animal matter into soil

GO ON TO THE NEXT PAGE.

Questions 16–17 refer to the following diagram.

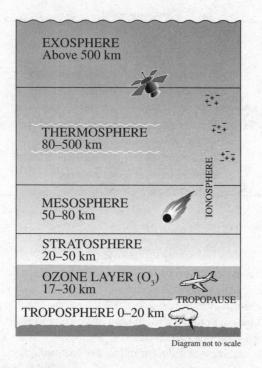

Diagram not to scale

ECOLOGICAL FOOTPRINT, PER PERSON BY COUNTRY, 2003

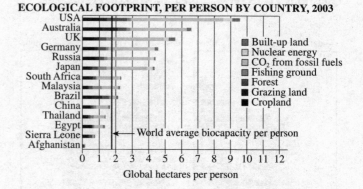

18. According to the graph above, which of the following countries' ecological footprints lay below the world average biocapacity per person in 2003?

   (A) Afghanistan, Sierra Leone, and South Africa

   (B) United States, Australia, and the UK

   (C) China, Egypt, and Thailand

   (D) Brazil, Malaysia, and Japan

16. Which of the following can be found in the tropopause?

   (A) The jet streams

   (B) The ozone layer

   (C) Aurora borealis and australis

   (D) The gases responsible for the greenhouse effect

17. Which of the following is the most common gas in the troposphere?

   (A) Oxygen

   (B) Nitrogen

   (C) Water vapor

   (D) Carbon dioxide

GO ON TO THE NEXT PAGE.

Questions 19–21 refer to the following information.

Because soil conservation is of major interest in sustainable agriculture, methods of farming that slow erosion and help preserve soil are an integral part of sustainable farming technique. One such method, strip cropping, involves the partitioning of a farmed field into long, narrow strips and using a crop rotation system by alternating what is planted in them. The strips can be contoured to follow the landscape, and the strips are planted alternately with cover crops and row crops. The strips of cover crop serve to slow the flow of water off the other strips and thus prevent erosion on a large scale. If some areas are particularly eroded, permanent protective vegetation can be grown there. In addition, the crop rotation helps improve soil fertility by periodically changing which nutrients are being leached from the soil and which returned.

19. Which of the following is a potential benefit of strip cropping?

    (A) Minimizing runoff

    (B) Providing shelter from wind erosion

    (C) Lessening the damage of overgrazing

    (D) Reducing the erosion caused by tillage

20. A small farmer has purchased a new plot of land that she intends to cultivate using sustainable practices. The major issues she wishes to address on the plot include excessive wind erosion, low soil fertility after long-term monoculture, and soil loss due to runoff. Would using strip cropping help her address these issues?

    (A) Yes, because strip cropping protects against wind erosion.

    (B) No, because strip cropping is mainly used to slow soil erosion.

    (C) No, because strip cropping will damage the soil fertility further.

    (D) Yes, because strip cropping will slow runoff and improve soil fertility.

21. Which of the following sustainable agricultural practices has benefits similar to those of strip cropping?

    (A) Terracing

    (B) Windbreaks

    (C) No-till farming

    (D) Rotational grazing

22. Which of the following best explains the outcome in the scenario below?

A wildfire damages a forest habitat and kills the majority of several animal populations living there. All populations begin to recover, but a generalist animal species outcompetes a few specialist species in overlapping niches, and recovers much more quickly, changing the balance of the local ecosystem permanently.

    (A) The specialist species have a disadvantage because there are several of them, and thus they recovered more slowly.

    (B) The specialist species have an advantage in changing conditions that allowed them to recover more quickly.

    (C) The generalist species has an advantage in changing conditions that allowed it to recover more quickly.

    (D) The generalist species has the advantage of being invasive, which allowed it to recover more quickly.

23. Which of the following is NOT a reason why Earth's Northern Hemisphere is hotter during its summer months?

    (A) More direct light from the sun means the light has to pass through less atmosphere before it reaches Earth's surface.

    (B) More direct light from the sun means a given amount of light is concentrated into a smaller area.

    (C) Shorter distance from the sun means less heat lost before it reaches Earth.

    (D) Longer daylight hours mean more sunlight for a given area per day.

24. Which of the following best describes the type of symbiotic relationship labeled "Type I" in the table below?

Interaction	Species A	Species B
Type I	receives benefit	receives benefit
Type II	receives benefit	not affected
Type III	receives benefit	harmed

    (A) Commensalism

    (B) Mutualism

    (C) Parasitism

    (D) Predation

GO ON TO THE NEXT PAGE.

25. The diagram below shows the projected recovery of stratospheric ozone levels over the course of the current century according to several models.

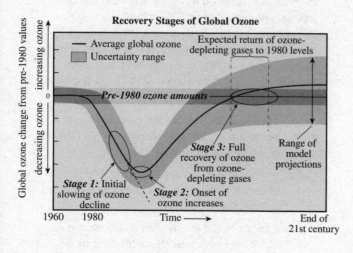

Recovery Stages of Global Ozone

What stage of this model is happening currently, according to the diagram?

(A) Stage 0: Rapid ozone decrease

(B) Stage 1: Initial slowing of ozone decline

(C) Stage 2: Onset of ozone increases

(D) Stage 3: Full recovery of ozone from ozone-depleting gases

26. A number of sheep are introduced to an island with plenty of grazing available, and the population increases rapidly. When the sheep population reaches its carrying capacity, which of the following is likely to occur?

(A) The population growth will continue to accelerate.

(B) The population growth will slow until it becomes constant.

(C) The population growth will stop as sheep choose not to reproduce until more resources are available.

(D) The population growth will overshoot the carrying capacity and some dieoff will occur until the resources available are adequate to sustain the population.

27. Which of the following is a source of atmospheric CO_2 for which humans are almost entirely responsible?

(A) Volcanism

(B) Respiration

(C) Combustion

(D) Decomposition

28. One major effect of global climate change is a rise in sea levels. This can cause several negative effects; which of the following is a possible positive effect of this change?

(A) Loss of estuary and shoreline habitats due to flooding

(B) Newly created marine habitats on flooded continental shelves

(C) Change in photic level, causing a lack of sunlight in ocean layers that were previously sunlit

(D) Loss of the protection from storm surge, tidal waves, and tsunamis afforded by mangroves and tidal marshes

29. Which of the following explains why biomagnification is a bigger problem in ocean life than in land animals?

(A) Marine food chains are longer.

(B) Marine ecosystems are more complex.

(C) Land animals encounter fewer pollutants.

(D) Land animals are less likely to ingest pollutants they encounter.

GO ON TO THE NEXT PAGE.

30. The west Texas region known as the Permian Basin is rich in oil, natural gas, and potash reserves. Which of the following could explain why this region has such a wealth of petroleum resources compared to surrounding areas?

    (A) The basin is in a geographic region of long-term subsidence, the pressure of which has been brought to bear on large deposits of ancient sediment.

    (B) The basin is too far south to be subject to glaciation, which means that surface layers of rock have not been scraped away.

    (C) The basin is in a region of high tectonic activity, the pressure of which accounts for a large amount of metamorphic rock.

    (D) The basin was once the seabed of an ancient ocean, but geologic processes have lifted it up above sea level.

31. Which of the following is an advantage of concentrated animal feeding operations over free-range grazing?

    (A) More animals can be raised for meat in a smaller space, which is more cost-effective for producers and therefore less costly to consumers.

    (B) Less crowding means animals are less susceptible to disease and there is less need for antibiotics to treat livestock.

    (C) Organic waste from the animals can be used as fertilizer and tends to be free of contaminants.

    (D) Animals are fed grains or feed, which are not as suitable as grass and can contribute to disease.

32. Which of the following explains how soil protects water quality?

    (A) Soils are formed when rock is weathered, transported, and deposited by wind and water.

    (B) Soil with greater water holding capacity is more productive and fertile.

    (C) Water can erode soil, removing fertile layers and exposing bare rock.

    (D) Soils effectively filter and clean water that moves through them.

33. Which of the following approximates the amount of energy lost through two trophic levels?

    (A) 1%

    (B) 10%

    (C) 90%

    (D) 99%

34. Because of impervious surfaces such as roads, sidewalks, parking areas, and rooftops, an acre of city land generates about 5 times as much runoff as an acre of forest. Which of the following represents a method to help reduce urban runoff?

    (A) Solar panels

    (B) Carbon offsets

    (C) Contour plowing

    (D) Permeable pavement

35. Which of the following characteristics is most likely to protect a given species against becoming endangered or extinct?

    (A) Mobility

    (B) Limited diet

    (C) High number of competitors

    (D) Specific habitat requirements

36. Which of the following correctly describes the shift in pH involved in global ocean acidification?

    (A) A change from slightly acidic to slightly basic

    (B) A change from slightly basic to slightly acidic

    (C) A change from slightly basic toward pH-neutral

    (D) A change from pH-neutral toward slightly acidic

GO ON TO THE NEXT PAGE.

37. The drawings below show how temperature inversion works in cities surrounded by mountains.

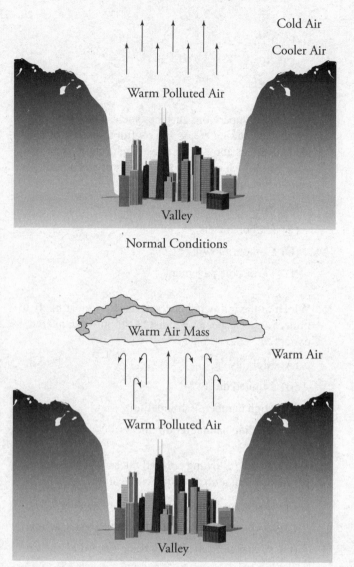

Normal Conditions

A Temperature Inversion

An environmental scientist wants to measure the effects of her city's temperature inversion on the yearly average levels of certain pollutants in the city's air. Which of the following is a variable she will need to control for when gathering her data?

(A) Regulations affecting the pollutants she's interested in

(B) Output levels for the pollutants she's interested in

(C) Seasonal temperature variations

(D) Number of cars in the city

38. Each of the following is a drawback of aquaculture EXCEPT

(A) Contamination of wastewater

(B) Density can increase the incidence of disease

(C) Fish that escape may compete or breed with wild fish

(D) Small requirements in terms of growing space and fuel

39. Which of the following does NOT constitute a natural disruption to an ecosystem?

(A) Seasonal flooding due to monsoons, high tides, or snowmelts

(B) Periodic sea level changes due to increased glaciation during ice ages

(C) Local climate change resulting from atmospheric interference of volcanic ash

(D) Forest fires used as part of forest management strategies leading to ecological succession

40. Which of the following species has a survivorship curve most similar to the one labeled A in the model below?

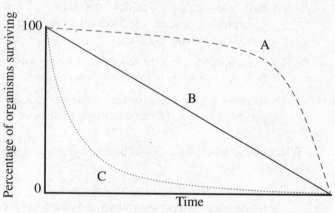

(A) Moose

(B) Squirrel

(C) Sparrow

(D) Tree frog

GO ON TO THE NEXT PAGE.

41. Which of the following describes Population *b* in the graph below?

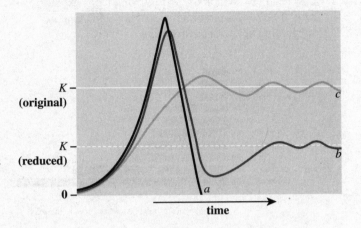

(A) Its population growth continues accelerating because the resources it needs are unlimited.

(B) Its population growth overshoots its carrying capacity; its environment is extremely damaged and the population dies out.

(C) Its population growth overshoots its carrying capacity; its environment is damaged and the carrying capacity is permanently lowered.

(D) Its population growth overshoots its carrying capacity; its environment is only lightly damaged and the resources recover, along with the population.

42. Geothermal energy is used for heating and electricity generation. Which of the following uses of geothermal energy can function without any infrastructure being built?

(A) Geothermal district heating

(B) Geothermal hot spring baths

(C) Enhanced geothermal systems

(D) Binary cycle geothermal power plants

43. Which of the following energy sources produces variable renewable energy?

(A) Biomass

(B) Wind power

(C) Geothermal energy

(D) Dammed hydroelectricity

44. Which fossil fuel burns the cleanest in terms of CO_2 and particulate emissions?

(A) Coal

(B) Crude oil

(C) Petroleum

(D) Natural gas

45. An area of protected wetland is under environmental threat because runoff from nearby agriculture is polluting water faster than the wetland ecology can purify it, and some activists in the area are calling for a ban on pesticides and fertilizers in soil that drains into the area. Which of the following is a potential disadvantage to this plan?

(A) Wetland animal populations may suffer health effects from the pollutants.

(B) An increase in the health of the wetland ecosystem may provide niches for more species to succeed.

(C) Floods may increase if the wetland area, which likely provides a buffer zone against them, decreases.

(D) Other types of development may move in, bringing new sources of pollution, if farms' financial stability is harmed.

GO ON TO THE NEXT PAGE.

46. The diagram below shows how a hydrogen fuel cell functions.

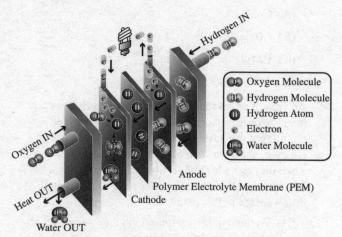

Oxygen Molecule
Hydrogen Molecule
Hydrogen Atom
Electron
Water Molecule

Hydrogen IN

Oxygen IN

Heat OUT

Water OUT

Anode
Polymer Electrolyte Membrane (PEM)
Cathode

Which of the following is a drawback of hydrogen fuel cells as a fuel source?

(A) Water is used as an input.

(B) Carbon dioxide, a greenhouse gas, is released.

(C) Energy is still needed to create the hydrogen gas used as input.

(D) Energy production from hydrogen fuel cells is highly inefficient.

47. The graph below shows the invasive species in Australia that have the greatest impact against native species there. Which of the following helps explain why the European rabbit has such an advantage in competing against native species?

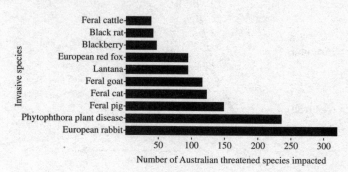

Invasive species

Feral cattle
Black rat
Blackberry
European red fox
Lantana
Feral goat
Feral cat
Feral pig
Phytophthora plant disease
European rabbit

50   100   150   200   250   300

Number of Australian threatened species impacted

(A) Rabbits are an *r*-selected, specialist mammal species; they reproduce rapidly and fit into a particular niche that tends to be rich in resources.

(B) Rabbits are an *r*-selected, generalist mammal species; they reproduce rapidly and use available resources at a faster rate than the varied native species with whom they compete.

(C) Rabbits are a *K*-selected, specialist mammal species; the greater investment of parents into offspring allows them to survive at a greater rate than the native species with whom they share a niche.

(D) Rabbits are a *K*-selected, generalist mammal species; the greater investment of parents into offspring allows them to survive at a greater rate than the varied native species with whom they compete.

GO ON TO THE NEXT PAGE.

Questions 48–50 refer to the following information and graph.

Some researchers tested the effects of an unknown substance on a population of mice. The graph below shows the percent mortality of the mouse population against the concentration of the substance.

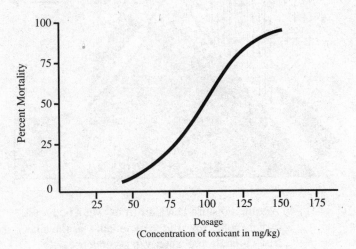

Dosage
(Concentration of toxicant in mg/kg)

48. According to the graph, what is the approximate LD_{50} of the substance on this population?

   (A) 5 mg/kg

   (B) 80 mg/kg

   (C) 100 mg/kg

   (D) 120 mg/kg

49. Would the unknown substance in this experiment be considered a poison to the mouse population?

   (A) No, because the mice may be able to detoxify the substance.

   (B) No, because the LD_{50} of this substance is not 50 mg or less per kg of body weight.

   (C) Yes, because the LD_{50} of this substance is 100 mg per kg of body weight.

   (D) Yes, because the threshold dose of this substance is below 50 mg per kg of body weight.

50. In a second phase of the experiment, the researchers introduced a second, known substance along with the unknown substance, and then measured the effect of the two substances combined on the mouse population. Assuming that the new substance served to decrease toxicity, what can be said about the LD_{50} of the two-substance combination?

   (A) It is less than 100 mg/kg.

   (B) It is more than 100 mg/kg.

   (C) It is now high enough that the substance will be considered a poison.

   (D) It is low enough that the two-substance combination is considered non-toxic.

51. Which of the following is a natural factor that contributes to stratospheric ozone depletion?

   (A) The El Niño–Southern Oscillation

   (B) Upwelling of cold water in the Arctic and Antarctic zones

   (C) An increase in the amount of UV rays that reach Earth's surface

   (D) Melting of ice crystals in the atmosphere at the beginning of the Antarctic spring

52. Which of the following explains why a population bottleneck may weaken the likelihood of a species' long-term survival?

   (A) Bottleneck events reduce species richness in a given area.

   (B) Fewer individuals left after a bottleneck event means that the population will never reach its former numbers.

   (C) Bottleneck events reduce genetic diversity, which makes a species less likely to recover from further disruptions.

   (D) Reduced genetic diversity after a bottleneck event means that the population will experience more genetic drift.

GO ON TO THE NEXT PAGE.

53. According to the soil texture triangle below, which type of soil is made up of 40% sand, 35% clay, and 25% silt?

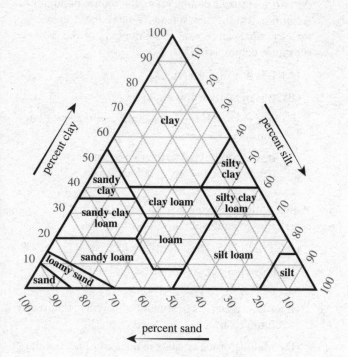

percent sand

(A) Clay

(B) Clay loam

(C) Sandy loam

(D) Silty clay loam

54. Which of the following describes the range of conditions (such as salinity, temperature, sunlight, etc.) in which an organism can live, and outside of which death or injury might occur?

(A) Ecological succession

(B) Ecological tolerance

(C) Species distribution

(D) Biodiversity

55. Which of the following gives the correct order of events leading to the rainshadow effect shown below?

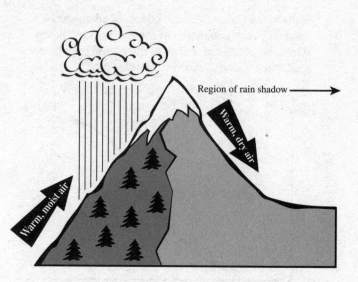

Region of rain shadow →

(A) Warm, moisture-laden air from over a body of water moves toward the mountain—as the air rises it cools and water vapor condenses—precipitation removes water vapor from the air—dry air remains on the other side of the mountain, creating a desert climate

(B) Cool, moisture-laden air from over a body of water moves toward the mountain—as the air rises it warms and water vapor condenses—precipitation removes water vapor from the air—dry air remains on the other side of the mountain, creating a desert climate

(C) Warm, moisture-laden air from over a body of water moves toward the mountain—as the air rises it cools and water vapor condenses, but since it does not fall a desert climate is produced—precipitation falls on the other side of the mountain

(D) Cool, moisture-laden air from over a body of water moves toward the mountain—as the air rises it warms and water vapor condenses, but since it does not fall a desert climate is produced—precipitation falls on the other side of the mountain

GO ON TO THE NEXT PAGE.

Questions 56–57 refer to the following passage and graph.

The theory of island biogeography describes the number of species found on an undisturbed island or other isolated area in terms of two factors: immigration and extinction.

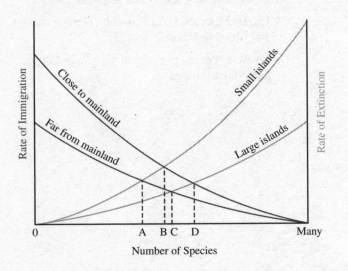

56. According to the graph, which of the following characterizes islands with the greatest species richness?

(A) Large islands close to the mainland

(B) Small islands close to the mainland

(C) Large islands far from the mainland

(D) Small islands far from the mainland

57. Which of the following could explain why invasive species tend to outcompete native ones when introduced to islands or isolated areas?

(A) Since limited resources on islands push local species to evolve to be generalists, more specialist mainland species have a short-term advantage in competition.

(B) Since limited resources on islands push local species to evolve to be specialists, more generalist mainland species have a short-term advantage in competition.

(C) Both island and mainland species tend to evolve to be specialists, but the greater distance traveled by mainland species leads island species to outcompete them.

(D) Both island and mainland species tend to evolve to be generalists, but the greater number of mainland species outcompetes the fewer island species.

58. One major source of noise pollution in urban environments is yard maintenance tools, especially the leaf blower. All of the following are more environmentally friendly alternatives to the use of this device EXCEPT

(A) Using lawn care services rather than individual homeowners being responsible

(B) Leaving leaves and other debris in place as mulch

(C) Using lawn space for gardens rather than grass

(D) Using a rake

59. Which of the following is associated with the cooler "La Niña" portion of the El Niño–Southern Oscillation?

(A) Air pressure high in the western Pacific and low in the eastern Pacific

(B) Air pressure high in the eastern Pacific and low in the western Pacific

(C) Air pressure high in the northern Pacific and low in the southern Pacific

(D) Air pressure high in the southern Pacific and low in the northern Pacific

60. Which of the following is an assumption made in the passage below?

Bisphenol A (BPA) is a known endocrine disruptor that is found in the linings of metal food cans, in plastic containers and bottles, and in the coating found on most receipt paper. Animal studies have shown that low levels of BPA are correlated with higher rates of mammary and prostate cancers, diabetes, low sperm count, early puberty, neurological problems, and obesity. Its effects seem to be greatest during early developmental stages, such as the prenatal stage. Therefore, BPA should be more actively banned from household and everyday products, especially baby formula packaging.

(A) BPA causes cancers after exposure during any stage of life.

(B) BPA's effects on animals are not similar to its effects on humans.

(C) Most obesity is caused by either BPA exposure or lifestyle factors.

(D) BPA has not recently been banned from use in household products and on receipt paper.

GO ON TO THE NEXT PAGE.

61. Which of the following greenhouse gases has the LEAST impact on global climate change?

    (A) CO_2

    (B) CFCs

    (C) Methane

    (D) Water vapor

62. The acronym HIPPCO describes the main factors leading to a decrease in biodiversity. Which of the following gives four of these factors?

    (A) Habitat destruction, invasive species, population movements, and coastal displacement

    (B) Habitat destruction, increase in greenhouse gases, pollution, and ocean acidification

    (C) Invasive species, polar warming, climate change, and ozone depletion

    (D) Invasive species, population growth, pollution, and climate change

GO ON TO THE NEXT PAGE.

Question 63 refers to the following graph.

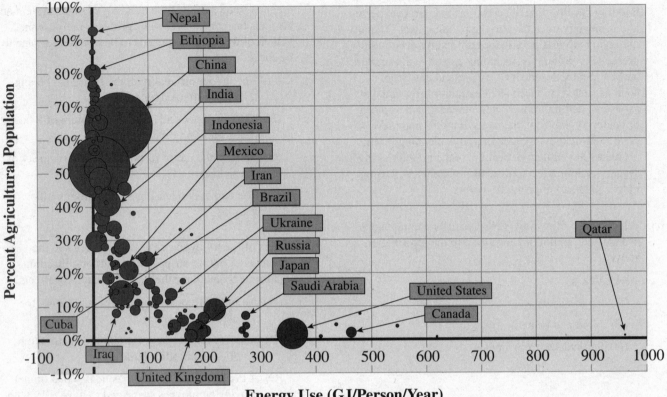

Energy Consumption and Agricultural Population for 205 countries, 2004

● Bubble area scaled to national population.

63. Which of the following best explains the trend in the graph above?

   (A) Countries with higher population use more energy.

   (B) Countries with lower population tend to use more energy per person.

   (C) Countries with lower agricultural populations are more industrialized and have higher energy demands.

   (D) Countries with higher agricultural populations are more mechanized and have higher energy demands.

64. Which of the following gives both an advantage and a disadvantage of the mechanization of farming practices that took place during the Green Revolution?

   (A) Mechanization increased profits but required reliance on genetically modified organisms.

   (B) Mechanization greatly increased efficiency on farms and also increased reliance on fossil fuels.

   (C) Mechanization increased the use of pesticides and the overuse of water resources for irrigation.

   (D) Mechanization made it easier to grow whole fields of the same crop, leaving farmers susceptible to disaster if pests or disease affected a specific crop too much.

GO ON TO THE NEXT PAGE.

Questions 65–67 refer to the following information.

Sick Building Syndrome (SBS) refers to a medical condition in which people who live or work in a given building develop shared symptoms that are unexpected and ultimately trace back to the building itself. Usually the main culprit is poor indoor air quality due to the presence of pollutants, including gases (such as radon, carbon monoxide, and VOCs), particulates (from sources such as tobacco smoke, indoor fires, and asbestos), and microbial contaminants (such as mold or bacteria). Ways to alleviate symptoms and prevent continued sickness include thorough cleaning; fixing problems in ventilation, heating, and air conditioning systems; eliminating sources of pollutants; and using indoor plants to improve air quality.

65. Which of the following is NOT usually a component of indoor air pollution, according to the information above?

    (A) VOCs

    (B) Hydrocarbons

    (C) Mold and bacteria

    (D) Particulates from smoke

66. A building manager suspects that a certain number of sicknesses in employees in the building are due to Sick Building Syndrome and takes certain measures to reduce the problem. She arranges for an expert diagnosis and repair of the building's HVAC system and also has the carpets and walls professionally cleaned. Which of the following needs to be true for her plan to be successful?

    (A) The indoor air quality in the building is being adversely affected by pollutants.

    (B) The employees all spend a minimum of 8 hours in the building per workday.

    (C) Mold is the factor responsible for a majority of the sickness.

    (D) The carpet is releasing VOCs through off-gassing.

67. Which of the following proposed solutions to the problems of indoor air pollution and Sick Building Syndrome would have the most far-reaching effects?

    (A) An initiative providing low-cost cleaning for affected buildings

    (B) A safety requirement for radon detectors to be installed in all office buildings

    (C) A tax leveraged against businesses that do not provide employees suspected to have SBS with paid sick leave

    (D) Building codes addressing multiple sources of indoor air pollution and requiring their prevention in all new construction

GO ON TO THE NEXT PAGE.

Questions 68–70 refer to the diagram below of the workings of an "earthship" passive house.

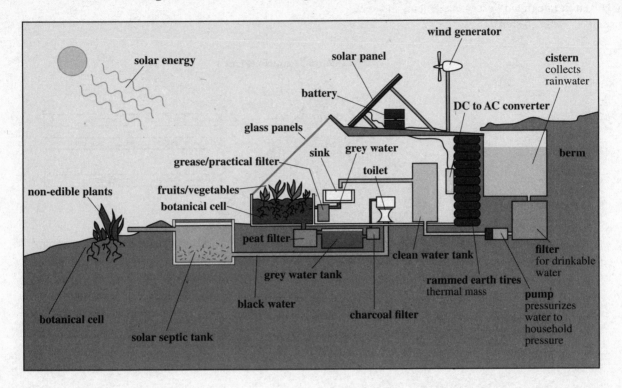

68. Which type(s) of solar energy does the house employ?

  I.   Passive solar energy system(s)
  II. Active solar energy system(s)
  III. Photovoltaic solar cells

(A) II only

(B) I and II only

(C) I and III only

(D) I, II, and III

69. The house uses each of the following methods to conserve energy and resources EXCEPT

(A) Hydroelectric power

(B) Rainwater collection

(C) Wind energy

(D) Solar energy

70. At the right side of the diagram, to the left of the labeled cistern, is a wall made of rammed earth tires labeled "thermal mass." With which of the following types of energy conservation does this feature help?

(A) Greywater reuse

(B) Rainwater collection

(C) Passive solar heating

(D) Solar and wind energy collection

GO ON TO THE NEXT PAGE.

71. The diagram below shows the hierarchy of steps involved in Integrated Pest Management, and some examples for each step.

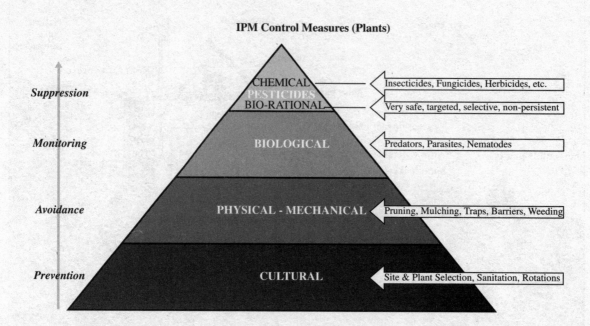

IPM Control Measures (Plants)

Suppression — CHEMICAL PESTICIDES / BIO-RATIONAL — Insecticides, Fungicides, Herbicides, etc. / Very safe, targeted, selective, non-persistent

Monitoring — BIOLOGICAL — Predators, Parasites, Nematodes

Avoidance — PHYSICAL - MECHANICAL — Pruning, Mulching, Traps, Barriers, Weeding

Prevention — CULTURAL — Site & Plant Selection, Sanitation, Rotations

Which of the following are potential disadvantages of Integrated Pest Management?

(A) It can be rigid and might be inadequate for certain environments.

(B) It can be complex and expensive to implement.

(C) It can lessen disruptions to the local biome.

(D) It can minimize threats to human health.

72. The graph below shows a trend in sea surface temperatures at the Great Barrier Reef over a period of 116 years.

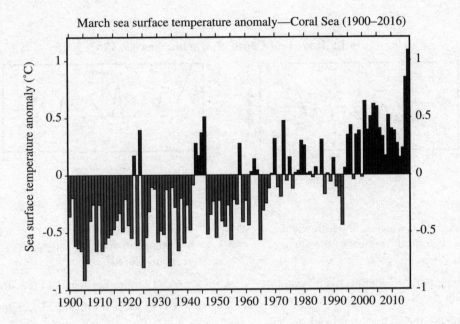

March sea surface temperature anomaly—Coral Sea (1900–2016)

Given that temperatures 1°C above average can cause bleaching, in which year did a mass bleaching event last long enough to still be occurring in March?

(A) 1942

(B) 1966

(C) 2008

(D) 2016

73. All of the following are ways global climate change can cause widespread habitat destruction EXCEPT

(A) Rise in sea level

(B) Increase in pollution

(C) Temperature changes

(D) Change in amounts of precipitation

GO ON TO THE NEXT PAGE.

Questions 74–76 refer to the following graphs.

Two graphs, showing the movement of the mean latitude and mean depth of a single given marine species over time, are shown below.

Change in Latitude and Depth of Marine Species, 1982–2015

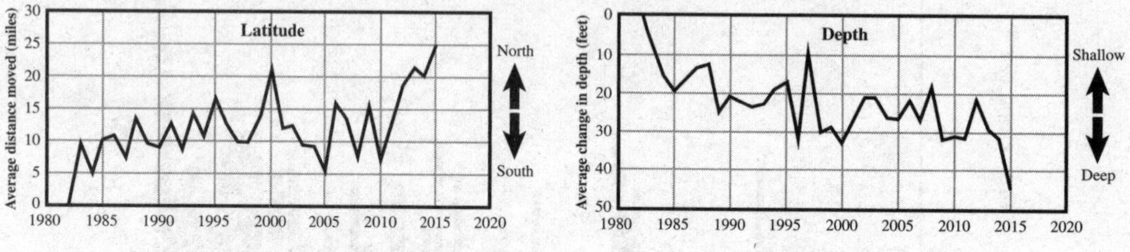

Data source: NOAA (National Oceanic and Atmospheric Administration) and Rutgers University. 2016. OceanAdapt.
http://oceanadapt.rutgers.edu.

74. According to the graphs, which of the following characterizes the average movement of the given species between 1999 and 2000?

   (A) Small amount of drift southward and small decrease in depth

   (B) Small amount of drift northward and small decrease in depth

   (C) Significant drift northward and small increase in depth

   (D) Significant drift southward and small increase in depth

75. Which of the following could explain the trends shown in the graphs—the species moving both northward and deeper over time?

   (A) Ocean warming has caused the species to move toward areas that were once warmer than the temperature range to which it is adapted.

   (B) Ocean cooling has caused the species to move toward areas that were once warmer than the temperature range to which it is adapted.

   (C) Ocean warming has caused the species to move toward areas that were once cooler than the temperature range to which it is adapted.

   (D) Ocean cooling has caused the species to move toward areas that were once cooler than the temperature range to which it is adapted.

76. Which of the following gives the approximate total net change in distance and depth of this species for the years measured?

   (A) 17 miles north and 44 feet deeper

   (B) 25 miles north and 23 feet deeper

   (C) 25 miles north and 44 feet deeper

   (D) 17 miles north and 23 feet deeper

77. Burning biomass releases CO_2. How can some biomass energy generation still be considered carbon-negative?

   (A) Photosynthesis cycles the CO_2 back into new crops.

   (B) Some biomass use falls below the allowable threshold for CO_2 emissions.

   (C) Some plant and animal material used as biomass grows quickly enough to replace what's used.

   (D) Biomass burning produces air pollution in the form of carbon monoxide and volatile organic compounds.

78. Which of the following is NOT a characteristic of a population in the transitional state of the Demographic Transition Model?

   (A) High birth rate

   (B) Lowered death rate

   (C) High infant mortality

   (D) High level of education for women

GO ON TO THE NEXT PAGE.

79. Malaria is a widespread disease, particularly in the tropical and subtropical regions, that claims hundreds of thousands of lives each year worldwide. Which of the following is a factor that may cause an increase in malaria cases?

    (A) Climate change

    (B) Drug resistance

    (C) Development of a malaria vaccine

    (D) More widespread use of mosquito netting

80. Which of the following is NOT likely to be true about an r-selected species?

    (A) They tend to have small body size.

    (B) Their life expectancy tends to be long.

    (C) They may reproduce only once in their lifespans.

    (D) Competition for resources in their habitat is relatively low.

END OF SECTION I

ENVIRONMENTAL SCIENCE
SECTION II
Time—70 minutes
3 Questions

Directions: Answer all three questions, which are weighted equally; the suggested time is about 23 minutes for answering each question. Where calculations are required, clearly show how you arrived at your answer. Where explanation or discussion is required, support your answers with relevant information and/or specific examples.

1. A state has a 10-year plan to reduce its CO_2 emissions by half, which involves the construction of wind farms. The graph below shows a comparison of several energy sources in terms of their CO_2 emissions.

Life cycle emissions from electricity generation, gCO_2/KWh

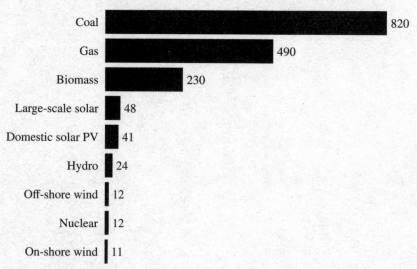

Coal — 820
Gas — 490
Biomass — 230
Large-scale solar — 48
Domestic solar PV — 41
Hydro — 24
Off-shore wind — 12
Nuclear — 12
On-shore wind — 11

(a) Using the graph, **identify** the three energy sources with the least CO_2 emissions and **explain** why the construction of wind farms is a good choice for the state in terms of meeting its goal.

(b) **Explain** how wind farms convert the kinetic energy of wind into electricity.

(c) The energy commission has tasked a team with predicting the success of the 10-year plan based on a preliminary trial of two years after the construction of the first wind farm. The team plans to monitor energy generation at the plant over the course of the two years and make predictions based on that and the data they already have about the amount of energy used and emissions generated while building the farm.

   i. **Describe** TWO other pieces of data the team must have in order for the commission to be able to accurately predict what difference each comparable wind farm will make in terms of the goal of reduction of CO_2 emissions.

   ii. If the majority (assume 100%) of the state's energy production prior to this trial was based on coal, **calculate** the approximate difference in CO_2 emissions per kilowatt-hour it will see if the new (on-shore) wind farms are able to supply 60% of the state's energy budget. **Show** your work.

   iii. The commission claims that if the new wind farms are able to supply 60% of the state's energy budget, then the goal of 50% CO_2 emissions reduction will have been met. **Justify** this claim using the calculation above.

   iv. **Identify** ONE real-world reason why a given wind farm might produce less energy than expected.

(d) The state's plan to reduce CO_2 emissions is focused solely on energy production. **Identify** TWO ways to reduce CO_2 emissions that are based instead in conservation (reducing energy use).

GO ON TO THE NEXT PAGE.

2. Below is a chart showing the stages of sewage treatment at a typical city water treatment facility.

Stage	Process	Byproduct	Filtered out
Physical Treatment	Filtering through screens		Larger debris
Primary Treatment	Chemical treatment and settling	Sludge	Suspended solids: 60% Organic waste: 30%
Secondary Treatment	Treatment with aerobic bacteria and settling	Sludge	Suspended solids: 97% Organic waste: 96% Toxic metals: 70% Organic chemicals: 70% Nitrogen: 50% Dissolved salts: 5%
Chlorination	Treatment with chlorine to remove any remaining living cells		

(a) **Identify** ONE pollutant that is NOT removed by this process.

(b) **Explain** how secondary treatment works to remove the wastes and pollutants it does.

(c) Sewage treatment of this type has byproducts that can be used.
   i. **Identify** TWO possible byproducts that are usable.
   ii. **Explain** how each can be used beneficially.

(d) The final stage identified in the table above, chlorination, sometimes causes the formation of trihalomethanes.
   i. **Explain** why this byproduct is undesirable.
   ii. **Propose** an alternate method that might reduce or avoid this negative effect, and **give one reason** this method has not been preferred to chlorination.

(e) Normally, after the final stage above, treated water is discharged into a city's streams, the ocean, or the city's gray water supply. Some cities prefer to discharge back into the groundwater. **Describe** what additional steps are necessary to make this acceptable under U.S. regulations.

GO ON TO THE NEXT PAGE.

3. Emeka wants to use the roof of the small house he built to collect rain. He intends to use it to irrigate his garden, and to supplement the municipal water source he is hooked up to for household use. He installs a rain collection barrel in his yard, and a downspout from the roof to fill it. Over the course of the following spring, his area gets about 25 centimeters of rain. The dimensions of the roof are 2.5 meters by 8 meters.

(a) **Describe** ONE environmental benefit of urban rainwater collection.

(b) Emeka's collection system is in operation without any changes for the length of one spring.

   i. **Calculate** the volume of rain he collects over that whole season, in cubic meters, by finding the volume of an imaginary rectangular prism with the dimensions of his roof as length and width, and the amount of rainfall as height (remember that the volume of a rectangular prism is length × width × height). Assume that he is using the water at a rate that assures his collection barrel does not overflow and no water is wasted. **Show** your work.

   ii. **Calculate** the amount of rainwater collected in liters. Convert the total volume of water Emeka collects into liters using the fact that a cube of side length 10 cm has a volume of 1 liter (remember that the volume of a cube is its side length cubed). Then convert the total volume to gallons (recalling that a gallon is about 3.785 liters), rounding to the nearest gallon. **Show** your work.

   iii. **Calculate** the average rate of his water collection to the nearest tenth in gallons per week if the spring lasts 13 weeks. **Show** your work.

(c) Rainwater is a relatively clean water source, but using a roof to collect it can introduce pollutants from the air (as well as bird droppings, moss, lichens, and dust) and make it non-potable. **Identify** TWO human-made pollutants likely to be found in rainwater collected this way in an urban setting.

(d) In order to use the water for dishes, bathing, and drinking, Emeka installs a filter system to remove contaminants and make it potable. Before he installed his rainwater collection system, his normal water use (for only himself in his small house) averaged 175 gallons per week. He pays for his municipal water use at a flat rate of 0.75¢ per gallon. At this rate, **calculate** how much money will he save per week now that he is supplementing with rainwater at the rate you found in part (b) (iii). **Show** your work.

(e) Emeka's longer-term goal is to collect enough water to be able to disconnect himself from the municipal water supply. To do this, he plans to install another rainwater collection barrel on the roof of his garden patio, which is 3 meters wide by 5 meters long. He claims this will meet his goal by making up for what he's been using from the municipal supply. **Calculate** the amount of water he'll collect per week (assuming the same average amount of rainfall) in this barrel as you did in part (b). **Show** your work. Then use it to **justify** Emeka's claim.

STOP

END OF EXAM

Practice Test 1: Answers and Explanations

PRACTICE TEST 1 ANSWER KEY

1.	C	21.	A	41.	C	61.	D
2.	A	22.	C	42.	B	62.	D
3.	A	23.	C	43.	B	63.	C
4.	B	24.	B	44.	D	64.	B
5.	D	25.	C	45.	D	65.	B
6.	C	26.	D	46.	C	66.	A
7.	B	27.	C	47.	B	67.	D
8.	C	28.	B	48.	C	68.	C
9.	D	29.	A	49.	B	69.	A
10.	A	30.	A	50.	A	70.	C
11.	D	31.	A	51.	D	71.	B
12.	D	32.	D	52.	C	72.	D
13.	B	33.	D	53.	B	73.	B
14.	C	34.	D	54.	B	74.	C
15.	A	35.	A	55.	A	75.	C
16.	A	36.	C	56.	A	76.	C
17.	B	37.	B	57.	B	77.	A
18.	C	38.	D	58.	A	78.	D
19.	A	39.	D	59.	B	79.	A
20.	D	40.	A	60.	D	80.	B

PRACTICE TEST 1 EXPLANATIONS

Section I—Multiple-Choice Questions

1. **C** Choices (B), (C), and (D) are all problems associated with clearcutting, while (A) is an unrelated practice, so eliminate (A). Choices (B) and (D), however, are problems not as closely tied to climate change as greenhouse gases, so (C) is the best answer.

2. **A** The melting of polar icecaps does release methane from inside the ice, so eliminate (C). It certainly causes a loss of habitat for those species who depend on the ice for habitat and food, so eliminate (B). It's also true that melting ice decreases the surface albedo of the region, meaning that less heat from the sun is reflected back to space. This results in a positive feedback cycle and more warming, so eliminate (D). However, ocean acidification is mainly caused by absorption of CO_2 from the atmosphere by ocean water, and isn't tied to polar melting, so (A) is correct.

3. **A** When stress overcomes a locked fault, a sudden slip occurs along the fault line: in other words, an earthquake. The other choices are all examples of geological events that can occur along fault lines and plate boundaries but are not caused by stress overcoming locked faults. Choice (A) is correct.

4. **B** Loss of areas for recreation represents disruption of a cultural service; fewer pollinators leading to decreased crop yield is an example of the impact of the disruption of a supporting service; and decreasing carbon sequestration interferes with a regulating service. Eliminate (A), (C), and (D). Since the loss of potential medicinal plant species is an example of the disruption of medicinal resources, a provisioning service, (B) is correct.

5. **D** The Coriolis effect causes deflection in movement of air and water (to the right in the Northern Hemisphere and to the left in the Southern Hemisphere). This affects large-scale phenomena in the movement of air and water, such as cyclones and hurricanes, as well as Hadley cells and the resulting trade winds. The El Niño–Southern Oscillation is an irregularly periodic variation in winds and sea temperatures caused by a change in the pressure-gradient force over the Pacific Ocean and is not related to the Coriolis effect. Choice (D) is correct.

6. **C** A *watershed* is the land area that drains into a particular stream, so (C) is correct. An *estuary* is a part of the wide lower course of a river where its current is met by the tides, while *headwaters* are the source of a river or stream. A *delta* is the location where a river drops its sedimentary load as it slows down to meet the ocean.

7. **B** Access to clean air and water, disease transmission, and food availability all depend on the density of population: with more people shared resources are spread thinner, and disease spreads more easily. However, natural disasters do not occur more frequently depending on the number of people, nor do they affect different numbers of people differently. Choice (B) is correct.

8. **C** The model shows the names of biomes found at different average temperatures and amounts of rainfall. The type of biome found where average temperatures are higher than 20°C and average precipitation is less than 50 cm is subtropical desert, the hottest and driest biome. Choice (C) is correct.

9. **D** The relationship between temperature, precipitation, and biome is that, generally, different biomes exist for different ranges of temperature and precipitation. Some biomes are very wet, which (A) supports; some are drier, and have fewer species, as (C) suggests. The coldest and hottest biomes also have fewer species, which makes (B) relevant. However, while it is true that there are temperatures that do not support life, these are not biomes, so (D) does not help explain the relationship shown and is the correct choice.

10. **A** The type of biome found where average temperatures are below 10°C and average precipitation is above 100 cm is taiga. Since this biome is generally characterized by conifer trees, with fungi, mosses, and lichens also common, (A) is correct.

11. **D** Use the age-structure pyramids to check the validity of each answer choice. In the year shown, the U.S. had a faster-growing population, so eliminate (A). The U.S. also had a higher proportion of its population under 45: eliminate (B). Be careful with (C): while the relative amount of the population in this age range does look greater on Germany's diagram, the absolute number is much smaller (look at the scales on the two diagrams). Germany's diagram is the classic shape for negative growth, whereas the diagram for the U.S. looks pretty close to what zero growth should look like. Choice (D) is correct.

12. **D** Based on the fact that Germany's population already showed negative growth several years ago, you should place Germany solidly in the postindustrial state, so (D) is correct.

13. **B** The total fertility rate is the number of children an average woman in a given population will bear during her lifetime. Since the shape of the age-structure pyramid for the U.S. shows close to zero growth, the TFR should be close to replacement levels. That means the number should be greater than 2, since the replacement rate would be exactly 2 if factors such as infant mortality didn't have to be accounted for. Eliminate (A). In a developing population, the replacement birth rate may be above 3 because of higher infant mortality rates; however, both common sense and the shape of the pyramid tell you that the U.S. was not in the transitional stage, so eliminate (C) and (D). Its replacement rate should lie between 2 and 3, so (B) is correct.

14. **C** The calculation required is: since E85 requires 34% more fuel than regular gasoline, for which the catalytic converter needs a ratio of 14.7 air to 1 fuel, the new ratio is 14.7 air to 134% of 1 fuel, or 14.7 air to 1.34 fuel. You can simplify that ratio to 10.97 to 1 using your calculator, but it isn't strictly necessary: you could also estimate that $14.7 \div 1.34$ will be less than 14.7, but not by too much ($14.7 \div 2$ is around 7.4, so you need a ratio greater than 7.4 to 1). The only choice that's in the correct range is (C), so it has to be correct.

15. **A** The phosphorous cycle is the only biogeochemical cycle that does not largely involve the atmosphere, since phosphorus does not exist in the atmosphere outside of dust particles. Therefore, no significant amount of phosphorous exists in precipitation, whereas uplift, decomposition, and leaching/runoff are all significant factors in the phosphorous cycle: (A) is correct.

16. **A** The tropopause is the buffer between the troposphere and the stratosphere. The gases responsible for the greenhouse effect are found in the troposphere, so eliminate (D). The ozone layer is found in the stratosphere, so eliminate (B). And the auroras are found in the thermosphere, so eliminate (C). The jet streams, powerful air currents that drive weather patterns across the globe, are found in the tropopause, so (A) is correct.

17. **B** The troposphere contains the air we breathe. The vast majority (78%) of it is nitrogen, so (B) is correct. About 21% of it is oxygen, and the remaining 1% includes all the greenhouse gases, such as carbon dioxide and water vapor.

18. **C** First, look at the graph to determine where world average biocapacity per person is represented. It's the label on the red line running vertically across the whole graph; the value shown is a little less than 2 global hectares per person. Next, since each answer choice lists three countries, find the one for which the bars representing total ecological footprint for *all three* fall under that value. Look at the bottom of the graph: the only countries whose total ecological footprint falls under the world average biocapacity value for 2003 are Afghanistan, Sierra Leone, Egypt, Thailand, and China. The only answer choice that chooses exclusively from this group is (C), so that choice is correct.

19. **A** Strip cropping is a sustainable farming technique meant to reduce soil erosion by slowing water flow, or runoff. Choice (A) is correct. The other choices are benefits of other sustainable farming techniques, such as windbreaks (B), no-till farming (D), and rotational grazing (C).

20. **D** Strip cropping reduces soil erosion by slowing water flow, or runoff. The passage also states that it involves crop rotation, and that that helps improve soil fertility. Both of these are issues the farmer wants to address, so eliminate (B) and (C). Choice (A) addresses the wrong issue, so eliminate it as well. Choice (D) is correct.

21. **A** Strip cropping's benefits are: reduction in soil erosion by slowing water flow, or runoff; and improvement in soil fertility, due to crop rotation. Choice (A), terracing, shares the benefit of reducing runoff, so keep it for now. Choice (B), windbreaks, are mainly used to prevent wind erosion, so eliminate it. No-till farming (C) is used to prevent the erosion that results from tillage (and, to some degree, to improve the impact of irrigation), so eliminate it as well. Rotational grazing (D) is a method of pasturing involving moving grazing animals to different paddocks and allowing the ones not in use to rest and recover. While it may impart some benefit to the soil in terms of reducing water erosion, its main benefits are to herd health and reduced input in terms of feeding the animals, so (A) is a better choice. Choice (A) is correct.

22. **C** In the scenario, the generalist species recovers more quickly and ends up with a greater share of the resources compared to several specialist species, meaning it must have had an advantage. Since (B) gives the specialists the advantage, eliminate it first. The question asks for an explanation of the outcome—in other words, *why* did the generalist species have the advantage? There is no information given about which species are native and which are invasive, so eliminate (D). There is no reason why there being several specialists to only one generalist would necessarily give the generalist the advantage, so eliminate (A). The key factor here is the wildfire: it caused a *change* in conditions, and that's what gave the generalist species the advantage in recovery. Choice (C) is correct.

23. **C** The underlying reason for the temperature difference between Earth's seasons is the axial tilt of the planet. This tilt means that the sunlight hits one hemisphere more directly than the other for each half of each year. This increases summer temperatures in two ways: by allowing light to pass through less of the atmosphere before it hits the surface, meaning less heat is absorbed along the way (A); and by concentrating the sunlight into a smaller area because of the angle at which it's hitting the planet (B). It also means that during the spring and summer in either hemisphere, days are longer than nights (while nights are longer than days during autumn and winter). This means more time during which the sun's rays are warming Earth (D). However, it's not true that Earth is closer to the sun during the northern summer. In fact, Earth's perihelion (the moment during its orbit when it's closest to the sun) takes place in early January each year, during the northern winter. Choice (C) is correct.

24. **B** The three types of relationship in the table are all described as symbiotic. Since predation is not a type of symbiosis, eliminate (D). The type of symbiosis in which both species receive benefit is mutualism, so (B) is correct.

25. **C** The current year is closest to the zone marked *Stage 2: Onset of ozone increases* in the diagram. This is consistent with the fact that depletion of ozone has already slowed considerably since the passing of the Montreal Protocol on Substances that Deplete the Ozone Layer in 1987. Choice (C) is correct.

26. **D** When the sheep population reaches its carrying capacity, its population growth will be checked in some way, so eliminate (A) and (B). The sheep are unlikely to see the problem and make conscious decisions to control their population, so eliminate (C). Rather, overshoot and dieback are likely until an equilibrium is reached: (D) is correct.

27. **C** Respiration and decomposition are biological processes that involve a wide variety of living things, including, but not limited to, humans. Volcanism is a geological process in which humans aren't generally implicated. However, combustion, or burning, is usually a process started by humans, so (C) is correct. All of these processes release CO_2 into the atmosphere.

28. **B** Choices (A), (C), and (D) all represent actual, but negative, effects of rising sea levels. Only (B) can be seen as a positive effect, so that choice is correct.

29. **A** Biomagnification is especially problematic in marine ecosystems because the food chains there are longer, so (A) is correct. Choices (B), (C), and (D) are all hard to verify and likely to be untrue in any general sense and thus can't explain this broad difference.

30. **A** The question asks for a choice that could *explain* the wealth of petroleum resources found in the Permian Basin. You don't need to know the specifics of the geologic history of west Texas in order to answer this question. Look instead for an answer that could explain the formation of fossil fuel deposits. Remember, fossil fuels are formed from organic matter that is subject to the pressures of geologic forces. Choice (B) explains why glaciers haven't ripped away surface layers, but says nothing about organic deposits, so eliminate it. Choice (C) gives a reason why metamorphic rock, not organic matter or fossil fuels, would be found in the area: eliminate it as well. Choice (D) is closer, because the seabed of an ancient ocean is likely to have large deposits of organic matter. However, it doesn't explain any geologic forces that could

have formed that matter into petroleum. Choice (A) contains the elements you need: deposits of ancient sediment (which would include organic matter), and a geologic process (subsidence) that would put the matter under pressure and could form fossil fuels. Choice (A) is correct.

31. **A** The question asks for an advantage of concentrated animal feeding operations (CAFOs, or feedlots) over free-range grazing. Eliminate answer choices that are disadvantages of CAFOs—(D)—or advantages of free-range grazing—(B) and (C). Choice (A), cost-effectiveness, is the main advantage CAFOs have over more environmentally friendly methods of meat production.

32. **D** Each choice describes some aspect of the relationship between soils (part of the rock cycle) and water (the water cycle). The only choice that shows how soil protects water quality, however, is (D): by filtering and cleaning water, soils remove adulterants and pollutants and improve the quality of the water passing through them. The other three choices explain ways in which water affects soil. Choice (D) is correct.

33. **D** Think of the amount of energy invested as 100 units. (What the units are does not matter to the calculation.) Through one trophic level, the 10% rule tells you that 10% of the energy will be conserved (and 90% lost). That means that 10% of 100 units, or 10 units, remain. Through the second trophic level, 10% of this amount will be conserved: 10% of 10 units, or 1 unit, remains. In total, 99 units were lost, which is 99% of the original number. Choice (D) is correct.

34. **D** The answer choices are all possible solutions to environmental problems, so look for the one that specifically addresses urban runoff. Choice (A), contour plowing, addresses agricultural runoff instead, so eliminate it. Choices (B) and (C) are further from the mark: solar panels are a form of clean energy and carbon offsets are meant to address the carbon footprints of individuals or corporations. Choice (D), permeable pavement, would allow the absorption of precipitation into the ground underneath city structures and thus decrease runoff, so it is correct.

35. **A** Limited diet, high numbers of competitors, and specific habitat requirements are all factors that can make a species *more* susceptible to becoming endangered or extinct in the face of habitat destruction, fragmentation, and/or loss, so eliminate (B), (C), and (D). Mobility, on the other hand, can help a species adapt to these problems by allowing it to seek out new habitats, so (A) is correct.

36. **C** Seawater is slightly basic, so eliminate (A) and (D)—especially (A), since it does not represent *acidification*, but rather a change toward less acidic conditions. In the last couple of centuries, surface ocean pH has decreased from approximately 8.25 to 8.14. This means that the change is not from basic to acidic—eliminate (B)—but rather just from slightly basic to less basic, which matches with the wording "toward pH-neutral." Choice (C) is correct.

37. **B** The environmental scientist wants to measure the effects of the temperature inversion on levels of pollutants, so other things that might affect levels of pollutants are what she would most likely need to control for. Seasonal temperature variations do not matter to the *yearly average* levels, so eliminate (C). While the number of cars and regulations that affect the pollutants do matter, they do not matter as directly as the actual levels of pollution entering the air, so (B) is better either than (A) or (D). Choice (B) is correct.

38. **D**   Choices (A), (B), and (C) all state disadvantages associated with aquaculture (fish farming). Choice (D), however, states an advantage, and is correct.

39. **D**   A natural disruption to an ecosystem is the result of natural, not human-caused, processes. Flooding because of seasonal weather events like monsoons, tides, and snowmelts fits this criterion, as do the effects of a volcanic eruption; similarly, ice age events cause natural changes to sea level. Only forest management is a human activity; as such, it does not constitute a natural disruption, so (D) is correct.

40. **A**   The curve labeled A in the model is a Type I survivorship curve, which shows high survivorship among young individuals and greater mortality at advanced age. Large mammals tend to have Type I survivorship, so (A) is correct. Squirrels and sparrows have Type II survivorship (matching line B in the graph), while tree frogs have Type III (matching C).

41. **C**   The graph shows what happens to three populations over time. All three populations start out increasing rapidly. Populations *a* and *b* overshoot their carrying capacity more severely; while population *a* dies out completely, *b* eventually recovers. Population *c* also overshoots, but less drastically, and it recovers as well. The difference between *b* and *c* on the graph is what happens after their recovery: while *c* fluctuates around the population's original carrying capacity, *b* does the same around a new, reduced carrying capacity. The explanation is likely that enough resources were damaged that the environment can now only sustain the population at a lower level. This makes (C) correct.

42. **B**   District heating (A) requires at least the infrastructure to deliver heating to separate homes, while enhanced geothermal systems (C) and binary cycle plants (D) are types of plants (infrastructure) that use geothermal energy to generate electricity. While baths certainly can involve construction, natural hot springs have been used as baths without any modification since ancient times. Choice (B) is correct.

43. **B**   Variable renewable energy is a renewable energy source that is non-dispatchable (cannot be used on demand) because of its fluctuating nature, such as wind power—(B) is correct. Dammed hydroelectricity and biomass are controllable renewable energy sources, and geothermal energy is also a relatively constant source.

44. **D**   Natural gas has lower CO_2 and particulate emissions than all other fossil fuel types. Choice (D) is correct.

45. **D**   The question asks for a potential *disadvantage* to the plan (banning pesticides and fertilizers). Choice (A) is a disadvantage of the pesticides themselves, while (C) is a disadvantage of a potential decrease in the area of the wetlands. Choice (B), on the other hand, is an *advantage* of helping the wetlands ecosystem. Only (D) posits a potential disadvantage to the plan: if the ban causes financial harm to the farms in question, then it's possible they will close and other development will move in, posing a whole new threat to the wetland area.

46. **C**   The diagram shows that oxygen and hydrogen gases are inputs, and the only outputs are heat, water, and electricity. You can eliminate (A) because water is an output, not an input, and (B) because no carbon dioxide is produced. Choice (D) is also incorrect: hydrogen fuel cells are relatively efficient. However, energy must still be used to make hydrogen gas: (C) is correct.

47. **B** Rabbits reproduce quickly, put little investment into their young, and have small body size, so they fit the profile of *r*-selected species rather well, for mammals. Eliminate (C) and (D). The graph shows that they outcompete the *largest number* of native species; in order to compete with so many other species, they must be generalists. Eliminate (A). Choice (B) explains why rabbits, as *r*-selected generalists, succeed against so many native species, and is therefore correct.

48. **C** The LD_{50} of a substance is the dosage it takes to kill 50% of the test population. On the graph, use a piece of paper or a pencil to find a straight horizontal line from the 50% mark on the *y*-axis to the curve. From that point on the curve, find a straight vertical line down to the *x*-axis, and note the place of intersection. It should intersect at just about 100 mg/kg, so that's the approximate LD_{50}. The correct choice is (C).

49. **B** A *poison* is any substance with an LD_{50} of 50 mg or less per kg of body weight, so this substance is not a poison. Eliminate (C) and (D). Since (B) correctly explains why, it is the best choice.

50. **A** Since *toxicity* just means the degree to which something is biologically harmful, if the toxicity is reduced, then the new combination should have a lower LD_{50} than the original unknown substance. The only choice you can prove with that information is (A). Both (B) and (C) are false. What constitutes "non-toxic" is a matter of some controversy and certainly not something you can prove with only this information. Choice (A) is correct.

51. **D** Since neither the El Niño–Southern Oscillation nor the upwelling of cold water in the Arctic and Antarctic zones has any effect on stratospheric ozone, eliminate (A) and (B). An increase in the amount of UV rays that reach Earth's surface, on the other hand, is an *effect* of stratospheric ozone depletion rather than a cause, so eliminate (C) as well. The melting of ice crystals in the atmosphere at the beginning of the Antarctic spring does provide a surface for the conversion of unreactive chlorine compounds into reactive ones: since this is a phenomenon driven by temperature and the Antarctic climate, not by humans, it is a natural factor. Choice (D) is correct.

52. **C** A bottleneck event is one in which the total number of individuals in a population is drastically reduced. This results in less species richness, but that is not relevant to the likelihood of survival for a single species, so eliminate (A). It also results in reduced genetic diversity. While that may increase genetic drift, there is no way to know whether increased genetic drift will help or hurt a given population, so eliminate (D). Additionally, remember that a decreased number of individuals never limits that population's eventual growth; limitations on population growth are imposed by availability of resources. Eliminate (B). Reduced genetic diversity does, however, leave a population more susceptible to the negative effects of further disruptions. Choice (C) is correct.

53. **B** To read a soil texture triangle, use the line indicated on each axis for a given percentage and find where the three lines intersect. The diagram indicates that you should use the left set of lines for the sand axis, the upper set for the clay axis, and the lower set for the silt axis. Following these for 40%, 35%, and 25%, respectively, you should find the intersection lies in the clay loam region. Choice (B) is correct.

54. **B**   *Ecological succession* means the establishment of a biological community in an area virtually barren of life, and the transitions in species composition of that biological community afterwards. *Species distribution* refers to the manner in which the individuals of a species are spatially arranged. The term *biodiversity* describes the number and variety of organisms found within a specified geographic region, or ecosystem, as well as the variability among living organisms, including the variability within and between species and within and between ecosystems. *Ecological tolerance* refers to the range of conditions in which an organism or species can live, so (B) is correct.

55. **A**   The rainshadow effect occurs where high-elevation areas lie near the coasts of warm bodies of water. When the prevailing winds push water-laden air toward land, the air must rise because of the terrain, which cools the air, causing water to condense and fall as precipitation. Thus, the windward side of the mountain or elevated area receives more precipitation and has a lush climate. When the air moves over the peak of elevation, it has been emptied of most of its water vapor, and the air moving down the other side is dry. That side receives less precipitation and has a desert climate. Choice (A) is correct.

56. **A**   Species richness refers to the number of different species represented in an ecological community. The islands with the greatest species richness are those with the most species; those for which the equilibrium point falls farthest to the right on the graph. It shows that large islands close to the mainland have an equilibrium point farthest to the right, so (A) is correct.

57. **B**   The limited resources on islands and in other isolated areas mean that competition for those resources is high. It benefits individuals to exploit niches, and thus specialization increases in these communities: eliminate (A) and (D). The species found on mainlands include some specialists, but the proportion of generalists is higher, and these generalists have a short-term advantage over specialists when they are introduced as invasive species, in that they can use different resources to thrive as long as those are available. Neither the number of species nor the distance traveled is relevant to the competition for resources the species face. Therefore, (B) is the best choice.

58. **A**   Leaving leaves in place, replacing lawns altogether with gardens, and using a simple rake are all good ways to avoid the use of leaf blowers. However, a lawn care service is just as likely to use these noisy tools as an individual homeowner is, so (A) is correct.

59. **B**   The El Niño–Southern Oscillation has a warmer phase—el Niño—and a cooler phase—la Niña. During the cooler phase, air pressure is high in the eastern Pacific and low in the western Pacific. During the opposite, warmer phase, air pressure is high in the western Pacific and low in the eastern Pacific, so (A) is describing the el Niño phase. Choice (B) is correct.

60. **D**   The passage gives evidence that BPA is correlated with a host of health problems and details everyday objects in which it can be found. It concludes that this endocrine disruptor should be banned. The necessary assumptions to a conclusion about what action should be taken are things like: *this course of action will help the problem* and *there is no reason it won't help or would make things worse*. Evaluate whether the choices given need to be true for the conclusion to make sense. The passage states that BPA is correlated with a higher incidence of cancer and that early developmental stages seem to be the most crucial: BPA doesn't

need to be shown to definitively cause cancer, or at all life stages, for banning it to be a helpful move. Eliminate (A). Choice (B) actually hurts the author's cause: if BPA's effects on humans are different from the animal effects cited, there is much less information to go on in terms of whether it should be banned or not. Eliminate (B). Similarly to (A), (C) contains information that is not necessary to the author's argument. There may be many more causes for obesity: as long as BPA is likely to be one of them, it makes sense to ban it. Choice (D), however, needs to be true for the author's point to make sense: if BPA has already been banned, then there's no need to do it anymore. Choice (D) is correct.

61. **D** CO_2 has a GWP (global warming potential) of 1 because other greenhouse gases' GWPs are defined using that gas as a reference point; CFCs and methane have higher GWPs than CO_2. However, while water vapor does contribute to global climate change, it doesn't have nearly the effect of these other gases. The hydrologic cycle keeps the amount of water vapor present in the atmosphere in balance. The small effect it has is due to the warming caused by other gases: once the atmosphere warms, it can hold more water vapor, and that does intensify the greenhouse effect some via positive feedback. Since its effect is less and does not occur independent of other causes, (D), water vapor, is the correct choice.

62. **D** The acronym HIPPCO stands for habitat destruction, invasive species, population growth, pollution, climate change, and over-exploitation. Choice (D) is correct.

63. **C** The graph shows that countries with lower agricultural populations have higher energy use per person. This fits with the trend that more industrialized countries have higher energy demands, since more industrialization corresponds with less agricultural population. Choice (C) is correct.

64. **B** Choice (B) gives both an advantage (increased efficiency) and a disadvantage (reliance on fossil fuels) of mechanization, so it is the best answer. Choice (A) is untrue: mechanization and genetic modification are separate innovations, and do not specifically require each other. Choice (C) gives two disadvantages (increased pesticide use and overuse of water), and choice (D) really only gives one disadvantage— monoculture—and explains its negative effects.

65. **B** The passage states that *pollutants, including gases (such as radon, carbon monoxide, and VOCs), particulates (from sources such as tobacco smoke, indoor fires, and asbestos), and microbial contaminants (such as mold or bacteria)* are often responsible for poor indoor air quality. Choices (A), (C), and (D) are mentioned in this list. The correct choice is (B), hydrocarbons, as these usually result from vehicle exhaust and manufacturing, not indoor sources.

66. **A** In order for the building manager's plan to be successful, the steps she takes (cleaning and repairing) need to make a difference to the health of the employees. If (A) were untrue, then neither cleaning nor repairing would make any difference—it would mean that the sickness was due to some other problem. Choice (A) is necessary to the plan and is correct. Choices (B), (C), and (D) are all possible, but none of them is necessary to the building manager's plan—any one of them could be untrue, but as long as the main problem is still *something* to do with air quality, her plan will work.

67. **D** Since the question asks for the proposed solution with the most far-reaching effects, look for an answer that would affect all or nearly all cases of SBS in the future. Radon is only one in a long list of pollutants

listed that relate to the syndrome, so eliminate (B). While providing cleaning for affected buildings would certainly help in plenty of cases, it's only one of several solutions, and won't help in cases in which the problem is caused by bad ventilation or problems with HVAC systems, so eliminate (A). A tax against businesses not providing sick leave will help the employees to cope with symptoms, but won't relieve the cause of those symptoms, so eliminate (C). If building codes require the prevention of several sources of indoor air pollution in *all* new construction, many future cases of SBS will be prevented entirely. Choice (D) is correct.

68. **C** Remember to use Process of Elimination on I/II/III questions. The diagram shows a solar panel, which is another name for a photovoltaic solar cell. Statement III is true, so eliminate (A) and (B). Since I is present in both remaining choices, statement I must be true: the house uses passive solar energy (you can tell it's true because the window is angled and only present on one side of the house—it's meant to capture solar energy through the greenhouse effect, and to capture more energy in winter than in summer). Now address statement II. Active solar energy systems use solar energy to heat a liquid that passes through mechanical and electric equipment to collect and store the energy captured. Since this type of system is not present in the diagram, statement II is false. Eliminate (D); the correct choice is (C).

69. **A** The solar panels and passive solar heating you noticed in Question 68 mean that (D) is used, so eliminate it. Look for evidence of the other methods. The labeled wind generator on the roof is evidence of the use of wind energy, so eliminate (C). The cistern (labeled "collects rainwater") at the right side of the diagram is evidence of rainwater collection, so eliminate (B). There is no indication of a hydroelectric power system in the house, so (A) is correct.

70. **C** The label "thermal mass" is a clue to what the wall of tires is used for. It must have something to do with heat, so you can eliminate (A) and (B). Although it relates to solar energy, it's not connected or related to the wind energy generator, so eliminate (D). Choice (C) is correct: a thermal mass is a mass that will collect and store heat for a time, allowing the passive solar radiation the house collects to last during non-daylight hours.

71. **B** Integrated Pest Management is a system used to effectively control pests while minimizing disruption to the environment. As the diagram shows, it involves a combination of methods and a hierarchy for when to use what. The complexity of the system compared to methods involving a singular technique should be evident, and complexity often leads to expense: (B) is correct. Rigidity is the opposite problem, so eliminate (A); and (C) and (D) both give advantages, not disadvantages, so eliminate them as well.

72. **D** The only year shown that has a March temperature more than 1°C above average is 2016, so (D) is correct.

73. **B** Since pollution is a separate problem from climate change, (B) is not a way climate change can cause habitat destruction, so it is the correct choice. Rise in sea level, temperature changes, and precipitation changes are all effects of climate change that can indeed cause habitat destruction.

74. **C** Between the years 1999 and 2000, the first graph shows an average distance of approximately 7 miles northward. Eliminate (A) and (D). The second graph shows an increase in depth of about 3 feet, so eliminate (B) as well. Choice (C) is correct.

75. **C** Since this is a marine species, think about what would make it move both northward and deeper. Since water is cooler in the depths than the shallows, the creature probably is in water that is warming, resulting in a movement toward cooler water. In the northern hemisphere, the northward movement would have the same result. Choice (C) is correct.

76. **C** Subtract the greatest from the least value on each graph: 25 miles – 0 miles = 25 miles north; 44 feet – 0 feet = 44 feet deeper. The reason both minimum values are zero is that these graphs both show the change *from initial position*, which is represented as 0 change. Choice (C) is correct.

77. **A** Carbon neutrality means achieving net zero carbon dioxide emissions, either by eliminating CO_2 emissions entirely or by balancing emissions with carbon removal; so the term "carbon-negative" means that the balance of carbon removal with emissions is on the side of removal. Since you know that burning biomass involves some CO_2 release, look for an answer that involves some sort of carbon removal. Choice (A) shows how photosynthesis can remove CO_2, so it is correct. Choice (B) focuses only on emissions, so it can't explain how you get to a carbon-negative balance; choice (C) focuses on renewability; and choice (D) simply brings up a separate problem.

78. **D** In the Demographic Transition Model, the transitional state (Phase 2) is the phase during which the greatest population increase occurs due to a high birth rate and a lowered death rate—eliminate (A) and (B). Since infant mortality and child labor tend to still be high in this phase, you can also eliminate (C). However, the level of education for women doesn't tend to increase until later stages in the DTM; conversely, higher levels of education for women tend to lower the total fertility rate and push a population toward later stages in the DTM. Choice (D) is correct.

79. **A** The question asks for a factor causing an *increase* in malaria cases. Eliminate (C) and (D): if these have an effect, it will be toward *decreasing* numbers of malaria cases. Choice (B) looks appealing, but keep in mind that drug resistance matters to *treatment*, and therefore *outcome*, of malaria cases; it should not affect the base number of cases that occur. Climate change, however, is likely to have a negative effect on this worldwide health problem; malaria is a vector disease carried by mosquitos, and as global temperatures increase the range of these insects widens and the potential for infection therefore grows. The correct choice is (A).

80. **B** An *r*-selected species is one that reproduces early in life and has a high capacity for reproductive growth. This normally corresponds with bountiful resources or low competition (D). Usually *r*-selected species produce many offspring each time they reproduce, but they may not reproduce very many times— perhaps even only once (C). They tend to be small (A) and mature quickly, but their lifespans tend to be relatively short. Thus (B) is NOT true and is the correct answer.

Section II—Free-Response Questions

Remember, you must write your responses in paragraph form!

Question 1

A state has a 10-year plan to reduce its CO_2 emissions by half, which involves the construction of wind farms. The graph below shows a comparison of several energy sources in terms of their CO_2 emissions.

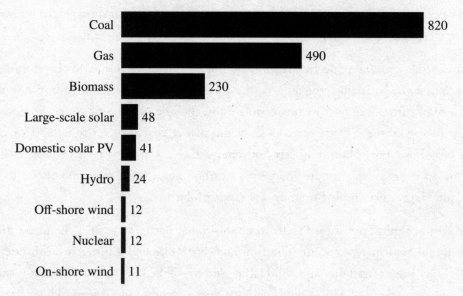

Life cycle emissions from electricity generation, gCO₂/KWh

Coal 820
Gas 490
Biomass 230
Large-scale solar 48
Domestic solar PV 41
Hydro 24
Off-shore wind 12
Nuclear 12
On-shore wind 11

(a) Using the graph, **identify** the three energy sources with the least CO_2 emissions and **explain** why the construction of wind farms is a good choice for the state in terms of meeting its goal.

The three energy sources with the least CO_2 emissions are off-shore wind, nuclear, and on-shore wind. Since both off-shore and on-shore wind are in this top three, wind farms are one of the best ways to meet the goal of reducing CO_2 emissions, as long as they are being used to replace other methods of energy generation that have higher CO_2 emissions (which most do).

(2 points maximum—1 point for identifying the three sources with the least emissions, and 1 point for explanation of why wind is a good choice)

(b) **Explain** how wind farms convert the kinetic energy of wind into electricity.

Wind turbines use the kinetic energy of moving air (wind) to spin a turbine, which in turn converts the mechanical energy of the turbine into electricity.

(1 point maximum for explanation)

(c) The energy commission has tasked a team with predicting the success of the 10-year plan based on a preliminary trial of two years after the construction of the first wind farm. The team plans to monitor energy generation at the plant over the course of the two years and make predictions based on that and the data they already have about the amount of energy used and emissions generated while building the farm.

i. **Describe** TWO other pieces of data the team must have in order for the commission to be able to accurately predict what difference each comparable wind farm will make in terms of the goal of reduction of CO_2 emissions.

Other pieces of data needed include: sources of energy the wind farm will replace, emissions from those sources (including during use and those involved in building/maintaining facilities), and amount of energy those sources generate.

(2 points maximum—1 point for each additional piece of data needed with a maximum of two)

ii. If the majority (assume 100%) of the state's energy production prior to this trial was based on coal, **calculate** the approximate difference in CO_2 emissions per kilowatt-hour it will see if the new (on-shore) wind farms are able to supply 60% of the state's energy budget. **Show** your work.

If the new wind farms supply 60% of the state's energy budget, then assume the remaining 40% is still provided by coal. Using the graph, this means that 60% of the total energy budget has emissions of about 11 gCO_2/KWh, and 40% has about 820 gCO_2/KWh. Thus the new emissions will be 0.6 × 11 + 0.4 × 820 = 334.6 gCO_2/KWh. Compare this to the emissions before the trial: assuming 100% of the budget was provided by coal, they were simply 820 gCO_2/KWh. The difference is 820 − 334.6 = 485.4 gCO_2/KWh.

(1 point maximum for correct calculation showing work)

iii. The commission claims that if the new wind farms are able to supply 60% of the state's energy budget, then the goal of 50% CO_2 emissions reduction will have been met. **Justify** this claim using the calculation above.

A reduction of 50% in CO_2 emissions means the previous rate of (approximately) 820 gCO_2/KWh should be reduced by half or more. Since half would be 410 gCO_2/KWh, and the rate calculated in part (ii) was 334.6 gCO_2/KWh, this does represent a reduction of more than half, and the goal would be met.

(1 point maximum for correct justification)

iv. **Identify** ONE real-world reason why a given wind farm might produce less energy than expected.

Possible answers:

- Estimates of weather (number of windy days) turn out to be incorrect
- Technical problems or natural disaster damage result in periods of non-functionality while repairs are made or problems are corrected
- Government or protestors impede the opening or functioning of the wind farm, resulting in delays or periods of non-functionality

(1 point maximum for an acceptable reason)

(d) The state's plan to reduce CO_2 emissions is focused solely on energy production. **Identify** TWO ways to reduce CO_2 emissions that are based instead in conservation (reducing energy use).

Possible answers:
- Improving vehicle fuel economy
- Increasing relative use of public transportation
- Requiring green building design for new buildings
- Increasing participation in energy-efficiency programs for homes and/or businesses
- Increasing participation in home/business energy use reduction programs
- Converting traditional yards to conservation landscaping

(2 points maximum—1 point for each acceptable conservation-based strategy)

Question 2

Below is a chart showing the stages of sewage treatment at a typical city water treatment facility.

Stage	Process	Byproduct	Filtered out
Physical Treatment	Filtering through screens		Larger debris
Primary Treatment	Chemical treatment and settling	Sludge	Suspended solids: 60% Organic waste: 30%
Secondary Treatment	Treatment with aerobic bacteria and settling	Sludge	Suspended solids: 97% Organic waste: 96% Toxic metals: 70% Organic chemicals: 70% Nitrogen: 50% Dissolved salts: 5%
Chlorination	Treatment with chlorine to remove any remaining living cells		

(a) **Identify** ONE pollutant that is NOT removed by this process.

Possible answers:
- Persistent organic pollutants (POPs)
- Radioactive isotopes
- Pharmaceuticals or Environmental persistent pharmaceutical pollutants (EPPPs)

(1 point maximum for one pollutant that is not removed)

(b) **Explain** how secondary treatment works to remove the wastes and pollutants it does.

In secondary treatment, either trickling filters or sludge processors are used to introduce aerobic bacteria to the sewage. These digest organic waste, and then the solids (including the bacteria) are settled out and removed as sludge.

(1 point maximum for correct explanation of secondary treatment process)

(c) Sewage treatment of this type has byproducts that can be used.

   i. **Identify** TWO possible byproducts that are usable.

   ii. **Explain** how each can be used beneficially.

Possible answers to this question are listed in the following table.

Possible byproduct	Beneficial use
Dried sludge (cake)	Can be sold as fertilizer
Methane	Can be used/sold as fuel

(4 points maximum—1 point for each possible byproduct with a maximum of two, 1 point for the beneficial use of each)

(d) The final stage identified in the table above, chlorination, sometimes causes the formation of trihalomethanes.

   i. **Explain** why this byproduct is undesirable.

Trihalomethanes are undesirable because they are environmental pollutants and possible carcinogens.

   ii. **Propose** an alternate method that might reduce or avoid this negative effect, and **give one reason** this method has not been preferred to chlorination.

Possible answers to this question are listed in the following table.

Alternate method	Disadvantage(s)
Ozonation	More expensive
UV Radiation	Needs multiple treatments, more expensive

(3 points maximum—1 for explanation of undesirability, 1 for alternate method, and 1 for disadvantage)

(e) Normally, after the final stage above, treated water is discharged into a city's streams, the ocean, or the city's gray water supply. Some cities prefer to discharge back into the groundwater. **Describe** what additional steps are necessary to make this acceptable under U.S. regulations.

Normally, the water must be further treated by tertiary treatment, which involves passing the secondary treated water through a series of sand and carbon filters and then further chlorination.

(1 point maximum for correct description of tertiary treatment process)

Question 3

Emeka wants to use the roof of the small house he built to collect rain. He intends to use it to irrigate his garden, and to supplement the municipal water source he is hooked up to for household use. He installs a rain collection barrel in his yard, and a downspout from the roof to fill it. Over the course of the following spring, his area gets about 25 centimeters of rain. The dimensions of the roof are 2.5 meters by 8 meters.

(a) **Describe** ONE environmental benefit of urban rainwater collection.

Possible answers: Reduces stormwater runoff (which can carry pollutants) to lakes and streams; reduces load on/need for municipal water treatment; reduces the chance that stormwater surge will overwhelm municipal sewers; helps even out water supplies over rain/drought cycles

(1 point maximum for a benefit)

(b) Emeka's collection system is in operation without any changes for the length of one spring.

i. **Calculate** the volume of rain he collects over that whole season, in cubic meters, by finding the volume of an imaginary rectangular prism with the dimensions of his roof as length and width, and the amount of rainfall as height (remember that the volume of a rectangular prism is length × width × height). Assume that he is using the water at a rate that assures his collection barrel does not overflow and no water is wasted. **Show** your work.

The volume of an imaginary rectangular prism with dimensions 2.5 meters by 8 meters by 25 centimeters (or 0.25 meters) is 2.5 m × 8 m × 0.25 m = 5 m^3.

ii. **Calculate** the amount of rainwater collected in liters. Convert the total volume of water Emeka collects into liters using the fact that a cube of side length 10 cm has a volume of 1 liter (remember that the volume of a cube is its side length cubed). Then convert the total volume to gallons (recalling that a gallon is about 3.785 liters), rounding to the nearest gallon. **Show** your work.

To find the volume of a cube of side length 10 cm, first convert to the units at hand (meters): the cube has side length 0.1 meter. Then find the volume: $(0.1 \text{ m})^3 = 0.001 \text{ m}^3$. Now divide the volume you found in part (i) to convert: $5 \text{ m}^3 \div 0.001 \text{ m}^3$ per liter = 5000 L.

To convert to gallons, use the conversion factor given: 5000 L $\div$ 3.785 L per gallon $\approx$ 1321.0039 gal. Round to the nearest gallon: 1321 gallons.

iii. **Calculate** the average rate of his water collection to the nearest tenth in gallons per week if the spring lasts 13 weeks. **Show** your work.

Emeka collects 1321 gallons over the course of 13 weeks. Divide to find the rate: 1321 gallons $\div$ 13 weeks $\approx$ 101.615 gallons per week. Round to the nearest tenth: 101.6 gallons per week.

(4 points maximum—1 point for correct calculation of total volume in cubic meters, 1 point each for conversions to liters and gallons, and 1 point for calculation of rate)

(c) Rainwater is a relatively clean water source, but using a roof to collect it can introduce pollutants from the air (as well as bird droppings, moss, lichens, and dust) and make it non-potable. **Identify** TWO human-made pollutants likely to be found in rainwater collected this way in an urban setting.

Possible answers: Particulates (including lead), pesticides, CO, CO_2, NO_x, SO_x, and ozone

(2 points maximum—1 point for each possible pollutant)

(d) In order to use the water for dishes, bathing, and drinking, Emeka installs a filter system to remove contaminants and make it potable. Before he installed his rainwater collection system, his normal water use (for only himself in his small house) averaged 175 gallons per week. He pays for his municipal water use at a flat rate of 0.75¢ per gallon. At this rate, **calculate** how much money will he save per week now that he is supplementing with rainwater at the rate you found in part (b) (iii). **Show** your work.

Using only municipal water, Emeka was paying about 175 gallons per week × 0.75¢ per gallon = 131.25¢ (or \$1.31) per week. Since Emeka is collecting 101.6 gallons of rainwater per week, his municipal usage should drop to 175 gallons – 101.6 gallons = 73.4 gallons per week. This should only cost him 73.4 gallons per week × 0.75¢ per gallon = 55.05¢ (or \$0.55) per week. The savings is 131.25¢ – 55.05¢ = 76.2¢ (or \$0.76) per week.

(1 point maximum for correct calculation of savings)

(e) Emeka's longer-term goal is to collect enough water to be able to disconnect himself from the municipal water supply. To do this, he plans to install another rainwater collection barrel on the roof of his garden patio, which is 3 meters wide by 5 meters long. He claims this will meet his goal by making up for what he's been using from the municipal supply. **Calculate** the amount of water he'll collect per week (assuming the same average amount of rainfall) in this barrel as you did in part (b). **Show** your work. Then use it to **justify** Emeka's claim.

First, find the volume of an imaginary rectangular prism with dimensions 3 meters by 5 meters by 25 centimeters (or 0.25 meters): 3 m × 5 m × 0.25 m = 3.75 m³. Next, divide the volume by the volume of a cube of side length 10 cm to convert to liters: 3.75 m³ ÷ 0.001 m³ per liter = 3750 L. To convert to gallons, use the conversion factor: 3750 L ÷ 3.785 L per gallon ≈ 991 gallons. Finally, divide by the same number of weeks (13) to find the rate: 991 gallons ÷ 13 weeks ≈ 76.2 gallons per week. In part (d) you found that Emeka's first collection barrel covered all but 73.4 gallons per week of his regular use: since 76.2 gallons per week is more than that, this additional collection barrel should give him just enough to not use the municipal supply.

(2 points maximum—1 point for correct calculation and 1 point for justification)

HOW TO SCORE PRACTICE TEST 1

Section I: Multiple-Choice

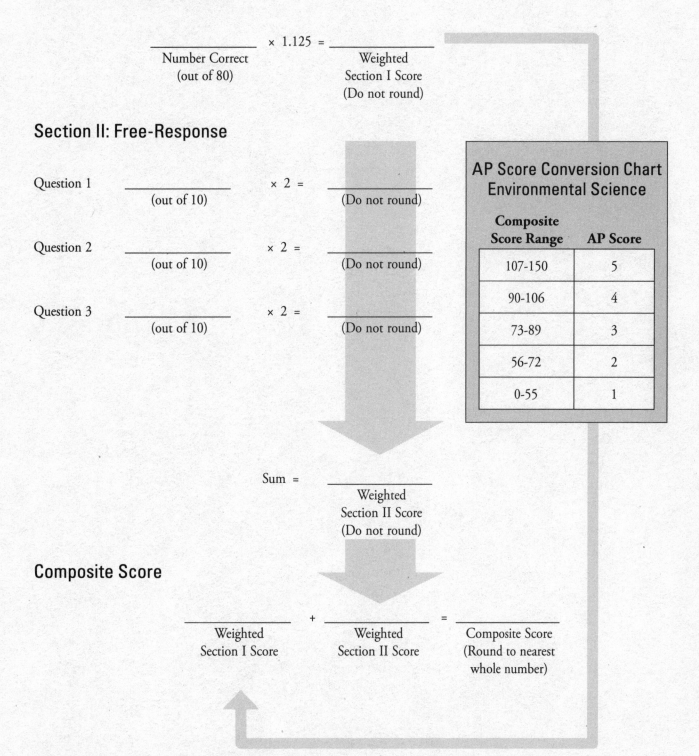

_____ × 1.125 = _____
Number Correct   Weighted
(out of 80)   Section I Score
                                (Do not round)

Section II: Free-Response

Question 1   _____   × 2 =   _____
                (out of 10)   (Do not round)

Question 2   _____   × 2 =   _____
                (out of 10)   (Do not round)

Question 3   _____   × 2 =   _____
                (out of 10)   (Do not round)

**AP Score Conversion Chart
Environmental Science**

Composite Score Range	AP Score
107-150	5
90-106	4
73-89	3
56-72	2
0-55	1

Sum =   _____
         Weighted
         Section II Score
         (Do not round)

Composite Score

_____   +   _____   =   _____
Weighted   Weighted   Composite Score
Section I Score   Section II Score   (Round to nearest
                                                  whole number)

Note: This score sheet is to help you estimate your approximate score for the official exam, not your actual score.

Part III
About the AP Environmental Science Exam

- The Structure of the AP Environmental Science Exam
- How the AP Environmental Science Exam Is Scored
- Overview of Content Topics
- How AP Exams Are Used
- Other Resources
- Designing Your Study Plan

THE STRUCTURE OF THE AP ENVIRONMENTAL SCIENCE EXAM

The AP Environmental Science Exam will be 2 hours and 40 minutes long. You will now have 90 minutes to answer 80 multiple-choice questions and 70 minutes to answer three free-response questions. This is a major change from previous exams, so below is a chart outlining the changes from the old and the new exam:

AP Update

The revised AP Environmental Science exam was scheduled to debut in May 2020, but because of the COVID-19 outbreak, the testing format was changed. The first administration of the new test is now planned for May 2021, so please refer to your free online student tools to see if there have been any breaking updates regarding the wording of questions or the representation of each content area within the test.

Previous Exam	Current Exam
100 Multiple-Choice Questions • Each MC question has five answer choices (A–E) • 1 hour, 30 minutes	**80** Multiple-Choice Questions • Each MC question will have **four** answer choices (A–D) • 1 hour, 30 minutes • **Five questions will have a text-based stimulus.**
Four Free-Response Questions • 1 hour, 30 minutes • One data set, one document-based, and two synthesis questions	**Three** Free-Response Questions • **1 hour, 10 minutes** • **Q1: Design and analyze an investigation** • **Q2: Analyze an environmental issue and propose a solution using models and representation** • **Q3: Analyze an environmental issue and propose a solution using calculations**

Section I: The Multiple-Choice Section

Multiple-choice questions will come in two types: individual questions and set-based questions referring to the same diagram or data presentation. This first type of question that we are going to discuss is the set-based question, which you may already be familiar with.

Questions 1–3 refer to the diagram of the atmosphere below:

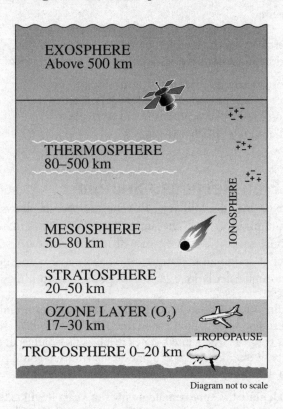

Diagram not to scale

1. Which layer contains the Earth's daily weather?

   (A) Troposphere
   (B) Stratosphere
   (C) Thermosphere
   (D) Mesosphere

2. The highest layer of the atmosphere heated by the IR radiation from the Earth is the

   (A) Stratopause
   (B) Mesosphere
   (C) Thermosphere
   (D) Troposphere

3. The approximate distance of the Mesosphere is which of the following?

   (A) 17 to 30 km
   (B) 50 to 80 km
   (C) 30 to 70 km
   (D) 20 to 50 km

This set of questions above are all related to the one diagram presented, so it is important that you reference the image to answer these particular questions. Within these types of set-based questions on the exam, you can expect to see stimulus material including diagrams, models, maps, data tables, charts, and graphs. We'll discuss the science underlying the questions later on, but in case you're interested—the answers are (A), (C), and (B).

The second type of multiple-choice question is more like the traditional multiple-choice questions you are used to seeing:

7. Salt intrusion into freshwater aquifers, beach erosion, and disruption of coastal fisheries all might occur as a result of

(A) rising ocean levels as global warming proceeds
(B) more solar ultraviolet radiation on the Earth
(C) more chlorofluorocarbons in the atmosphere
(D) reduced rates of photosynthesis

Section II: The Free-Response Section

Environmental Science is interdisciplinary. This means that it draws from several sciences (biology, chemistry, and physics) and the humanities (government, economics, and social studies). The free-response section of the exam will include three questions that will test your ability to do the following: design an investigation, analyze an environmental problem, propose a solution, and work through calculations. A free-response question may look like this:

3. According to the United States Energy Information Administration, the consumption of natural gas in the United States increases 8 percent per year. The United States receives its supplies of gas from a variety of international and domestic locations. Natural gas is used in the home, for industry, and for power generation.

(a) **Calculate** the approximate number of years it will take to double the consumption of natural gas. Show all work.
(b) **Describe** one method for the recovery and transportation of natural gas.
(c) **Describe** two benefits to the environment that would occur by switching from coal to natural gas-fired electric power generation.
(d) Some people advocate increasing the use of coal instead of natural gas for the production of electricity. **Explain** one argument that the proponents of coal might use to justify their position.

As you can see, for this multipart free-response question, your answers should not be one-dimensional; they will need to encompass many different subjects and areas of thought.

We'll talk more about how to go about writing your responses to these questions in Chapter 2, but for now, remember that you have only 70 minutes to answer all three questions. This translates to about 23 minutes per question, so any practice you can get before test day will be invaluable! Fortunately for you, we have put sample free-response questions and answers in each chapter of this book to give you that practice.

HOW THE AP ENVIRONMENTAL SCIENCE EXAM IS SCORED

What Will My Score Look Like, and What Will It Mean?

After taking the test in early May you will receive your score sometime around the first week of July, right around the time when you've just started to forget about the entire experience. Your score will be a single number from 1 to 5. Here's what those numbers mean.

May 2019 Score Results			
Score	What this score means	Approximate % of all test-takers receiving this score	Will a student with this score receive credit?
5	Extremely Qualified	9.5%	Yes
4	Well Qualified	25.9%	Most Likely
3	Qualified	14.2%	Maybe
2	Possibly Qualified	25.5%	Very Rarely
1	Not Qualified	24.9%	No

Those percentages are pretty intimidating, huh? Quite a few students get either a 1 or 2 on this exam. Why is this? Well, it's probably because these students were not willing to put the time and energy into studying and reviewing the necessary topics.

However, by purchasing this book, you've already proven that you aren't one of those students. No one said this test would be easy, but it is definitely manageable.

How Much Will Each Section Count Toward My Final Score?

As we mentioned, a computer will grade the multiple-choice section of your exam. Your final grade will be made up of the two sections, which will be given different weights: 60 percent of the grade will come from the 80 multiple-choice questions, and the remaining 40 percent will come from the three free-response questions, with about 13 percent coming from each of your responses.

Your Multiple-Choice Score

For this section of the test, you will be scored only on the number of questions you answer correctly. That's right, only the ones you get right are counted! Since there is no penalty for wrong answers, it is in your best interest to answer as many multiple-choice questions as possible. Although random guessing is better than nothing, you should use Process of Elimination (POE) to narrow down the choices first. We'll get into the details of POE in Chapter 1. If you are running out of time, remember, you need to fill in all the bubbles before time is up. If you don't have time for POE, just choose any letter and move on.

Your Free-Response Score

Each AP free-response question is scored on a scale from 0 to 10, with 10 being the best score. The scores of all three free-response questions are added together to obtain your free-response score.

At the beginning of the free-response question grading, rubrics are formulated by the chief reader, a university professor who also attends development committee meetings, and other AP Environmental Science leaders. These rubrics are refined again and again to provide the best possible scoring rubric for each question. Leaders for each question will direct readers and reread tests to make sure that every test is graded accurately and fairly according to the rubric. Computer-generated statistics aid the chief reader and the other leaders in this quality assurance, and each reader (either high-school AP Environmental Science teachers or university professors) is assigned to read only one free-response answer and becomes an expert on that question.

As you can imagine, reading a thousand or more test papers in eight days is an awesome task; therefore, anything you can do to make your free-response answers easier to read is appreciated! A well-organized and readable exam may receive a higher score because the reader can actually find the information he or she needs in order to award the earned points.

There are several different ways to earn those 10 points. However, you can never earn all 10 points unless you answer all parts of the free-response question. So, if the free-response question has 4 parts, you cannot score 10 points unless you answer all 4 parts of that question correctly!

We've said that the highest possible score on each question is a 10. Answers that receive a 10 demonstrate a clear and thorough knowledge of the material and include a superb response to all parts of the question. If the answer requires a calculation, the student performs the calculation showing all work and labeling all units. If it includes placing points on a graph or graphing an answer, the graph is drawn correctly, and the points are placed in the correct place, with all elements (including the *x*- and *y*-axes) labeled correctly.

Answers that receive a 9 also demonstrate a thorough knowledge of the material, but are docked a point due to sloppy work, such as not labeling graphs or not showing work when doing calculations. Answers that score a 7 or 8 do not give a satisfactory response to one part of the question, while answers that score a 3, 4, or 5 give an unsatisfactory response for more than one part.

In all cases, even if you don't think you know the answer to one or more questions, you should thoroughly read the question, complete the rest of the free-response questions in the section, and then come back to it.

To look at real questions from past exams and get late-breaking information directly from the College Board, visit apstudents.collegeboard.org.

Your Final Score

Your final 1-to-5 score is a combination of your section scores. Remember that the multiple-choice section counts for 60 percent of the total and that the free-response questions count for 40 percent. However, the bottom line is that both sections are very important, and you must concentrate on doing your best on both parts.

Keep in mind that even if you do not get college credit for this course, you will not have wasted your time. Research shows that just taking an AP course helps your college performance. Small consolation, perhaps, but as we mentioned earlier, because you're making this effort to prepare properly, you'll most likely get due credit.

But How Do I Get Credit If (I Mean, When) I Score a 4 or 5?

While your score will be sent to schools, you should keep a copy of your score and take it when you register for classes at the college or university you attend.

Also, as this is one of the newer AP courses—and an interdisciplinary course—you should probably keep the syllabus and your laboratory notebook from your AP Environmental Science course. You may need to show it to college or university counselors in order to get college credit for a science laboratory course.

OVERVIEW OF COURSE UNITS, PRACTICES, AND BIG IDEAS

As you may know, College Board has updated the AP Environmental Science Course and Exam Description. Below outlines the nine units and topics the College Board suggest to be covered in your AP course. While this book may not follow this exact outline, it does in fact cover all subjects necessary to prepare for the exam.

Unit 1: The Living World: Ecosystems (6–8%)
Introduction to Ecosystems
Terrestrial Biomes
Aquatic Biomes
The Carbon Cycle
The Nitrogen Cycle
The Phosphorus Cycle
The Hydrologic (Water) Cycle
Primary Productivity
Trophic Levels
Energy Flow and the 10% Rule
Food Chains and Food Webs

Unit 2: The Living World: Biodiversity (6–8%)
Introduction to Biodiversity
Ecosystem Services
Island Biogeography
Ecological Tolerance
Natural Disruptions to Ecosystems
Adaptations
Ecological Succession

Unit 3: Populations (10–15%)
Generalist and Specialist Species
K-Selected, r-selected Species
Survivorship Curves
Carrying Capacity
Population Growth and Resource Availability
Age Structure Diagrams
Total Fertility Rate
Human Population Dynamics
Demographic Transition

Unit 4: Earth Systems and Resources (10–15%)
Plate Tectonics
Soil Formation and Erosion
Soil Composition and Properties
Earth's Atmosphere
Global Wind Patterns
Watersheds
Solar Radiation and Earth's Seasons
Earth's Geography and Climate
El Niño and La Niña

Unit 5: Land and Water Use (10–15%)
The Tragedy of the Commons
Clearcutting
The Green Revolution
Impacts of Agriculture Practices
Irrigation Methods
Pest Control Methods
Meat Production Methods
Impacts of Overfishing
Impacts of Mining
Impacts of Urbanization
Ecological Footprints
Introduction to Sustainability
Methods to Reducing Urban Runoff
Integrated Pest Management
Sustainable Agriculture
Aquaculture
Sustainable Forestry

Unit 6: Energy Resources and Consumption (10–15%)
Renewable and Nonrenewable Resources
Global Energy Consumption
Fuel Types and Uses
Distribution of Natural Energy Resources
Fossil Fuels
Nuclear Power
Energy from Biomass
Solar Energy
Hydroelectric Power

Geothermal Energy
Hydrogen Fuel Cell
Wind Energy
Energy Conservation

Unit 7: Atmospheric Pollution (7–10%)
Introduction to Air Pollution
Photochemical Smog
Thermal Inversion
Atmospheric CO_2 and Particulates
Indoor Air Pollutants
Reduction of Air Pollutants
Acid Rain
Noise Pollution

Unit 8: Aquatic and Terrestrial Pollution (7–10%)
Sources of Pollution
Human Impacts on Ecosystems
Endocrine Disruptors
Human Impacts on Wetlands and Mangroves
Eutrophication

Thermal Pollution
Persistent Organic Pollutants (POPs)
Bioaccumulation and Biomagnification
Solid Waste Disposal
Waste Reduction Methods
Sewage Treatment
Lethal Dose 50%
Dose Response Curve
Pollution and Human Health
Pathogens and Infectious Diseases

Unit 9: Global Change (15–20%)
Stratospheric Ozone Depletion
Reduction Ozone Depletion
The Greenhouse Effect
Increases in the Greenhouse Gases
Global Climate Change
Ocean Warming
Ocean Acidification
Invasive Species
Endangered Species
Human Impacts on Biodiversity

In addition to the course units, the College Board also emphasizes four Big Ideas and seven Science Practices connecting the many topics you will learn throughout each unit. These are:

Big Ideas	Science Practices
BIG IDEA 1: Energy Transfer	Practice 1: *Concept Explanation*
	Practice 2: *Visual Representations*
BIG IDEA 2: Interactions between Earth Systems	Practice 3: *Text Analysis*
	Practice 4: *Scientific Experiments*
BIG IDEA 3: Interactions between different Species and the Environment	Practice 5: *Data Analysis*
	Practice 6: *Mathematical Routines*
BIG IDEA 4: Sustainability	Practice 7: *Environmental Solutions*

As you go through your AP course during the school year, make sure you are paying very close attention to how your AP teacher ties in each topic with the Big Ideas and Science Practices.

HOW AP EXAMS ARE USED

Different colleges use AP Exam scores in different ways, so it is important that you go to a particular college's website to determine how it uses AP Exam scores. The three items below represent the main ways in which AP Exam scores can be used:

Are You Preparing for College?

Check out all of the useful books from The Princeton Review, including *SAT Prep, ACT Prep, The Best 386 Colleges,* and more!

- **College Credit.** Some colleges will give you college credit if you score well on an AP Exam. These credits count toward your graduation requirements, meaning that you can take fewer courses while in college. Given the cost of college, this could be quite a benefit, indeed.

- **Satisfy Requirements.** Some colleges will allow you to "place out" of certain requirements if you do well on an AP Exam, even if they do not give you actual college credits. For example, you might not need to take an introductory-level course, or perhaps you might not need to take a class in a certain discipline at all.

- **Admissions Plus.** Even if your AP Exam will not result in college credit or allow you to place out of certain courses, most colleges will respect your decision to push yourself by taking an AP course or even an AP Exam outside of a course. A high score on an AP Exam shows mastery of more difficult content than is taught in many high-school courses, and colleges may take that into account during the admissions process.

OTHER RESOURCES

There are many resources available to help you improve your score on the AP Environmental Science Exam, not the least of which are your teachers. If you are taking an AP class, you may be able to get extra attention from your teacher, such as obtaining feedback on your free-response essays. If you are not in an AP course, reach out to a teacher who teaches science and ask if he or she will review your essays or otherwise help you with content.

Another wonderful resource is **AP Students**, the official student site of the AP Exams. The scope of the information at this site is quite broad and includes:

- the updated course description, which provides details on what content is covered and sample questions

- the latest AP Environmental Science Free-Response and Scoring Guidelines

- access to AP Classroom if you are enrolled in a course (teacher assistance required)

- free-response prompts from previous years and exam tips

The AP Students home page address is https://apstudent.collegeboard.org.

Finally, **The Princeton Review** offers Homework Help for the AP Environmental Science Exam. Our expert instructors can help you refine your strategic approach and add to your content knowledge. For more information, call 1-800-2REVIEW.

DESIGNING YOUR STUDY PLAN

In Part I, you identified some areas of potential improvement. Let's now delve further into your performance on Practice Test 1, with the goal of developing a study plan appropriate to your needs and time commitment.

Read the answers and explanations associated with the multiple-choice questions (starting on page 35). After you have done so, respond to the following questions:

- Review the Overview of Content Topics on pages 64–65, and, next to each one, indicate your rank of the topic as follows: "1" means "I need a lot of work on this," "2" means "I need to beef up my knowledge," and "3" means "I know this topic well."

- How many days/weeks/months away is your AP Environmental Science Exam?

- What time of day is your best, most focused study time?

- How much time per day/week/month will you devote to preparing for your AP Environmental Science Exam?

- When will you do this preparation? (Be as specific as possible: Mondays and Wednesdays from 3 P.M. to 4 P.M., for example.)

- Based on the answers above, will you focus on strategy (Part IV) or content (Part V) or both?

- What are your overall goals in using this book?

Study Breaks Are Important
Don't burn yourself out before test day. Remember to take breaks every so often— go for a walk, listen to a favorite album, or get some fresh air.

Part IV
Test-Taking Strategies for the AP Environmental Science Exam

Chapter 1
How to
Approach
Multiple-Choice
Questions

THE PRINCETON REVIEW APPROACH

There are basically two ways to prepare for the AP Environmental Science Exam.

- Know absolutely everything about everything. Bad idea.

- Review only what you need to know and tackle the test strategically. Good idea.

This is The Princeton Review's way—and the best way—to improve your score.

Rather than trying to teach you everything there is to know about environmental science, we at The Princeton Review focus on test-taking strategies. Naturally, we'll review some hard science as well. But rather than cluttering your brain, we'll look only at the environmental science you need to know for the test, explaining and highlighting key concepts along the way.

First, we'll give you some simple, straightforward strategies for tackling multiple-choice questions and for writing free-response answers. Let's now take a closer look at how to approach the multiple-choice section.

The Two-Pass System

Proven Techniques
The two-pass system allows you to pick up easy points from the start!

The AP Environmental Science Exam covers a broad range of topics. There's no way, even with our extensive review, that you will know everything about every topic in environmental science. So, what should you do?

Adopt a two-pass system. The two-pass system entails going through the test and answering the easy questions first. Save the more time-consuming questions for later. (Don't worry—you'll have time to do them later!) First, read the question and decide if it is a "now" or "later" question. If you decide this is a "now" question, answer it in the test booklet. If it is a "later" question, come back to it. Once you have finished all the "now" questions on a double page, transfer the answers to your bubble sheet. Flip the page and repeat the process.

Once you've finished all the "now" questions, move on to the "later" questions. Start with the easier questions first. These are the ones that require calculations or that require you to eliminate the answer choices (in essence, the correct answer does not jump out at you immediately). Transfer your answers to your bubble sheet as soon as you answer these "later" questions.

Watch Out for Those Bubbles!

Because you're skipping problems, you need to keep careful track of the bubbles on your answer sheet. One way to accomplish this is by answering all the questions on a page and then transferring your choices to the answer sheet. If you prefer to enter them one by one, make sure you double-check the number beside the ovals before filling them in. We'd hate to see you lose points because you forgot to skip a bubble!

Process of Elimination (POE)

It makes sense to assume that you need to know your material backward and forward in order to get the right answer. In other words, if you don't know the answer beforehand, you probably won't answer the question correctly. This is particularly true of fill-in-the-blank and essay questions. We're taught to think that the only way to get a question right is by knowing the answer. However, that's not the case on Section I of the AP Environmental Science Exam. You can get a perfect score on this portion of the test without knowing a single right answer—provided you know all the wrong answers!

What are we talking about? This is perhaps the most important technique to use on the multiple-choice section of the exam. Let's take a look at an example.

41. The long-term storage of phosphorus and sulfur occurs in which of the following?

   (A) Bacteria
   (B) Rocks
   (C) Water
   (D) Plants

Applied Strategies
It is often easier to identify a wrong answer than a right answer. Use POE to get rid of bad answers!

Now, if this were a fill-in-the-blank-style question, you might be in a heap of trouble. But let's take a look at what we've got. You see the elements phosphorus and sulfur in the question, which leads you to conclude that we're talking about elements. Right away, you can probably remember that these aren't normally components of water, so you can eliminate (C). Also, plants don't live a long time, so sulfur and phosphorus can't be stored for the long-term in plants, right? Get rid of (D). The same goes for bacteria, so lose (A). You're left with (B), the correct answer.

We think we've illustrated our point: Process of Elimination is the best way to approach the multiple-choice questions. Even when you don't know the answer right off the bat, you'll surely know that two or three of the answer choices are not correct. What then?

Aggressive Guessing

As mentioned earlier, you are scored only on the number of questions you get right, so we know guessing can't hurt you. But can it help you? It sure can. Let's say you guess on five questions; odds are you'll get one right. So, you've already increased your score by one point. Now, let's add POE into the equation. If you can eliminate as many as two answer choices from each question, your chances of getting them right increase and so does your overall score. Remember, don't leave any bubbles blank on test day!

Word Associations

Another way to rack up the points on the AP Environmental Science Exam is by using word associations in tandem with your POE skills. Make sure that you memorize all of the words in the Glossary, which is Chapter 12 of this book. Know them backward and forward. As you learn them, make sure you group them by "association," as you're bound to be tested on them on the AP Environmental Science Exam. What do we mean by "word associations"?

Let's take the example of air pollution. You'll soon see from our review, and possibly your course study, that there are several compounds associated with various types of air pollution. For example, ozone, VOC, and nitrogen oxides are all terms associated with air pollution. Now, take a look below at a typical question about pollution.

2. All of the following are important in smog production EXCEPT

(A) photochemical reactions
(B) stratospheric ozone
(C) tropospheric ozone
(D) volatile organic compounds

This might seem like a difficult question, but let's think about the associations we just discussed. The question asks us about smog. Choices (C) and (D) are all terms that we've associated with air pollution. Therefore, we can eliminate them. Maybe you're unsure about whether or not photochemical reactions are part of air pollution, but since you know for sure that stratospheric ozone has nothing to do with smog production (or for that matter, air pollution), you might guess that (A) is the correct answer (and you'd be right!).

We'll explain what these words mean in Chapter 9, in which we discuss pollution, but the point is that without even racking your brain, you've managed to get this down to two answer choices—not bad! You would have a fifty-fifty chance of guessing correctly on this question.

By combining the associations we'll offer throughout this book with aggressive POE techniques, you'll be able to rack up points on problems that might have seemed particularly difficult at first.

Mnemonics—or the Environmental Science Name Game

One of the big keys to simplifying biology is to organize terms into a handful of easily remembered packages. The best way to accomplish this is by using mnemonics. A mnemonic, as you may already know, is a convenient device, such as a rhyme or phrase, for remembering something. Environmental science is all about names: the names of chemicals, processes, theories, and more. How are you going to keep them all straight without a little help?

For example, the major components of air pollution are:

- **S**ulfur dioxide—SO_2

- **P**articulates

- **L**ead—Pb

- **O**zone—O_3

- **N**itrogen dioxide—NO_2

- **C**arbon monoxide—CO

The first letter of each component spells SPLONC, which is otherwise known as Some Pollution Lands On Nature Constantly. Learn the mnemonic and you'll never forget the science!

Mnemonics can be as goofy as you like, so long as they help you remember. Be creative! Remember, the important thing is that you remember the information, not how you remember it.

Identify Question Types

Many of the traps on the AP Environmental Science Exam deal with the way in which the question is asked. Here's information about a few types of multiple-choice questions you may see on the updated exam.

EXCEPT/NOT/LEAST Questions

Some of the multiple-choice questions in Section I may be EXCEPT/NOT/LEAST questions. With this type of question, you must remember that you're looking for the wrong (or the least correct) answer. The best way to approach these is by using POE.

More often than not, the correct answer is a true statement, but is wrong in the context of the question. Cross off the four that apply, and you're left with the one that does not. Here's an example of this type of question.

27. All of the following are components of integrated waste management EXCEPT

   (A) using canvas bags that can be reused rather than disposable bags
   (B) using old appliances for construction of artificial reefs
   (C) using disposable diapers instead of cloth diapers
   (D) using reused glass bottles

If you don't remember anything about integrated waste management, you should at least understand that the question is asking about waste. So, which of the choices does *not* deal with a way to reduce or reuse waste? Well, (C) would result in more, and not less, waste; and it is the correct answer. Remember, the best way to answer these types of questions is to spot all the right statements and cross them off. You'll wind up with the wrong statement, which happens to be the correct answer.

Unspecified One-or-More

Another type of multiple-choice question that might appear on the updated exam is called the Unspecified One-or-More question. These questions are designed to have you select all of the correct answers, though they do not prompt you on how many might be correct. In this case, you need to carefully analyze each answer, independent of the other answers. Be sure to consider each choice carefully, determine which ones are correct, and then look at the answer options to see which one corresponds with the selection of answers you have determined is correct. Here's an example of this type of question.

1. The rain shadow effect may cause which of the following?

   I. Drier conditions on the leeward side of mountain ranges
   II. Warmer conditions on the windward side of mountain ranges
   III. More light on the leeward side of mountain ranges

   (A) I only
   (B) II only
   (C) III only
   (D) I and III only

The correct answer is (A). Mountains cause warm, moist air to rise and compress, ultimately creating rain on the windward side of mountains. As a result, the leeward sides of mountains are much dryer. Therefore, the correct answer must have something to do with precipitation or moisture. In considering each answer independently, the first answer is clearly correct but the second two are not.

Chapter 1 Drill

Let's practice some of the different types of multiple-choice questions that you just learned about. For answers and explanations, see Chapter 13.

1. Once numbering more than 20,000 birds, the Hawaiian goose (or "Nene") was reduced to 30 individuals by 1918. After listing on the Endangered Species List in 1967, the population increased from 400 birds in 1980 to 1,600 in 2008. Calculate the annual growth rate from 1980 to 2008.

   (A) 13%
   (B) 7%
   (C) 5%
   (D) 2.5%

2. Why is the barrier island such a fragile habitat?

   I. Climate change may cause sea level rise that can cover the island.
   II. Coastal development may remove protective dunes and mangrove colonies that exist on some barrier islands.
   III. Storms can cause erosion of barrier islands.

   (A) I only
   (B) II only
   (C) I and II only
   (D) I, II, and III

3. Residents of suburbia may notice a decline in the population of house cats at the same time that they hear the yip of coyotes more frequently in their neighborhood. What type of population control mechanism is exhibited in this ecosystem?

   (A) Top-down control
   (B) Bottom-up control
   (C) Keystone succession control
   (D) Competitive advantage control

4. Marine hypoxic or "dead zones" caused by humans are

   (A) primarily caused by nutrient pollution
   (B) areas of such low oxygen concentration that animal life mostly dies
   (C) often attributed to agricultural runoff
   (D) all of the above

5. The greatest cause for fishery depletion and collapse is

   (A) invasive species
   (B) aquaculture technologies
   (C) pollution
   (D) overexploitation and overharvesting

6. All of the following sources produce non-point sources of pollution EXCEPT

   (A) large diesel shipping trucks
   (B) smokestack of a chemical manufacturing company
   (C) urban runoff from parking lots
   (D) agricultural waste from a cattle stockyard

7. In the year 2000, a country had a population of 10 million people with a birth rate of 6.3% and a death rate of 1.3%. If these rates remain constant and there is no migration, the population of that country will be close to 40 million in _____.

   (A) 2005
   (B) 2014
   (C) 2018
   (D) 2028

8. The Dust Bowl

   (A) was caused solely by a particularly bad drought
   (B) only affected farmers
   (C) was largely a result of deep-till farming practices on the Great Plains
   (D) was prevented because of the poor price of wheat in the 1920s

9. The nitrogen cycle may include all of the following processes EXCEPT

   (A) respiration
   (B) assimilation
   (C) fixation
   (D) ammonification

Chapter 2
How to
Approach
Free-Response
Questions

THE ART OF THE FREE-RESPONSE ESSAY

You're given three free-response questions to answer in 70 minutes. That's only about 23 minutes per question. Each of these three questions will present a scenario and ask you to answer several smaller questions (generally 3 to 4). You can get a maximum of 10 points per free-response question, and you need to answer every part of the question to get all 10 points. Each question has a certain grading rubric assigned to it, which is what the free-response readers use to give you points for your responses. The best way to rack up points on this section is to give the graders what they're looking for. Fortunately, we know precisely how to do this.

Now or Later?

For the updated exam, one of the free-response questions will ask you to design an investigation, and one will ask you to propose a solution to an environmental problem using models. The third question will ask you the same as the second, but this time using calculations, so a calculator can be very useful for this question.

While you do have to answer each of these questions, you do *not* need to answer them in order. The best strategy is to read the scenario (not the sub-questions) and decide if this is a question you want to attempt now or later. Do this before reading the next scenario. If you decide to do it later, move on to the next question.

If you decide to do it now, look at the questions and start answering them. Remember the grading rubric, and make it easy for the grader to give you points. If the question asks for two solutions, label the solutions (for example, "a" and "b") so the grader can easily find them. If the question asks for a calculation, show all your work, including any formulas you are using. If the question asks you to plot something, clearly label the *x*- and *y*-axes, and any relevant point or area on the chart.

New Calculator Policy!

Good news! You are now allowed to use a four-function, scientific, or graphing calculator on the entire AP Environmental Science exam. This is a great resource to help you on both the multiple-choice section and the free-response section. Just make sure to use it only when necessary!

Calculate the Math

Yes, there is math in environmental science and you will be allowed to use a calculator on the exam. So even if you're able to do the math in your head, it helps to know you can now double-check your work on a calculator when solving challenging questions!

Hot-Button Terms

The AP essay graders have a checklist of key terms and concepts that they use to assign points. We like to call these "hot-button" terms. Simply put, for each hot button that you include in your essay, you will receive a predetermined number of points. For example, if the essay question deals with photochemical smog, the AP graders are instructed to give students two points for writing: "In the presence of sunlight and heat, VOCs (volatile organic compounds), NO_x, and ozone combine to form smog"—or something very similar to that. So where do you find these key terms? Funny you should ask—they are at the end of each chapter in this book and in the Glossary in Chapter 12. Make sure you have a grasp of all of the words in those lists, and use them as hot buttons in your essays.

Make an Outline

As you read the question, brainstorm a list of terms and concepts you want to cover. Use the sub-questions to help you with your list of terms and concepts. Next, draft an outline that will help you organize them into some logical order. While you do not get points for organization, a well-organized essay is easier to write (and more importantly, easier to grade). The best way to organize your response is to write a clear, simple outline (just a couple of bullets per section). Outlining should take no more than about two to three minutes.

Of course, if you just composed a list of key scientific terms, you wouldn't be writing an essay. It is important to remember that the three free-response questions are essay questions, and they need to be written in paragraph style. An answer that's written as a list or an outline is not acceptable and will not be scored. On average, you will need to write no more than one or two paragraphs for each question.

If the question asks for two examples, give just that—two examples. If you present more than two examples, the grader may not even count them toward your score. Make sure you read carefully and provide just what the question asks for.

Label All Diagrams and Figures

Sometimes it's easier to present a diagram or figure as part of your essay. You may illustrate your answer, but all illustrations should be labeled and discussed in the verbiage of your answer. Remember to properly label your diagram or figure; otherwise the AP graders will give you no more than partial credit for your work.

Know the Labs Covered in Your AP Course

At least one of the four essay questions will be experimentally based. Sometimes the questions will refer back to a laboratory experiment conducted in your AP class. Consequently, the laboratory component of your course is an integral part of this exam. In Chapter 11, we'll review some of the laboratory experiments you may have performed in your AP Environmental Science class.

Outline Before You Write
You have about 23 minutes to answer each free-response question. Use the first 2 to 3 minutes to make an outline of your answer. This process will help you organize your thoughts and direct your writing.

Review Your Answers

After answering all parts of a question, give yourself a couple of minutes to review your answers before moving to the next question. Remember, once you are done with a question, you are DONE. Do not go back to a question you have completed—even if you have time at the end of the test.

Practice, Practice, and More Practice!

The only way to get good at writing an essay in 23 minutes is to keep at it. Try out the strategies you've learned on the practice questions found on the next page and in the free-response questions found at the end of each content chapter in Part V.

Chapter 2 Drill

Keeping in mind the things that we just covered when tackling free-response questions, try out some of these questions for practice. For answers and explanations, see Chapter 13.

1. Before 1880, the North American black-tailed prairie dog population thrived in the hundreds of millions and shaped the Great Plains temperate grassland ecosystem for over 200 other species of plants and animals observed living on or near prairie dog burrow colonies. But, by 1972, the population was estimated to be approximately 3,100 with the projection of becoming extinct by 2000. Since 1972, the population has increased from 3,100 individuals to approximately 12,400 in 2012.

    (a) **Calculate** the population growth rate from 1972 to 2012.
    (b) **Identify** and **describe** TWO major causes for the original decline of these species.
    (c) **Identify** and **describe** one likely ecological impact of the loss of black-tailed prairie dogs in the grassland biome.
    (d) Make ONE economic or ONE ecological **argument for** protecting the black-tailed prairie dog or another endangered species that you identify and one economic or one ecological **argument against** it.
    (e) **Identify** and **describe** one piece of United States or international legislation that is in place to prevent the decline of species or encourage the regrowth of species population.

2. The diagram below is of diversity downstream from the point of discharge from a sewage treatment plant.

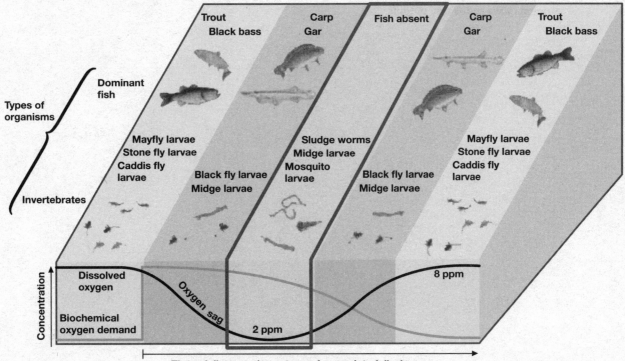

    (a) **Explain** the limiting factor that is responsible for these different zones, creating different biodiversity levels in each area.
    (b) **Explain** the health of the stream as you move downstream from the point of discharge based on what you notice about changes in biodiversity.
    (c) Sewage wastewater treatment plants mimic some of the same elements of wetlands in improving water quality before it reaches streams. **Explain** how wastewater treatment plants perform these equivalent responsibilities in the primary and secondary treatments.
    (d) **Describe** one possible pollutant to target during disinfection of wastewater before its final release back into nature and describe a method of sewage wastewater treatment that is meant to disinfect water.

3. Your growing city of 500,000 people has realized they need to produce another waste stream destination. You are appointed the chairman of the committee to make the decision of which is the best facility for your community. The two options open for discussion are either the development of a new sanitary landfill or an incineration facility. You must select only one of these options.

    (a) **Provide** TWO environmental reasons why your recommendation is better than the alternative.
    (b) **Identify** and **describe** ONE economic reason why your recommendation is better than the alternative.
    (c) A local recycling company has concerns about how your recommendation may affect their business. Use the assumptions below to answer the questions that follow. For each calculation, **show** all work.

Recycling company pay out	5 cents
Recycling income from wholesaler	7 cents
Mileage of recycling truck	10 miles per gallon
Fuel cost	$2 per gallon
Recycling truck daily distance travel	100 miles
Truck driver pay	$100 per day

    (i) **Calculate** the gross income the company would need to make weekly in order to earn a $350 profit.
    (ii) **Calculate** how many cans must be recycled to make a profit of $50 per day.

    (d) **Describe** TWO conservation measures (other than recycling) that the city could take to reduce the total amount of waste the city outputs.

Chapter 3
Using Time
Effectively
to Maximize
Points

BECOMING A BETTER TEST-TAKER

Very few students stop to think about how to improve their test-taking skills. Most assume that if they study hard, they will test well, and if they do not study, they will do poorly. Most students continue to believe this even after experience teaches them otherwise. Have you ever studied really hard for an exam, and then blew it on test day? Have you ever aced an exam for which you thought you weren't well prepared? Most students have had one, if not both, of these experiences. The lesson should be clear: factors other than your level of preparation influence your final test score. This chapter will provide you with some insights that will help you perform better on the AP Environmental Science Exam and on other exams, as well.

PACING AND TIMING

A big part of scoring well on an exam is working at a consistent pace. The worst mistake made by inexperienced or less savvy test-takers is that they come to a question that stumps them, and, rather than just skip it, they panic and stall. Time stands still when you're working on a question you cannot answer, and it is not unusual for students to waste five minutes on a single question (especially a question involving a graph or the word EXCEPT) because they are too stubborn to cut their losses. It is important to be aware of how much time you have spent on a given question and on the section you are working on. There are several ways to improve your pacing and timing for the test.

- **Know your average pace.** While you prepare for your test, try to gauge how long you take on 5, 10, or 20 questions. Knowing how long you spend on average per question will help you identify how many questions you can answer effectively and how best to pace yourself for the test.

- **Have a watch or clock nearby.** You are permitted to have a watch or clock nearby to help you keep track of time. It is important to remember, however, that constantly checking the clock is in itself a waste of time and can be distracting. Devise a plan. Try checking the clock after every 15 or 30 questions to see if you are keeping the correct pace or need to speed up. This will ensure that you are cognizant of the time but will not permit you to fall into the trap of dwelling on it.

- **Know when to move on.** Since all questions are scored equally, investing appreciable amounts of time on a single question is inefficient and can potentially deprive you of the chance to answer easier questions later on. If you are able to eliminate answer choices, do so, but don't worry about picking a random answer and moving on if you cannot find the correct answer. Remember, tests are like marathons; you do best when you work through them at a steady pace. You can always come back to a question you don't know. When you do, very often you will find that your previous mental block is gone, and you will wonder why the question perplexed you the first time around (as you gleefully move on to the next question). Even if you still don't know the answer, you will not have wasted valuable time you could have spent on easier questions.

- **Be selective.** You don't have to do any of the questions in a given section in order. If you are stumped by an essay or multiple-choice question, skip it or choose a different one. Select the questions or essays that you can answer and work on them first. This will make you more efficient and give you the greatest chance of getting the most questions correct.

- **Use Process of Elimination on multiple-choice questions.** Many times, one or more answer choices can be eliminated. Every answer choice that can be eliminated increases the odds that you will answer the question correctly. Review the section on this strategy in Chapter 1 to find these incorrect answer choices and increase your odds of getting the question correct.

Remember, when all the questions on a test are of equal value, no one question is that important. Your overall goal for pacing is to get the most questions correct. Finally, you should set a realistic goal for your final score. In the next section, we will break down how to achieve your desired score and ways of pacing yourself to do so.

GETTING THE SCORE YOU WANT

Depending on the score you need, it may be in your best interest *not* to try to work through every question. Check with the schools to which you are applying.

Years ago, AP Exams eliminated the "guessing penalty" of a quarter of a point for every incorrect answer. Instead, students are assessed only on the total number of correct answers. It is really important to remember that if you are running out of time, you should fill in all the bubbles before the time for the multiple-choice section is up. Even if you don't plan to spend a lot of time on every question and even if you have no idea what the correct answer is, it's to your advantage to fill something in.

TEST ANXIETY

Everybody experiences anxiety before and during an exam. To a certain extent, test anxiety can be helpful. Some people find that they perform more quickly and efficiently under stress. If you have ever pulled an all-nighter to write a paper and ended up doing good work, you know the feeling.

However, too much stress is definitely a bad thing. Hyperventilating during the test, for example, almost always leads to a lower score. If you find that you stress out during exams, here are a few preemptive actions you can take.

- **Take a reality check.** Evaluate your situation before the test begins. If you have studied hard, remind yourself that you are well prepared. Remember that many others taking the test are not as well prepared, and (in your classes, at least) you are being graded against them, so you have an advantage. If you didn't study, accept the fact that you will probably not ace the test. Make sure you get to every question you know something about. Don't stress out or fixate on how much you don't know. Your job is to score as high as you can by maximizing the benefits of what you do know. In either scenario, it is best to think of a test as if it were a game. How can you get the most points in the time allotted to you? Always answer questions you can answer easily and quickly before you answer those that will take more time.

- **Try to relax.** Slow, deep breathing works for almost everyone. Close your eyes, take a few slow, deep breaths, and concentrate on nothing but your inhalation and exhalation for a few seconds. This is a basic form of meditation, and it should help you to clear your mind of stress and, as a result, concentrate better on the test. If you have ever taken yoga classes, you probably know some other good relaxation techniques. Use them when you can. (Obviously, anything that requires leaving your seat and, say, assuming a handstand position won't be allowed by any but the most free-spirited proctors.)

- **Eliminate as many surprises as you can.** Make sure you know where the test will be given, when it starts, what type of questions are going to be asked, and how long the test will take. You don't want to be worrying about any of these things on test day or, even worse, after the test has already begun.

The best way to avoid stress is to study both the test material and the test itself. Congratulations! By buying or reading this book, you are taking a major step toward a stress-free AP Environmental Science Exam.

Work Hard, Play Hard
Remember to give yourself small rewards as you prepare for the AP Environmental Science Exam.

Part V
Content Review for the AP Environmental Science Exam

HOW TO USE THE CHAPTERS IN THIS PART

You may need to come back to the following chapters more than once. Your goal is to obtain mastery of the content you are missing, and a single read of a chapter may not be sufficient. At the end of each chapter, you will have an opportunity to reflect on whether you truly have mastered the content of that chapter.

Guess What?

You're about to embark on a comprehensive content review of AP Environmental Science. It's a lot to remember, and we can help! Check out the study guides in your Student Tools, which you can access by registering your book. Just follow the steps on the Get More (Free) Content page (page viii).

Chapter 4
Earth's Interdependent Systems

In this first chapter, we review planet Earth's structure and the materials from which it is made, especially those used by humans as resources for our needs and economic gain. This overview serves as a foundation for the chapters that follow, so pay close attention! According to the College Board, about 10 to 15 percent of the test is based directly on the content covered in this chapter. If you are unfamiliar with a topic presented here, consult your textbook for more in-depth information.

> The College Board calls this section "Earth Systems and Resources" in the new Course and Exam Description.

The chapter starts with a brief Welcome to Planet Earth providing basic information on the planet's age and location in the solar system. The remaining topics correspond to the four physical spheres that make up our planet and regulate life on Earth—the materials and structure of these spheres as well as the mechanics of how each one works. We also discuss the importance of each sphere to humans. The four spheres are:

- The **Solid Earth**—Earth's solid, rocky outer shell.
  - Topics: the movement of tectonic plates, volcanoes and earthquakes, and the different types of rock.
  - The upper shell of the Solid Earth is called the **Lithosphere** and is the part that interacts most with the other spheres.

- The **Atmosphere**—the envelope of gases that surrounds Earth.
  - Topics: the greenhouse effect, climate, and weather events.

- The **Hydrosphere**—Earth's oceans and freshwater bodies.
  - Topics: ocean zones, ocean currents, and water conservation issues.

- The **Pedosphere**—more commonly known as soil.
  - Topics: soil's makeup, soil development, and soil conservation issues.

We've broken down discussion of the four spheres into neat sections for review, and we'll go through everything you need to know about these systems for test day.

The four physical spheres provide resources that support the Biosphere, the fifth of Earth's spheres. The biosphere is comprised of all the living organisms that inhabit the planet and draw on the physical resources of the other four spheres. We'll discuss the biosphere in depth in the next chapter.

As illustrated below, all five spheres interact to shape the variety of landforms, biomes, and phenomena that make environmental science such an endlessly fascinating area of study!

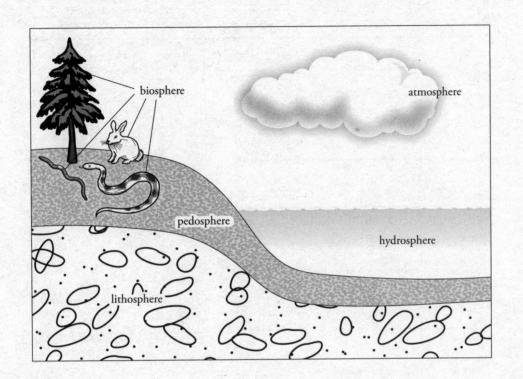

WELCOME TO PLANET EARTH

The first thing you should know about Earth is its history. Earth is thought to be between 4.5 and 4.8 billion years old. That amount of time is pretty inconceivable to humans, but the following **geologic time scale** will help you get a sense of the vast amount of time that has gone by since Earth was formed. You will not be responsible for memorizing all of the eons, eras, periods, and epochs for this exam, but you should be familiar with the major ones—they will come in handy.

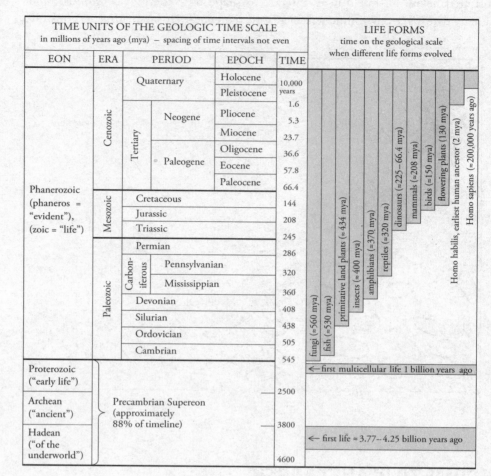

Figure 1: Geologic Scale

Here are some important takeaways from this table.

- We are currently in the Holocene Epoch.

- The Quaternary and Tertiary are the two most recent geologic periods.

- Non-avian dinosaurs lived during the Mesozoic Era. (Note that evolutionary biologists consider birds to be avian, or flying, dinosaurs. Pterosaurs such as pterodactyls are a group of reptiles less closely related to dinosaurs and birds.)

- The Precambrian eons represent the vast majority of the geologic time scale.

- The next epoch will be called the Anthropocene, to recognize humankind's accelerating effects on the planet's physical resources, climate, and life forms. Many geologists propose that we have already entered this new, post-Holocene epoch.

Where Is Earth in the Solar System?

Earth is the third planet from the sun in our solar system, which contains a total of eight currently known and recognized planets. From the sun outward, the planets are Mercury, Venus, Earth, Mars, Jupiter, Saturn, Uranus, and Neptune.

Each planet has its own orbit around the sun in the shape of an ellipse (a "stretched" circle). And you probably already know that it takes Earth about $365\frac{1}{4}$ days, or 1 year, to complete its orbit of the sun.

The Solid Earth

Planet Earth is made up of three concentric zones of rocks that are either solid or liquid (molten). The innermost zone is the core. The core has two parts: a solid inner core and molten outer core. The inner core is composed mostly of nickel and iron and is solid due to the tremendous pressure from overlying matter. The outer core is composed mostly of iron, also mixed with nickel as well as some lighter elements, and is semi-solid due to lower pressure. Surrounding the outer core is the **mantle**, which is made mostly of solid rock. Near the top of the mantle lies a layer of slowly flowing rock called the **asthenosphere**. The **lithosphere**, a thin, rigid layer of rock, is the Earth's outer shell. The lithosphere includes the rigid upper mantle above the asthenosphere and the **crust**, the solid surface of the Earth. Think of it as floating atop the asthenosphere like a cracker atop a thick layer of hot pudding.

Memorize the order and characteristics of Earth's layers; they will be asked about on test day!

The following diagrams show the chemical and physical properties of Earth's layers and a detail of the lithosphere.

Earth's Layers

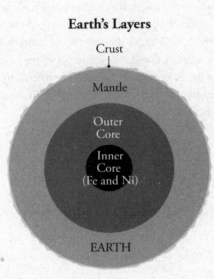

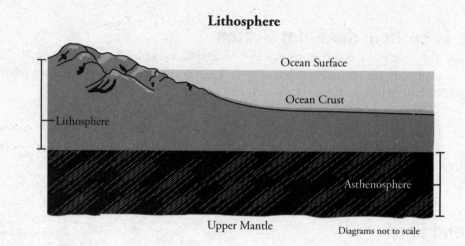

Lithosphere

Diagrams not to scale

Tectonic Plates

Scientists theorize that during the Paleozoic and Mesozoic Eras, the continents were joined together, forming a supercontinent known as **Pangaea**. Roughly 200 million years ago, Pangaea began to break apart.

> **Fast Fact**
> In ancient Greek, the word *pan* meant whole, while the word *gaia* meant Earth. Thus, Pangaea means the whole Earth!

Today, it is believed that the Earth's crust is composed of several large pieces of lithosphere—called **tectonic plates**—that move slowly over the mantle of the Earth. There are a total of a dozen or so tectonic plates that move independently of one another. The majority of the land on Earth sits above six giant plates; the remainder of the plates lie under the ocean as well as the continents.

Some plates consist only of ocean floor, such as the Nazca plate, which lies off the west coast of South America, while others contain both continental and oceanic material. One example of the latter is the North American plate, where the United States is located; this plate extends out to the mid-Atlantic ridge. There is even a plate that is located exclusively within the Asian continent; its boundaries nearly coincide with those of Turkey. The largest plate is the Pacific plate—it primarily consists of ocean floor, but also includes Mexico's Baja Peninsula and southwestern California. The major plates of Earth are shown on the map on the following page.

Earth's Plates

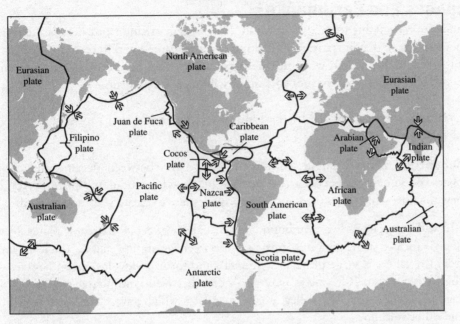

The edges of the plates are called **plate boundaries**, and the places where two plates abut each other are where events like sea floor spreading and most volcanoes and earthquakes occur. There are three types of plate boundary interactions:

- **Convergent boundary:** Two plates are pushed toward and into each other. One of the plates slides beneath the other, pushed deep into the mantle.

- **Divergent boundary:** Two plates move away from each other. This creates a gap between plates that may be filled with rising magma (molten rock). When this magma cools, it forms new crust.

- **Transform fault boundary:** Two plates slide against each other in opposite directions—as when you rub your hands back and forth to warm them up. These are also called simply transform boundaries.

So, what happens when plates collide? It depends on whether the collision happens between two oceanic boundaries, between two continental boundaries, or at an oceanic-continental boundary. Converging ocean-ocean and converging ocean-continent boundaries often result in **subduction,** in which a heavy ocean plate is pushed below the other plate and melts as it encounters the hot mantle. Converging continent-continent boundaries result in orogeny, the uplifting of plates that form large mountain chains as they crunch into each other. Examples include the Himalayas (which were created by a collision between the plate carrying India and the Asian plate), the Urals of western Russia, the Alps of southern Europe, and the Appalachian Mountains of the Eastern U.S.

One important result of plate movement is the creation of volcanoes and earthquakes. Let's examine those next.

Volcanoes and Earthquakes

Volcanoes are mountains formed by pressure from magma rising from Earth's interior. **Active volcanoes** are those that are currently erupting or have erupted within recorded history (that is, within the last 10,000 years), while **dormant volcanoes** have not been known to erupt during this period. It's thought that **extinct volcanoes** will never erupt again.

Active volcanoes are categorized by the kind of tectonic event that produces them. They are associated with the following:

- **Subduction zones** occur at convergent boundaries between oceanic and continental plates, or sometimes between two oceanic plates. The subducting plate is recycled into new magma, which rises through the overlying plate to create volcanoes inland.

- **Rift valleys** occur at divergent boundaries, usually between two oceanic plates. New ocean floor is formed as magma fills in the gap between separating plates. Thick magma rising from rift valleys is made of basaltic minerals and forms pillow lava upon contact with the cold ocean water. Rift valleys may also occur between continental plates; a prominent example is the Great Rift Valley of eastern Africa, which gave rise to Mount Kilimanjaro and other volcanoes.

- **Hot spots** do not form at plate boundaries. Instead, they are found in the middle of tectonic plates, in locations where columns of unusually hot magma melt through the mantle and weaken the Earth's crust. The Hawaiian islands continue to form over a hot spot beneath the Pacific plate. Volcanoes over oceanic hot spots are basaltic, resulting in milder eruptions; while volcanoes over continental hot spots are characterized by rhyolitic rocks, which produce more violent eruptions.

There are four types of volcanoes:

- **Shield volcanoes** have a broad base and are tall with gentle slopes. They generally form over oceanic hot spots and usually have mild eruptions with slow lava flow. Sometimes, however, when water enters the vent, they can be very explosive, forming pyroclastic flows, a fluidized mixture of hot ash and rock.

- **Composite volcanoes** have a broad base and are also tall but with steeper slopes. They are formed at subduction zones and are associated with violent eruptions that eject lava, water, and gases as superheated ash and stones.

- **Cinder volcanoes** are small, short, and steeply sloped cones. They form when molten lava erupts and cools quickly in the air, hardening into porous rocks (called cinders or scoria) that fracture as they hit Earth's surface. Cinder volcanoes generally form near other types of volcanoes.

- **Lava domes** are small and short with steep slopes and a rounded top. They are formed from lava that is too viscous to travel far but instead hardens into a dome shape. This type of volcano occurs near or even inside other types of volcanoes.

Earthquakes are the result of vibrations (often due to sudden plate movements, such as stress overcoming a locked fault) deep in the Earth that release stored energy. Earth's tectonic plates move slowly all the time, at about the same pace as fingernail growth, but earthquakes are very sudden movements. They often occur as two plates slide past one another at a transform boundary. The **focus** of the earthquake is the location at which it begins within the Earth, and the initial surface location of the event is the **epicenter.** The size, or magnitude, of earthquakes is measured by using an instrument known as a **seismograph,** which was devised by Charles Richter in 1935. The **Richter scale** measures the amplitude of the highest S-wave of an earthquake. Observed values range from 0 to 9.5, although theoretically there is no maximum value. Each increase in Richter number corresponds to an increase of approximately 33 times the energy of the previous number.

> **Fast Fact**
> An **S-wave**, also known as a shear **wave**, is a seismic body **wave** that shakes the ground up and down or side to side, perpendicular to the direction the **wave** is moving.

In January 2010, an earthquake of magnitude 7.0 struck the nation of Haiti. The quake's epicenter occurred in the boundary region separating the Caribbean plate and the North American plate. Official estimates from the U.S. Geological Service had 222,570 people killed, 300,000 injured, 1.3 million displaced, 97,294 houses destroyed, and 188,383 damaged across the Port-au-Prince area and much of southern Haiti. This includes at least four people killed by a local tsunami in the Petit Paradis area near Léogâne. **Tsunamis** are very large waves, or chains of waves, caused by the movement of the Earth during an earthquake or volcanic eruption and can be extremely destructive.

In March 2011, Japan suffered an earthquake of 9.0 magnitude off the eastern coast, near Sendai. This quake caused a massive tsunami wave that was 33 feet high. Both the earthquake and tsunami caused extensive damage: many buildings, roads, and railways were destroyed; major fires occurred; villages, thousands of homes, and people were washed away; and at least three nuclear power plants experienced dangerous explosions.

The Rock Cycle

Rocks are all around us, in the soil, our buildings, and the ore used in industry. So, where do all those rocks come from? The answer is: other rocks. The oldest rocks on Earth are 3.8 billion years old, while others are only a few million years old. This means that rocks are recycled. These transformations are described by the **rock cycle**. In the rock cycle, time, pressure, and the Earth's heat interact to create three basic types of rocks.

- **Igneous** rock results when rock is melted (by heat and pressure below the crust) into a liquid and then resolidifies when cooled. The molten rock (**magma**) comes to the surface of the Earth, and when it emerges it is called lava; cooled lava becomes solid igneous rock. An example of an igneous rock is basalt.

- **Sedimentary** rock is formed as sediment (eroded rocks and the remains of plants and animals) builds up and is compressed. Sedimentary rock forms under water as sediments or dissolved minerals deposit on a stream bed or ocean floor. They are compressed as more material is deposited and then cemented together. An example of a sedimentary rock is limestone.

- **Metamorphic** rock is formed as a great deal of pressure and heat produces physical and/or chemical changes in existing rock. This can happen as sedimentary rocks sink deeper into the Earth and are heated by the high temperatures found in the Earth's mantle. An example of a metamorphic rock is slate which results from the metamorphosis of shale.

The diagram below illustrates the rock cycle. Make sure you are familiar with it before the exam!

The Rock Cycle

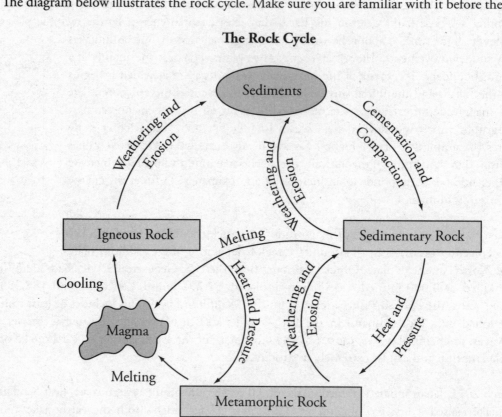

THE ATMOSPHERE

In the broadest definition, the atmosphere is a layer of gases that's held close to Earth by the force of gravity. The inner four layers of the atmosphere reach an altitude that's just about 12.5% of Earth's radius. The layer of gases that lies closest to Earth is the **troposphere;** it extends from Earth's surface to about, on average, 12 km (7.5 miles) at the poles and 20 km (12.4 miles) at the equator. The troposphere is where all the weather that we experience takes place. The layer also contains 99% of the atmosphere's water vapor and clouds. Generally, the troposphere is well-mixed from bottom to top—with the exception of periodic temperature inversions. The troposphere gets colder with altitude, decreasing 6.5°C for every kilometer of altitude (or 3.5°F for every thousand feet).

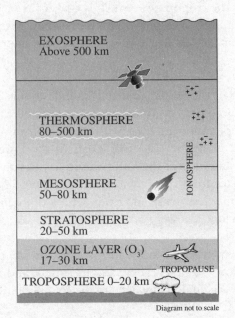

EXOSPHERE
Above 500 km

THERMOSPHERE
80–500 km

MESOSPHERE
50–80 km

STRATOSPHERE
20–50 km

OZONE LAYER (O$_3$)
17–30 km

TROPOPAUSE

TROPOSPHERE 0–20 km

IONOSPHERE

Diagram not to scale

Make sure you have a solid grasp of the order and characteristics of the layers of Earth's atmosphere!

Because of its density, the troposphere contains about 75–80% of Earth's atmosphere by mass. You've probably heard about the troposphere before in the news because of the **greenhouse effect.** The troposphere contains the air we breathe, which is made up of 78% nitrogen and 21% oxygen. The remaining 1% includes the so-called "greenhouse" gases (GHGs). The proportion of these gases in the troposphere is minuscule, but their effects on conditions on Earth are disproportionately significant; the most important of which are water vapor (H_2O), carbon dioxide (CO_2), and methane (CH_4). As the sun's rays strike Earth, some of the solar radiation is reflected back into space; however, greenhouse gases in the troposphere intercept and absorb a lot of this radiation. This warming effect of greenhouse gases was a good thing, until their concentration in the atmosphere shot up after the Industrial Revolution. We'll further investigate the greenhouse effect in Chapter 9, but for now take a look at the following figure.

The Greenhouse Effect

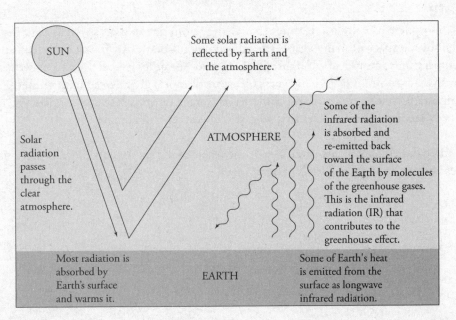

SUN

Some solar radiation is reflected by Earth and the atmosphere.

Solar radiation passes through the clear atmosphere.

ATMOSPHERE

Some of the infrared radiation is absorbed and re-emitted back toward the surface of the Earth by molecules of the greenhouse gases. This is the infrared radiation (IR) that contributes to the greenhouse effect.

Most radiation is absorbed by Earth's surface and warms it.

EARTH

Some of Earth's heat is emitted from the surface as longwave infrared radiation.

Crowning the troposphere is the **tropopause,** which is a layer that acts as a buffer between the troposphere and next layer up, the stratosphere. This buffer zone is where the jet streams travel, air currents that are important drivers of weather patterns and important factors in planning airline routes.

The **stratosphere** sits on top of the tropopause and extends about 20–50 km (7.5–31 miles) above Earth's surface. As opposed to conditions in the troposphere, gases in the stratosphere are not well mixed and temperatures increase with distance from the Earth. This warming effect is due to a thin band of **ozone** (O_3) that exists in the lower half of this layer. The ozone traps the high-energy radiation of the sun, holding some of the heat and protecting the troposphere and Earth's surface from this radiation. The stratosphere is similar to the troposphere in gas composition, only less dense and drier, with a thousand times less water vapor. Commercial jets may also fly in the lower part of this layer.

Above the stratosphere are two layers called the mesosphere and the thermosphere. The **mesosphere** extends to about 80 km (50 miles) above Earth's surface and is the area where meteors usually burn up. Temperatures again decrease here, to the coldest point in the atmosphere at the top of this layer, around –90°C (–130°F).

The **thermosphere** extends from 80 to around 500 km above the Earth (50–435 miles). Gases are very thin (rare) and it's in this layer that the spectacular and colorful auroras (northern lights and southern lights) take place. The furthest layer is the exosphere, extending to 10,000 km (6,200 miles) or more above the Earth, although the upper limit of this layer is not definitively settled. The concentration of gases is thinnest here. Human-made satellites orbit in the exosphere and in the upper thermosphere. The **ionosphere** is not a distinct layer but dispersed throughout the upper mesosphere, the thermosphere, and the lower exosphere. The ionosphere comprises regions of ionized gases that absorb most of the energetic charged particles from the sun—the protons and electrons of the solar wind. Interestingly, the ionosphere also reflects radio waves, making long-distance radio communication possible. You'll need to know how the climates that we experience on Earth are created by the atmosphere, so let's go into this next.

Climate

Earth's atmosphere has physical features that change from day to day as well as patterns that are consistent over a space of many years. The day-to-day properties such as wind speed and direction, temperature, amount of sunlight, pressure, and humidity are referred to as **weather**. The patterns that are constant over many years (30 years or more) are referred to as **climate**. The two most important factors in describing climate are average temperature and average precipitation amounts. **Meteorologists** are scientists who study weather and climate.

The weather and climate of any given area is the result of the sun unequally warming Earth (and the gases above it) as well as Earth's rotation.

Air Circulation in the Atmosphere

The motion of air around the globe is the result of solar heating, the rotation of Earth, and the physical properties of air, water, and land. There are three major reasons why Earth is unevenly heated.

- More of the sun's rays strike the Earth at the equator in each unit of surface area than strike the poles in the same unit area. This is because the angle of the sun's rays strikes the Earth more directly at the equator.

- The tilt of Earth's axis points regions toward or away from the sun. When pointed toward the sun, those areas receive more direct or intense light than when pointed away. This causes the seasons.

- Earth's surface at the equator is moving faster than at the poles, because the circumference is larger but the rotation time is the same. Because an object or air mass nearer to the equator is moving more rapidly (from east to west), it will maintain this eastward momentum as it moves away from the equator where the surface is moving more slowly, winding up further east. Therefore, winds moving north from the equator near the surface are deflected to the right or east, and winds moving south from the equator are deflected to the left or east. Conversely, winds blowing toward the equator will be deflected to the west because they are moving eastward more slowly than is the surface in the lower latitudes where they are moving to. So, in the Northern Hemisphere, this westward deflection will be to the right, and in the Southern Hemisphere, it will be to the left. This deflection pattern is known as the **Coriolis effect**. The resulting wind patterns are known as the **prevailing winds**, belts of air that distribute heat and moisture unevenly around the globe.

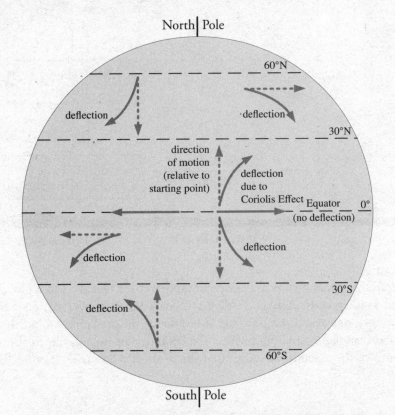

Deflection of wind due to the Coriolis Effect. Deflection is to the right in the Northern Hemisphere and to the left in the Southern Hemisphere. Winds heading due east and due west are still deflected, except at the equator. This is due to the interaction of rotational momentum in a circle, centrifugal force (directed outward from the Earth's axis), and gravity (directed inward toward Earth's center).

Solar energy warms Earth's surface. The heat is transferred to the atmosphere by radiation heating. The warmed gases expand, become less dense, and rise, creating vertical air flow called **convection currents**. The warm currents can also hold a lot of moisture compared to the surrounding air. As these large masses of warm, moist air rise, cool air flows along Earth's surface to occupy the area vacated by the warm air. This flowing air or **horizontal airflow** is one way that surface winds are created.

As warm, moist air rises into the cooler atmosphere, it cools to the **dew point**, the temperature at which water vapor condenses into liquid water. This condensation creates clouds. If condensation continues and the water drops get bigger, they can no longer be held up by the convection in Earth's atmosphere and they fall as **precipitation** (which can be frozen or liquid). This cold, dry air is now denser than the surrounding air. This air mass then sinks to Earth's surface, where it is warmed and can gather more moisture, thus starting the **convection cell** rotation again.

Convection Cell

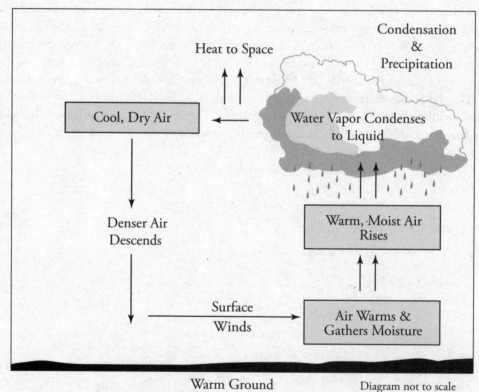

Diagram not to scale

On a local level, this phenomenon accounts for land and sea breezes. On a global scale, these cells are called **Hadley cells**. A large Hadley cell starts its cycle over the equator, where the warm, moist air evaporates and rises into the atmosphere. The precipitation in that region is one cause of the abundant equatorial rainforests. The cool, dry air then descends about 30 degrees north and south of the equator, forming the belts of deserts occurring around Earth at those latitudes.

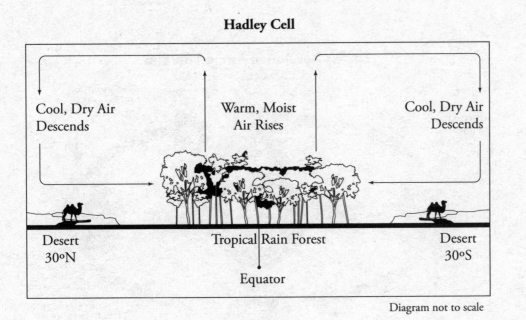

Hadley Cell

Cool, Dry Air Descends

Warm, Moist Air Rises

Cool, Dry Air Descends

Desert 30°N

Tropical Rain Forest

Desert 30°S

Equator

Diagram not to scale

Seasons

The motion of Earth around the sun and Earth's axial tilt of 23.5 degrees together create the seasons that we experience on Earth. It varies because Earth's tilt means that it hits most directly, and for the longest number of hours per day, on the parts of Earth that face the sun most directly, which changes across Earth's orbit. In other words, the main source of energy for a given place on Earth (the sun's rays) varies depending on latitude and season. Any place will receive the most solar radiation on its longest summer day and the least on its coldest winter day. The Earth's **revolution**, or trip around the sun, takes 365.25 days. The calendar that we use is based on the Earth's revolution around the sun: every four years, we have a leap year with one extra day to account for the accumulation of four quarter-days. As the Earth revolves around the sun, it is spinning on its **axis**; this spinning action is called **rotation**. On Earth, it takes 24 hours, or one day, to make a complete rotation. As you can see in the figure below, the Earth's revolution around the sun is the reason we have seasons.

Interestingly, because of the Earth's tilt, the sun rises and sets just once a year at the North and South Poles. Approximately six months of the year at the poles are daytime, while the other six months are dark and considered nighttime. (See the figure on the next page.)

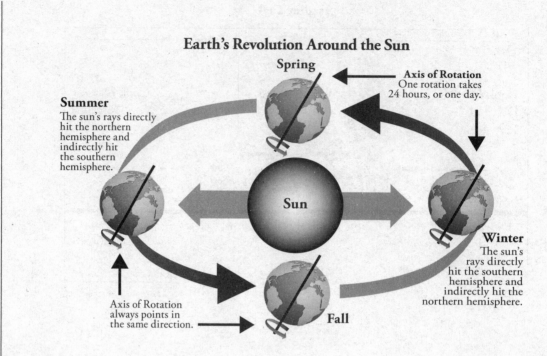

Earth's Revolution Around the Sun

Spring

Axis of Rotation
One rotation takes
24 hours, or one day.

Summer
The sun's rays directly
hit the northern
hemisphere and
indirectly hit
the southern
hemisphere.

Sun

Winter
The sun's
rays directly
hit the southern
hemisphere and
indirectly hit the
northern hemisphere.

Axis of Rotation
always points in
the same direction.

Fall

Albedo (Reflectance)

Another important property that affects the climate of different regions on the Earth is **albedo,** the percentage of insolation reflected by a surface. The lower the surface albedo, the more solar radiation that is absorbed. An albedo value of 0 corresponds to zero reflectance and absorption of all radiation, whereas an albedo value of 1 corresponds to reflection of all incoming radiation. Snow and ice have high albedo values, while land and trees have lower albedo values. Changes in the albedo can lead to alterations in the temperature. For example, snowfall may raise the albedo of an area, leading to an increase in the reflection of solar radiation and a decrease in temperature.

Let's move on to discuss wind. You might be thinking, "I know what wind is, so I can skip this section!" Don't skip it! The AP Environmental Science Exam may ask you about the specific types of winds and air movements coming up next—so it's better to be safe than sorry.

Types of Winds

So, what is "wind"? Why does everyone refer to wind when they're discussing weather? Well, the term "wind" is widely used to refer to air currents, and we already know that air currents tend to flow from regions of high pressure to regions of low pressure. But let's review some important details you'll need to know about wind before we move on to our review of the hydrosphere. Formally speaking, **wind** is air that's moving as a result of the unequal heating of Earth's atmosphere. It is part of Earth's circulatory system and moves heat, moisture, soil, and even pollution around the planet.

One crucial wind-related phenomenon that you'll need to know about for the AP Environmental Science Exam is trade winds. **Trade winds** were named for their ability to quickly propel trading ships across the ocean. The trade winds that blow between about 30 degrees latitude and the equator are steady and strong, and travel at a speed of about 11 to 13 mph. They are caused by the surface currents of the Hadley cells, described above, along with Earth's direction of rotation (counterclockwise if viewed toward the North Pole). In the Northern Hemisphere, the trade winds blow from the northeast and are known as the **Northeast Trade Winds;** in the Southern Hemisphere, the winds blow from the southeast and are called the **Southeast Trade Winds.**

Another important type of moving air mass, called a **westerly,** named for the direction from which it originates, travels north and east in the Northern Hemisphere and south and east in the Southern Hemisphere near the equator (between 30 degrees and 60 degrees). The movement of air that accounts for the westerlies, called the Ferrel cell, is the reverse of the Hadley cell but operates on the same thermodynamic principles. The eastward movement of westerlies are a result of the Coriolis effect. **Polar easterlies** are formed by similar forces: in polar easterlies, winds between latitudes 60 degrees and the North Pole blow from the north and east, and winds between 60 degrees and the South Pole blow from the south and east.

It's important that you can distinguish among the different types of wind and blow through the wind questions on test day!

Between the wind belts mentioned above, air movement is less predictable, and often no wind blows at all for days. For example, between about 30 degrees to 35 degrees north and 30 degrees to 35 degrees south of the equator lies the region known as the **horse latitudes** (or the subtropical high). Subsiding dry air and high pressure result in very weak winds in this region. Some people say that sailors gave the region of the subtropical high the name "horse latitudes" because ships relying on wind were unable to sail in these areas—so, afraid of running out of food and water, sailors would throw their horses (and other live cargo) overboard to save on food and water and to make the ship lighter and easier to move.

Similarly, the air near the equator is relatively still because air at these locations is constantly rising and not blowing. For this reason, early sailors called this region the **doldrums.** The region of the doldrums, occurring between 5 degrees north and 5 degrees south of the equator, is also known as the Intertropical Convergence Zone, or ITCZ for short. The trade winds converge in the region of the ITCZ, producing convectional storms that produce regions with some of the world's heaviest precipitation.

The last type of moving air system that you'll need to be familiar with for the exam is the **jet stream.** Jet streams are high-speed currents of wind that occur in the tropopause; these fast-moving air currents have a large influence on local weather patterns.

Winds Around the World

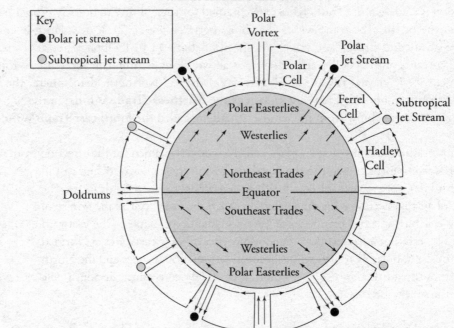

Now, let's move on to review the types of weather that result from all of these moving air masses.

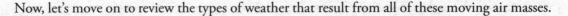

Weather Events

Weather and climate are affected not only by the insolation in a given area but also by geologic and geographic factors, such as terrain (mountains, plains, distance from the ocean) and ocean temperatures. **Monsoons**, or seasonal winds that are usually accompanied by very heavy rainfall, occur when land heats up and cools down more quickly than water does. In a monsoon, hot air rises from the heated land and a low-pressure system is created. The rising air is quickly replaced by cooler moist air that blows in from over the ocean's surface. As this air rises over land, it cools, and the moisture it carries is released in a steady seasonal rainfall. This process happens in reverse during the dry season, when masses of air that have cooled over the land blow out over the ocean. Monsoons primarily occur in coastal areas. Check out this helpful illustration.

How a Monsoon Forms

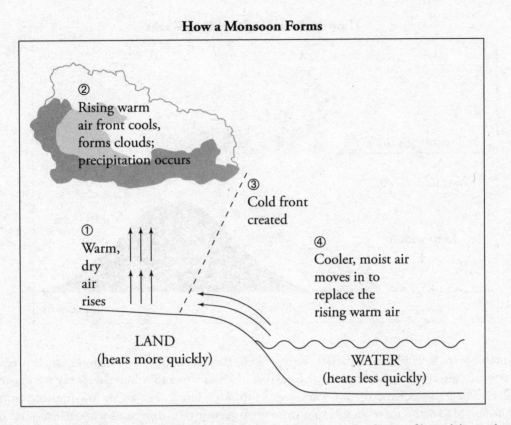

② Rising warm air front cools, forms clouds; precipitation occurs

③ Cold front created

① Warm, dry air rises

④ Cooler, moist air moves in to replace the rising warm air

LAND (heats more quickly)

WATER (heats less quickly)

On a smaller—local or regional—scale, this effect can be seen on the shores of large lakes or bays. In these areas, again the land warms faster than does the water during the day, so the air mass over the land rises. Air from over the lake moves in to replace it, and this creates a breeze. At night, the reverse happens: the land cools more quickly than the water, and the air over the lake rises. The air mass from the land moves out over the lake to replace the rising air, and this creates a breeze, as well. This small-scale monsoon effect is called the lake effect, something of a misnomer since the phenomenon isn't limited to inland lakes. If you live near one of the Great Lakes or in the Bay Area of San Francisco, you may have experienced the lake effect yourself!

As we mentioned above, the air that moves in from over the ocean or a large body of water contains large amounts of water. If an air mass is forced to climb in altitude—if, for instance, it encounters an obstruction such as a mountain—the air will be forced to rise. When the air mass rises, it will cool, and water will precipitate out on the ocean side of the mountain. By the time the air mass reaches the opposite side of the mountain, it will be virtually devoid of moisture. This phenomenon is known as the **rain shadow effect** and is responsible for the impressive growth of the Olympic rainforest (within Olympic National Park) on the coast of Washington State. Interestingly, Olympic National Park has a temperate rainforest on its west side where annual precipitation is about 150 inches (making it perhaps the wettest area in the continental United States) and forests on its much drier east side, where the annual precipitation is around 15 inches.

How the Rain Shadow Effect Works

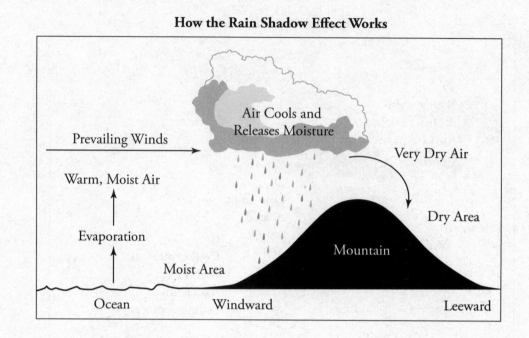

Remember trade winds from the last section? Well, they occur in steady and somewhat predict-able wind patterns, but they may cause local disturbances when they blow over very warm ocean water. When this occurs, the air warms and forms an intense, isolated, low-pressure system while also picking up more water vapor from the ocean surface. The wind will circle around this isolated low-pressure air area (counterclockwise in the Northern Hemisphere and the opposite in the Southern Hemisphere—once again due to the Coriolis effect!). The low-pressure system will continue to move over warm water, increasing in strength and wind speed; this will eventu-ally result in a tropical storm.

Certain tropical storms are of sufficient intensity to be classified as **hurricanes.** Hurricanes must have winds with speeds greater than 120 km/hr. The rotating winds of a hurricane remove water vapor from the ocean's surface, and heat is released as the water vapor condenses. This addition of heat energy continues to contribute to the increase in wind speed, and some hurricanes have winds traveling at speeds of nearly 400 km per hour! A major hurricane contains more energy than that released during a nuclear explosion, but since the force is released more slowly, the damage is generally less concentrated. Another important note about this type of storm is that they are referred to as hurricanes in the Atlantic Ocean, but they are called **typhoons** or **cyclones** when they occur in the Pacific Ocean. Go figure!

El Niño is a climate variation that takes place in the tropical Pacific about once every three to seven years, and it lasts for about one year. Under normal weather conditions, trade winds move the warm surface waters of the Pacific away from the west coast of Central and South America. As a result, the cold ocean water that lies under the displaced water moves to the surface (causing the thermocline, or line of demarcation between two layers of water to rise), bringing nutrients with it and keeping the temperature of the coastal water, relatively cool. This phenomenon is called upwelling.

During El Niño, the normal trade winds are weakened or reversed because of a reversal of the high and low pressure regions on either side of the tropical Pacific. This reversal of pressure systems is known as the **Southern Oscillation.** Without these regular trade winds off the Central and South American coast, the process of upwelling slows or stops, and the water off the coast becomes warmer and contains fewer nutrients. This means that during El Niño, the northern United States and Canada experience warmer winters and a less intense hurricane season; the eastern United States and regions of Peru and Ecuador that are typically dry have higher-than-average rainfall; and the Philippines, Indonesia, and Australia are drier than normal.

Differing Wind Patterns

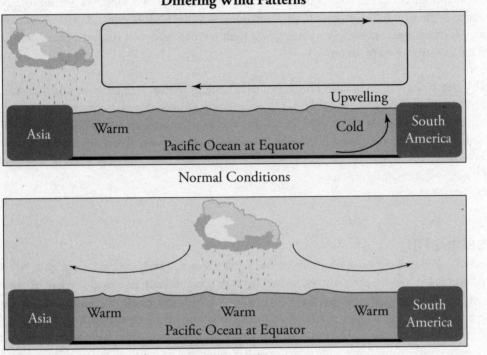

Normal Conditions

El Niño Conditions

One environmentally important effect that El Niño has on humans is that, because of the suppression of upwelling, the offshore fish populations of certain coastal areas decline. In countries like Peru, which relies heavily on fishing, El Niño has devastating economic effects.

The reverse of El Niño is known as **La Niña.** The Coriolis effect contributes to La Niña conditions. As air moves toward the equator to replace rising hot air, the moving air deflects to the west and helps move the surface water, allowing the upwelling. During La Niña, the surface waters of the ocean surrounding Central and South America are colder than normal. The term *El Niño* comes from the fact that traditionally these conditions were observed to begin around Christmas time (*"el niño"* means "infant boy" in Spanish). The alternations of atmospheric conditions that lead to El Niños or La Niñas are referred to as **ENSO** events. (This stands for El Niño and Southern Oscillation.)

THE HYDROSPHERE

Water covers about 71 percent of planet Earth. Most of the water on Earth's surface is salt water. On average, the salt water in the world's oceans has a salinity of about 3.5 percent. This means that for every 1 liter (1,000 ml) of sea water, there are 35 grams of salts (mostly, but not entirely, sodium chloride) dissolved in it. (1 ml of water weighs approximately 1 gram.) In fact, 1 cubic foot of seawater would evaporate to leave about 2 pounds of sea salt! However, **sea water** is not uniformly saline throughout the world. The planet's freshest sea water is in the Gulf of Finland, part of the Baltic Sea. The most saline open sea is the Red Sea, where high temperatures and confined circulation result in high rates of surface evaporation.

Freshwater is water that contains only minimal quantities of dissolved salts, especially sodium chloride. All freshwater ultimately comes from precipitation of atmospheric water vapor, which reaches inland lakes, rivers, and groundwater bodies directly, or after melting of snow or ice. Let's start with a discussion of freshwater before discussing the world's oceans. We will end this section with a review of the ways in which humans use water, global problems associated with water usage, access to water, and issues of water conservation.

Freshwater

Freshwater is deposited on the surface of the Earth through precipitation. Water that falls on the Earth and doesn't move through the soil to become groundwater moves along Earth's surface via gravity, forms small streams, and then eventually forms larger ones. The size of the stream will continue to increase as water is added to it, until the stream becomes a river, and the river will flow until it reaches the ocean. The land area that drains into a particular stream is known as a **watershed,** or drainage basin. A particular watershed has some characteristics that define it: its area, length, slope, soil and vegetation types, and how it's separated from adjoining watersheds.

As water moves into streams, it carries with it sediment and other dissolved substances, including small amounts of oxygen. Turbulent waters are especially laden with dissolved oxygen and carbon dioxide, such as those found at the source, or headwaters, of a stream. As a general rule, the more turbulent the water, the more dissolved gases it will contain.

Freshwater Bodies

As you probably know, freshwater that travels on land is largely responsible for shaping Earth's surface. Erosion occurs when the movement of water etches channels into rocks and soil. The moving water then carries eroded material farther downstream.

Because of obstructions on land, moving water does not move in a straight line. Instead, it follows the lowest topographical path, and as it flows continuously along the same path it cuts farther into its banks to eventually form a curving channel. As the water travels around these bends, its velocity decreases, and the stream drops some of its sedimentary load. Water always follows the path of least resistance as it travels from the highlands to the sea.

Rivers drop most of their sedimentary load as they meet the ocean because their velocity decreases significantly at this juncture. At these locations, landforms called **deltas**—which are made of deposited sediments—are created. Another important type of freshwater body that you should know about is the estuary. **Estuaries** are sites where the "arm" of the sea extends inland to meet the mouth of a river. Estuaries are often rich with many different types of plant and animals species, because the freshwater in these areas usually has a high concentration of nutrients and sediments. The waters in estuaries are usually quite shallow, which means that the water is fairly warm and that plants and animals in these locations can receive significant amounts of sunlight. Subcategories of estuarine environments that you should know for the exam include saltwater marshes, mangrove forests, inlets, bays, and river mouths.

Some of Earth's most important ecologically diverse ecosystems are **wetlands**— areas along the shores of fresh bodies of water, wet inland habitats fed only by rainwater, and ephemeral (seasonally temporary) water bodies. Types of wetlands include marshes, swamps, bogs, prairie potholes (which exist seasonally), and floodplains (which occur when excess water flows out of the banks of a river and into a flat valley). So, those are the main types of freshwater bodies you'll need to know. Let's review the stratification of freshwater bodies, get through oceans, and move on to the fun stuff—the impacts of water use on humans.

Vertical Stratification in Freshwater Biomes

In all natural bodies of water, there exist layers of water that vary significantly in their temperature, oxygen content, and nutrient levels. These layers are affected differently by seasonal changes and other disturbances, and this also contributes to how they are categorized.

In freshwater, the layers are the **epilimnion**, which is the uppermost and thus the most oxygenated, layer; and the **hypolimnion,** which is the lower, colder, and denser layer. The demarcation line between these two layers, at which the temperature shifts dramatically, is the **thermocline.**

The layers of freshwater bodies may also be categorized differently, according to the types of organisms that can live in them. You should definitely be familiar with the following terms for the AP Environmental Science Exam, so take note!

- **Littoral zone:** Begins with the very shallow water at the shoreline. Plants and animals that reside in the littoral zone receive abundant sunlight. The end of this zone is defined as the depth at which rooted plants stop growing.

- **Limnetic zone:** Surface of open water; the region that extends to the depth that sunlight can penetrate. Organisms that are residents in this zone are short-lived and rely on sunlight to carry out photosynthesis.

- **Profundal zone:** Water that is too deep for sunlight to penetrate. Because the profundal zone is aphotic (a place where light cannot reach), photosynthesizing plants or animals cannot live in this region.

- **Benthic zone:** The deepest layer in a body of water, characterized by very low temperatures and low oxygen levels.

The World's Oceans

Before we get into our review of the world's oceans, let's consider another aquatic ecosystem (besides wetlands and estuaries) that's an important source of biodiversity. This one is a salt-water ecosystem. Certain landforms that lie off coastal shores are known as **barrier islands.** Because barrier islands are created by the buildup of deposited sediments, their boundaries are constantly shifting as water moves around them. These spits of land are generally the first hit by offshore storms, and they are important buffers for the shoreline behind them.

In tropical waters, a very particular type of barrier island called a **coral reef** is quite common. These barrier islands are formed not from the deposition of sediments, but from a community of living things. The organisms responsible for the creation of coral reefs are cnidarians, which secrete a hard, calciferous shell; these shells provide homes and shelter for an incredible diversity of species, but they are also extremely delicate and thus very vulnerable to physical stresses as well as changes in light intensity, water temperature, ocean depth, and pH. The increase in ocean temperatures and dissolved CO_2 due to climate change is resulting in more acidic waters resulting in coral bleaching. Coral bleaching occurs when acidic conditions cause the coral to expel the colorful algae which provided them with food.

Like freshwater bodies, oceans are divided into zones based on changes in light and temperature. Study the following terms, and know them cold for the test!

- **Coastal zone**: This zone consists of the ocean water closest to land. Usually it is defined as being between the shore and the end of the continental shelf (the edge of the tectonic plate.)

- **Euphotic zone:** The photic, upper layers of water. The euphotic zone is the warmest region of ocean water; this zone also has the highest levels of dissolved oxygen.

- **Bathyal zone:** The middle region; this zone receives insufficient light for photosynthesis and is colder than the euphotic zone.

- **Abyssal zone:** This is the deepest region of the ocean. This zone is marked by extremely cold temperatures and low levels of dissolved oxygen, but high levels of nutrients because of the decaying plant and animal matter that sinks down from the zones above.

Ocean Zones

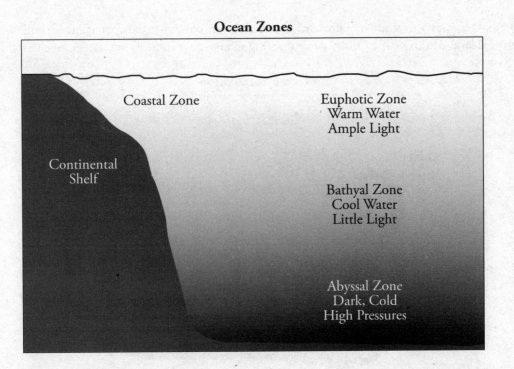

Coastal Zone

Euphotic Zone
Warm Water
Ample Light

Continental
Shelf

Bathyal Zone
Cool Water
Little Light

Abyssal Zone
Dark, Cold
High Pressures

Both freshwater and saltwater bodies experience a seasonal movement of water from the cold and nutrient-rich bottom to the surface. These **upwellings** provide a new nutrient supply for the growth of living organisms in the photic regions. Therefore, they are followed by an almost immediate exponential growth in the population of organisms in these zones, especially the single-cell algae, which may form blooms of color called algal blooms. These algae can also produce toxins that may kill fish and poison the beds of filter feeders such as oysters and mussels. One notorious recurring toxic algal bloom is referred to as **red tide**; this is caused by a proliferation of dinoflagellates.

Water is densest at 3.98°C, or 39°F. In non-tropical regions of the Earth, after ice melt in spring, the water-surface temperature of lakes and ponds will rise from 0°C to 4°C, whereupon this dense surface water will sink to the bottom of the lake or pond. This will displace water at the bottom of the lake or pond to the surface. This overturn brings oxygen to the bottom and nutrients to the top of the lake or pond and occurs during spring and fall as the temperature of the ecosystem changes from cold to warm or the reverse.

Ocean Currents

Ocean currents play a major role in modifying conditions around the Earth that can affect where certain climates are located. As the sun warms water in the equatorial regions of the globe, prevailing winds, differences in salinity (saltiness), and Earth's rotation set masses of ocean water in motion. For example, in the Northern Hemisphere, the **Gulf Stream** carries sun-warmed water northward along the east coast of the Unites States and across the Atlantic Ocean as far as Great Britain. This warm water displaces the colder, denser water in the polar regions, which can move south to be re-warmed by the equatorial sun. Northern Europe is kept 5° to 10°C warmer than it would be were the current not present.

Oceanographers also study a major current, the "**ocean conveyor belt**," that moves cold water in the depths of the Pacific Ocean while creating major upwellings in other areas of the Pacific.

Ocean Circulation

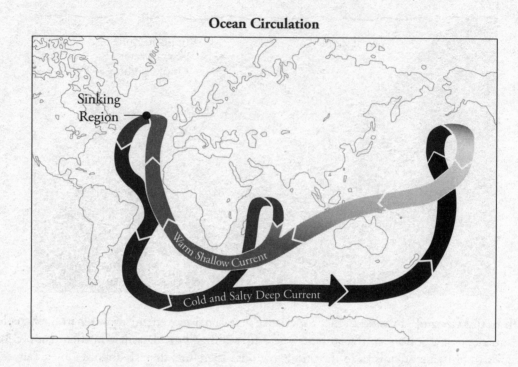

Water, Water, Everywhere…or Not?

As you know, we all need water in order to live. In particular, communities need water for many different industries, including fisheries, recreation, transportation, and agriculture. Agriculture is one of the biggest water-users of all—about 73 percent of the global demand for water is for crop irrigation. Industry accounts for about 21 percent of all water use, and domestic use accounts for about 6 percent.

Since the 1950s, global water use has tripled—mostly due to population growth and improvements in the global standard of living. One way that humans have recently dealt with potential water shortages in communities is through **interbasin transfer**. During interbasin transfer, water is transported very long distances from its source, through aqueducts or pipelines. An example of this type of engineering is the pipeline that now exists between the western and eastern slopes of the Rocky Mountains in Colorado. Known as the Big Thompson Project, 213,000 acre-feet of water is delivered annually to the eastern slope of Colorado. However, this method has several negative effects. It can result in different geographic areas arguing over water rights. It can also have serious environmental repercussions; interbasin transfer can increase the salinity of the water body being exploited and even change the local climate of an ecosystem.

In North America especially, humans rely on groundwater as a primary source of water for everyday use. **Groundwater** refers to any water that comes from below the ground—that is, from wells or from **aquifers,** which are underground beds or layers of Earth, gravel, or porous stone that hold water. Water found in an **unconfined aquifer** is free to flow both vertically and horizontally. A **confined aquifer**, however, has boundaries that don't readily transport water. Our reliance on and use of groundwater has several detrimental environmental effects; for example, it can result in a depressed water table and the drying up of local groundwater sources. In the late 1990s, a drought in Florida resulted in such a severe reduction in the aquifers that roads collapsed from lack of subterranean structural support. This subsidence (or sinking) of Earth's surface is another serious consequence of groundwater withdrawal.

Additionally, aquifers can become **compacted**—meaning that the mineral grains making up the aquifer collapse on each other and the material is unable to hold as much water. Furthermore, in most urban areas, humans have rendered the groundwater incapable of being replenished by building structures and roads that are impermeable to precipitation.

Global Water Needs

Scientists differentiate between countries that are water-stressed and those that are water-scarce. Countries that are **water-stressed** have a renewable annual water supply of about 1,000–2,000 m^3 per person, but countries that are **water-scarce** have less than 1,000 m^3 per person and lack sufficient freshwater resources to meet demand. Currently, approximately 4 billion individuals experience severe water scarcity for at least one month a year, and 500–700 million people in 43 countries experience severe water scarcity year-round. Many of the countries that are currently considered water-scarce are developing countries that have rapidly increasing populations—which means that their water-scarcity problems will grow over time.

Water scarcity is affected by national and regional politics, civil strife, and other issues affecting access and distribution; and lists of water-scarce countries differ by the source reporting. However, among the countries currently experiencing the most severe water scarcity are Yemen, Libya, Jordan, Western Sahara, and Djibouti. Unfortunately, more and more countries are expected to become water-scarce by the year 2050.

Water Use in the United States

The United States is not considered water-scarce, but certain regions of the United States are considered water-stressed. Additionally, water use in the United States is out of control—we use water more quickly than it can possibly be replenished, so water scarcity is definitely in our future if we continue to use water at our present, furious rate.

The hydrologic cycle supplies the water that we use for all of our activities. Water used in our home, manufacturing, cooling equipment that generates electricity, and irrigating croplands are a few examples. To give you a sense of water use in the United States, take a look at the chart on the next page that shows trends in water withdrawals and population from 1950–2010.

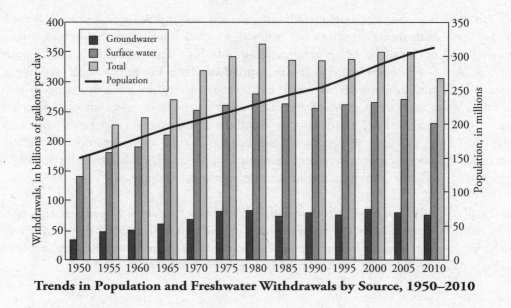

Trends in Population and Freshwater Withdrawals by Source, 1950–2010

As you see, the nation's water use peaked in about 1980 and has been fairly steady since then. Many of the stresses for greater water use have risen since 1980, such as population, irrigation of crops to feed this larger population, and more industry, yet total water use has not risen. The fact that water use has leveled off despite the increase in these stresses shows that water conservation efforts and greater efficiencies in using water have had a positive effect in the last 35 years. Nonetheless, we often hear environmental news stories covering droughts, proposed emergency water relief plans, and rules about lawn watering.

What Are We Doing About It?

Water is a tricky business. It's difficult for politicians and lawmakers to put restrictions on water use because many people think that water should be free. After all, it falls from the sky; we can take a bucket from the lake down the street and no one will arrest us for stealing. For the AP Environmental Science Exam, you should know about certain concepts of human water rights. The first is the idea of riparian right. **Riparian** means of, on, or relating to the banks of a natural course of flowing water, and **riparian right** is the right of people who have legal rights to use that area. Alternately, in **prior appropriation,** water rights are given to those who have historically used the water in a certain area. In other words, prior appropriation can be thought of as water squatters' rights!

It has been proposed that, in order to solve current global water crises, we simply take tons of ocean water and desalinate it—this is a fairly simple process physically, but unfortunately it isn't currently economically viable on a large scale, because it takes a great deal of energy to remove the salt through distillation or reverse osmosis. As water becomes scarcer globally, it will be important for countries to think of ways to regulate the use of water, as estimated global water consumption is set to continue rising. As global water crises become more common, research into the economic viability of desalination has increased.

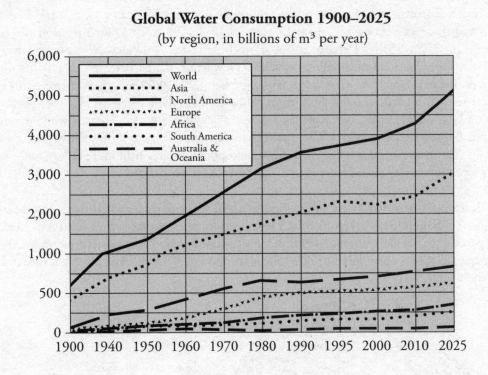

Global Water Consumption 1900–2025

(by region, in billions of m³ per year)

Legend:
— World
····· Asia
— — North America
▲▼▲▼ Europe
—·—· Africa
••••• South America
— — Australia & Oceania

SOIL

One very important but often underappreciated player in Earth's interdependent systems is soil. Soil plays a crucial role in the lives of the plants, animals, and other organisms and acts as an essential link between the **abiotic** (the nonliving components of the world) and the **biotic** (that's right, the living components of the world). Soil is sometimes called the pedosphere and exists at the interface of the other four systems or spheres: the lithosphere, atmosphere, hydrosphere, and biosphere. As we'll see in the next chapter, soil plays an active role in the cycling of nutrients. Let's take a moment to review the major characteristics of soil that you'll be expected to know for the test.

Soil Is More Than Just Dirt

Although we may be tempted to think of soil as simply "dirt," soil is actually a complex, ancient material teeming with living organisms. Many soils are tens of thousands of years old! A typical soil has the following composition: about 45% is mineral in the form of rocks broken down into tiny particles. About 25% is air and 25% water, which fill the pores between soil particles. About 5% is organic matter, both living and dead. In just one gram of soil, there may be as many as 50,000 protozoa, as well as bacteria, algae, fungi, and larger organisms such as earthworms and nematodes.

Soils can be categorized according to a number of physical and chemical features, including color and texture. There are many ways to test soil (looking at its chemical, physical, and biological properties) that can help people make decisions about its use (such as for irrigation and fertilization). The United States Department of Agriculture (USDA) divides soil particles into three size classes. The class with the smallest particles is **clay,** which has particles less than 0.002 mm in diameter. The next largest size class is **silt,** with particles 0.002–0.05 mm in diameter. **Sand** is the coarsest soil, with particles 0.05–2.0 mm in diameter. Sand particles are usually too spherical in shape and too large to easily stick together; sand particles have larger pore spaces, with more room for water but less ability to hold it. Clay particles more easily adhere to each other because they are small and flat in shape; there is less room for water but the water is held more tightly. Soils with a high proportion of clay particles are extremely compact.

A soil's texture is defined by its proportion of these three particle-size classes. A soil texture triangle is a diagram that allows for the identification and comparison of soil types based on their percentages of clay, silt, and sand. A soil's texture is called **loam** if it has a proportion of the three size classes considered optimal for plant growth: 7–27% clay, 28–50% silt, and less than 52% sand.

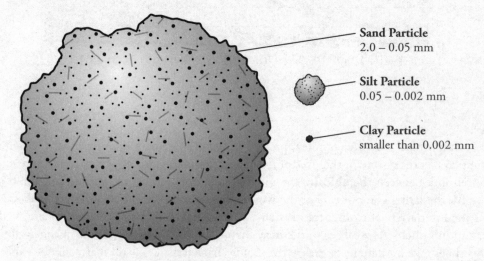

Sand Particle
2.0 – 0.05 mm

Silt Particle
0.05 – 0.002 mm

Clay Particle
smaller than 0.002 mm

pH Pop Quiz!
Q: Which of the following tastes is associated with acidic food?
(A) Sour
(B) Salty
(C) Sweet
(D) Bitter
Turn the page for the answer.

Another very important characteristic of soil types is soil **acidity** or **alkalinity.** Recall that the pH of a substance ranges from 0–14 and is a measure of the concentration of hydronium (H_3O^+) ions in aqueous solution. Therefore, soil pH is actually the pH of the soil's water component and, in the laboratory, is commonly measured by suspending soil in water or a dilute neutral salt solution. Most soils fall into a pH range of about 4–8, meaning that most soils range in pH from being neutral to slightly acidic. Soil pH is important because it affects the solubility of nutrients, and this in turn determines the extent to which these nutrients are available for absorption by plant roots. If the soil in a region is too acidic or basic, certain soil nutrients will not be able to be used by the regional plants. One last thing about soils that the College Board wants you to know: when the soil solution becomes more acidic, it more readily dissolves heavy metals such as mercury (Hg) or aluminum (Al) from soil minerals. Leached into the ground water, these ions can travel to streams and rivers and harm both plants and aquatic animal life. For example, aluminum ions can damage the gills of fish and cause them to suffocate.

Where Does Soil Come From?

Basically, soil is a combination of organic material and rock that has been broken down by chemical and biological **weathering.** Therefore, it should not be surprising to learn that the types of minerals found in soil in a particular region will depend on the identity of the base rock of that region.

Water, wind, temperature, and living organisms are all prominent agents of weathering, and all weathering processes are placed into the following three rather broad categories (which you should definitely know for exam day).

- **Physical weathering** (also known as **mechanical weathering**): Any process that breaks rock down into smaller pieces without changing the chemistry of the rock. The forces responsible for physical weathering are typically wind and water.

- **Chemical weathering:** Occurs as a result of chemical reactions of rock with water, air, or dissolved minerals. Chemical processes result in minerals that are broken down or restructured into different minerals. This type of weathering tends to dominate in warm or moist environments. One example of chemical weathering is rust, which forms when iron and other metallic elements come in contact with water.

- **Biological weathering:** Weathering that takes place as the result of the activities of living organisms, which may act through physical or chemical means. When tree roots enlarge the cracks in rocks as they grow, that is a physical process. When plant roots or lichens growing on rocks release organic acids that dissolve minerals, that is a chemical process.

Memorize the different types of weathering processes and the order and characteristics of soil layers. You will definitely be asked about these on test day!

Soil is made up of distinct layers with very different characteristics. Let's discuss those next.

Soil Layers

Soil comprises distinct layers known as **horizons,** each of which has distinct physical, chemical, and biological properties. Study the following diagram, which illustrates and describes the different horizons. Not every soil contains all of these horizons.

- **O horizon:** This layer is made up of organic matter at various stages of decomposition. It includes animal waste, leaves and other plant tissues (such as dead roots), and the decomposing bodies of organisms. The stable residue left after most organic matter has decomposed is a dark, crumbly material called **humus**.

- **A horizon:** This is the topsoil—the topmost mineral horizon and the most intensively weathered soil layer. Its dark color is due to accumulation of organic matter from the O horizon. In soils lacking an E horizon, this may also be called the zone of **leaching**.

- **E horizon:** This is the eluviated horizon. It is light in color and coarse in texture; no organic matter has traveled down from the A horizon, while clays and minerals like iron and aluminum oxides have been washed out by leaching and **eluviation**. The E horizon isn't found in all soils, but mostly in soils developed under forest.

Eluviation is the movement of water-borne minerals, humus, and other materials from higher soil layers to lower soil layers. This is due to the downward movement of water via gravity. **Illuviation** is the deposition of these materials in a lower soil horizon. **Leaching** is similar to eluviation but refers specifically to dissolved (not suspended) organic and chemical compounds, and implies loss of these substances from the soil profile by draining into the groundwater.

- **B horizon:** Sometimes called the subsoil, this is where organic matter, clay, and minerals washed out of the upper horizons accumulate. Thus, it is called the zone of accumulation or the zone of **illuviation**.

- **C horizon:** This layer is the parent material—unconsolidated material, loose enough to be dug up with a shovel. Weathering at this depth is minimal so the soil retains identifiable features of the parent material (rock from which the A and B horizons formed). This horizon has much less biological activity than the horizons above it.

- **R horizon:** Beneath the soil lies a layer of consolidated (cemented), unweathered rock. Because it hasn't been weathered, this isn't part of the soil, strictly speaking. If the material from which horizons A through C formed was not transported from elsewhere, the R horizon has the same source minerals as the parent material above. The R stands for regolith, a word meaning "blanket or surface rock" in ancient Greek.

Soil Horizons

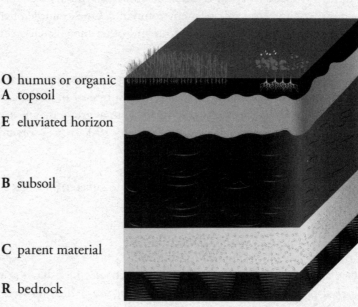

O humus or organic
A topsoil
E eluviated horizon

B subsoil

C parent material

R bedrock

Soil Development

pH Pop Quiz!
A: Acidic foods, such as vinegar and lemon juice, taste sour because of their acidic nature. The answer is (A).

- **Four Processes of Soil Development.** Soil development is an intricate business! The development of distinct soil horizons is accomplished by four basic processes that work on soil minerals and particles, organic matter, and soil chemistry. These four processes include **additions** of materials from off-site; **losses** through erosion or leaching or biological activity (plant uptake); vertical **translocations** or movement from one soil horizon to another; and **transformations** of materials in place by weathering or chemical activity.

- **Six Soil-Forming Factors**. The processes of soil development are influenced by six soil-forming factors. It takes hundreds to thousands to millions of years for a soil to develop its characteristic layers or profile. To wrap it all up in one big sentence: any soil you see is a dynamic formation produced by the effects of climate and biological activity (organisms), as modified by topography (relief) and human influences, acting on parent materials over time. The six soil-forming factors can be remembered by the awkward acronym Cl–O–R–P–T–H. Here is a summary of how each factor works.

- **Climate**: Climate involves differences in temperature and precipitation across the globe, and both heat and water facilitate chemical and biochemical reactions. Seasonal fluctuation of heat and moisture affects processes such as freeze-thaw cycles that weather rock. Climate also helps to determine what organisms grow in a particular location.

- **Organisms** (Biological Activity): Different local conditions support different organisms, which influence the soil in a multitude of ways. Microorganisms perform biochemical functions such as decomposition of organic matter and transformation of minerals into different forms. Animals move soils, consume vegetation, and add nutrients through waste and decomposing bodies. Plants perform physical weathering through root growth, take up soil nutrients and water, alter soil chemistry in various ways, and add nutrients when they die and decompose.

- **Relief** (Topography): Topographical relief affects where water moves on the landscape and also depth to the water table from location to location. Relief similarly affects erosion—which locations are likely to lose surface material through the action of wind and rain, and which locations are likely to accumulate eroded material. Topographical relief also leads to differences in how much sun different locations receive. Through these characteristics, relief influences which organisms grow in a particular location.

- **Parent Material**: This is the starting point for soil development. Its mineral properties, hardness, and topographical form affect how it is weathered into soil. As parent material varies from location to location, so will the soil that develops at each location. As an example, parent material rich in quartz, such as granite and sandstone, weathers into sandy soil. Shale weathers into soil richer in silt and clay.

- **Time**: More time equals more change! Hard parent material weathers more slowly and softer material more quickly. A flat, stable topographic position develops horizons more quickly than do slopes and depressions where material is lost and gained.

- **Human Influence**: The effects of human activity must increasingly be acknowledged as a factor in soil development. Use of fertilizer, pollution, and acid rain alter soil chemistry on a broad scale. Construction activities like digging and plowing tend to haploidize soils by mixing soils and blurring the distinctions between horizons. Human activities also lead to compaction (through the traffic of vehicles and machinery), erosion (through removal of stabilizing vegetation), and salinization (high salt content, through irrigation and depletion of groundwater).

Soil Problems for (and Caused by) Humans

In order to be able to grow all of the foods that humans consume, we must have enough **arable**—suitable for plant growth—soil to meet our agricultural needs. Soil fertility refers to soil's ability to provide essential nutrients, like nitrogen (N), potassium (K), and phosphorus (P), to plants. Humus (remember, it's in the O layer!) is also an extremely important component of soil because it is rich in organic matter.

Remember that soils composed of a balanced mixture of the three particle sizes (clay, silt, and sand) are described as **loamy,** and these types of soil are considered the best for plant growth. Another important characteristic of soil for agricultural purposes is its **structure**, or the extent to which it aggregates or clumps. Soil **aggregates** are formed and held together by such substances as clay particles and organic matter—plants and roots, the root-like filaments of fungi, and sticky substances released by bacteria and fungi. The most fertile soils have good structure.

Soil is considered a non-renewable resource due to the great length of time required to form arable soil. It takes 500 to 1,000 years to form a single inch of soil, and at least 3,000 years to form enough fertile soil to support crop growth.

Monoculture

Unfortunately, certain agricultural activities can change the texture and structure of soil; for example, repeated plowing tends to break down soil aggregates, leaving "plow pan" or "hard pan," which is hard, unfertile soil.

Whereas communities traditionally planted many different types of crops in a field, in modern agriculture the **monoculture,** or the planting of just one type of crop over a large area, predominates. Over the history of agriculture, a significant decrease in the biodiversity of crop species has taken place—both in the number of crop species and in the genetic makeup of individual species. This creates numerous problems. First of all, a lack of genetic variation makes crops more susceptible to pests and diseases. Secondly, the consistent planting of one crop in an area eventually leaches the soil in that area of the specific nutrients that the plant needs in order to grow. One way to prevent this phenomenon is to practice **crop rotation**, in which different crops are planted in the area in each growing season. Another practice that farmers will use to increase sustainability is **polyculture**, planting several crops on the same plot of land simultaneously. This increases biodiversity.

Other problems with modern agriculture include its reliance on large machinery (which can damage soil through compaction), and the fact that as an industry, agriculture is a huge consumer of energy. Energy is consumed both in the production of pesticides and fertilizers and in the use of fossil fuels to run farm machinery.

The past 50 years or so have seen a huge increase in worldwide agricultural productivity, which is largely due to the mechanization of farming that resulted from the Industrial Revolution. The boom in agricultural productivity is known as the **Green Revolution,** and, unfortunately, it has since had many detrimental effects on the environment. For example, the use of chemical pesticides resulted in the emergence of a new variety of pesticide-resistant insects. Recently, the introduction of genetically modified plants has enabled researchers to take steps in solving the problem of pesticide-resistant insect species.

Another drawback to the Green Revolution resulted from the dramatic increase in irrigation worldwide; the largest human use of freshwater, 70%, is for irrigation! Furrow irrigation, which involves cutting furrows between crop rows and filling them with water, is inexpensive but loses about 1/3 of the water used to evaporation and runoff. Flood irrigation, which involves flooding a field with water, can lead to waterlogging and loses about 20% of the water to evaporation and runoff. Spray irrigation involves pumping water into spray nozzles and spraying fields; it only loses about 1/4 of the water but requires energy to run and can be expensive. One problem is waterlogging: if too much water is left to sit in soil, it can raise the water table of the groundwater, causing plants to have trouble absorbing oxygen through their roots. Additionally, over-irrigated soils undergo salinization. In **salinization,** the soil becomes water-logged; when it dries out, salt forms a layer on its surface. This eventually leads to **land degradation.** In order to combat this problem, researchers have developed **drip irrigation,** which allots an area only as much water as is necessary and delivers the water directly to the roots using perforated hoses that release small amounts of water: this is far more efficient, with only about 5% of water lost to evaporation and runoff, but is more expensive.

Soil Erosion

The small rock fragments that result from weathering may be moved to new locations in the process of erosion, and bare soil (upon which no plants are growing) is more susceptible to erosion than soil covered by organic materials.

Because of the constant movement of water and wind on Earth's surface, the **erosion** of soil is a continual and normal process. However, when erosion removes valuable topsoil or deposits soil in undesirable places, it can become a problem for humans. Eroded topsoil usually ends up in bodies of water, posing a problem for both farmers, who need healthy soil for planting, and people in general, who rely on bodies of water to be uncontaminated with soil runoff.

The most significant portion of erosion caused by humans results from logging and from agriculture—especially slash-and-burn agriculture, which is characterized by cutting down and burning trees to clear land for agricultural purposes. The removal of plants in an area makes the soil much more susceptible to the agents of erosion.

Unfortunately, human activities—unsustainable agricultural practices, overgrazing, urbanization, and development deforestation—have significantly increased the levels of erosion in the upper layers of soil. These processes will continue to create problems for farmers searching for arable land until new techniques that preserve the integrity of soil are introduced and utilized.

Soil Conservation

In order to conserve soil resources, several best management practices have been developed. These practices return organic matter to the soil, slow down the effects of wind, and reduce the damage to the soil from tillage (plowing). Here are some of the more common methods.

- Use of animal waste (manure), compost, and the residue of plants to increase the amount of organic matter in the soil.

- The practice of organic agriculture, a method of farming that utilizes compost, manure, crop rotation, and non-chemical methods to enhance soil fertility and control pests. Organic producers avoid or strictly limit the use of chemical fertilizers and pesticides as well as genetically modified organisms.

- Modification of tillage practices to reduce the breakup of soil and to reduce the amount of erosion. These include no-till farming, contour plowing, and strip planting.

- Use of trees and other wind barriers to reduce erosion from wind.

Much of what we know about soil conservation was established relatively recently. The Soil Conservation Act was passed in 1935 and led to the creation of the Soil Conservation Service. These developments came in response to the Dust Bowl of the 1930s, which was a period of unprecedented dust storms caused by severe drought and ill-advised farming practices. The Soil Conservation Service was a federal agency founded by Hugh Hammond Bennett. Its mission was to promote sustainable soil conservation practices among farmers and other landowners and to help restore ecological balance across the nation's landscape. The agency is now called the Natural Resources Conservation Service.

Soil Laws

While the following soil laws are not required for the exam, they are relevant to the discussion of soil!

Date	Name of Legislation	What It Did
1977	Soil and Water Conservation Act	This act established soil and water conservation programs to aid landowners and users; it also set up conditions to continue evaluating the condition of U.S. soil, water, and related resources.
1985	Food Security Act	Nicknamed the Swampbuster, this act discouraged the conversion of wetlands to nonwetlands. In 1990, federal legislation denied federal farm supplements to those who converted wetlands to agriculture and provided a restoration of benefits to those who converted lands to wetlands.

CHAPTER 4 KEY TERMS

Use this list to review the key terms in this chapter. Keep these terms in mind when you are brainstorming hot buttons for your essays.

The Solid Earth

geologic time scale
core
mantle
asthenosphere
lithosphere
crust
Pangaea
tectonic plates
plate boundaries: convergent, divergent, transform fault
volcanoes: magma, active, dormant, extinct, subduction, rift valley, hot spot, shield volcano, composite volcano, cinder volcano, lava dome
earthquake: focus, epicenter, seismograph, Richter scale, S-wave
tsunami
rock cycle: sedimentary, metamorphic, igneous

Atmosphere

troposphere
greenhouse effect
tropopause
stratosphere
ozone
mesosphere
thermosphere
exosphere
ionosphere
weather
climate
meteorologists
prevailing winds
Coriolis effect
convection currents
horizontal airflow
dew point
precipitation
convection cell
Hadley cell
seasons: insolation, revolution, axis, rotation
albedo

winds: trade wind, westerly, polar easterly, horse latitude, doldrums, jet stream
monsoon
lake effect
rain shadow effect
hurricane
typhoon, cyclone
El Niño, Southern Oscillation, La Niña, ENSO events

Hydrosphere

sea water
freshwater
watershed
delta
estuary
wetland
epilimnion
hypolimnion
thermocline
freshwater zones: littoral, limnetic, profundal, benthic
barrier island
coral reef
saltwater zones: coastal, euphotic, bathyal, abyssal
upwelling
red tide
ocean currents: Gulf Stream, ocean conveyor belt
interbasin transfer
groundwater
aquifer: unconfined aquifer, confined aquifer, subsidence, water-stressed, water-scarce
riparian right
prior appropriation

Soil or Pedosphere

abiotic, biotic, clay, silt, sand, loam, acidity, alkalinity
weathering: physical, chemical, biological
soil horizons: humus, leaching, eluviation, illuviation

Want Printable Lists?
Head over to your Student Tools (your Princeton Review online companion for this book), where you can find printable versions of all Key Terms lists.

4 soil-forming processes: additions, losses,
   translocations, transformations
6 soil-forming factors (Cl-O-R-P-T-H):
   climate, organisms, relief, parent material,
   time, human influence
arable
soil structure: aggregates
monoculture
polyculture
crop rotation
soil compaction
Green Revolution
salinization
land degradation
drip irrigation
erosion
soil conservation: tillage, organic agriculture,
   wind barriers

Chapter 4 Drill

Directions: Each of the questions or incomplete statements below is followed by four suggested answers or completions. Select the one that is best in each case. For answers and explanations, see Chapter 13.

1. A seismograph is used to measure

    (A) tectonic plate size
    (B) the magnitude of an earthquake
    (C) the temperature of volcanic materials
    (D) the frequency of earthquakes in a region

2. The aurora borealis occurs in which of the following parts of the atmosphere?

    (A) Troposphere
    (B) Thermosphere
    (C) Mesosphere
    (D) Hydrosphere

3. Which of the following are the two most important factors in determining a habitat's climate?

    (A) Temperature and wind speed
    (B) Wind direction and precipitation
    (C) Wind speed and rate of evaporation
    (D) Temperature and precipitation

4. The atmosphere is warmed as gases such as water vapor and carbon dioxide absorb the infrared heat radiated from the Earth. This process is best described as

    (A) ozone depletion
    (B) the greenhouse effect
    (C) biomagnification
    (D) ionization

5. The hydrosphere includes all of the following EXCEPT

    (A) watershed
    (B) wetlands
    (C) parent rock
    (D) rivers

6. Imagine an area where there are cold waters, low oxygen levels, and bottom-dwelling fish. This description best fits the

    (A) benthic zone
    (B) littoral zone
    (C) limnetic zone
    (D) open water zone

7. The amount of Earth's surface that is covered by water is approximately

    (A) 12 percent
    (B) 36 percent
    (C) 50 percent
    (D) 75 percent

8. An area where salt and freshwater mix that has a very high level of productivity is correctly called

    (A) the open ocean
    (B) the abyssal zone
    (C) the headwaters
    (D) an estuary

9. Which of the following correctly describes the waters in an upwelling area?

    (A) Cold and nutrient rich
    (B) Warm and nutrient poor
    (C) Cold and nutrient poor
    (D) Heavily polluted by human waste

10. Which of the following best describes an unconfined aquifer? It is an area where

    (A) water always comes to the surface
    (B) water is free to flow in all directions
    (C) water is held in place by impenetrable rocks
    (D) pollutants enter the aquifer

11. Occurring primarily in coastal areas, the weather phenomenon that occurs when a low-pressure system develops as hot air rises from heated land is known as

    (A) a hurricane
    (B) a tornado
    (C) a monsoon
    (D) a typhoon

Free-Response Question

1. Scientists designed an experiment to learn about the functioning of the hydrologic cycle and the phosphorus cycle in a forest. Using two areas of the same size and geologic features, they cut all the trees down from one plot and did not disturb the other plot. They were able to accurately measure the amount of water that flowed out of the two plots as well as measure the amounts of phosphorus found in the runoff.

   (a) **Describe** what the differences would be in the volume of water running off the two plots and give ONE reason why. Assume that the two areas received the same amounts of precipitation.

   (b) **Describe** the differences in the levels of phosphate found in the runoff of the two plots. Assume that both plots started off with the same amount of phosphorus in the soil.

   (c) **Describe** ONE negative effect that might occur in a stream that receives the runoff water and sediment.

   (d) When a tropical rain forest is cut down and used as farmland, the fertility of the soil only lasts a few years. **Give** an explanation as to why there is little organic matter in rain forest soil and what would happen to that material after deforestation.

Summary

o The lithosphere is constantly changing due to plate tectonics, volcanoes, earthquakes, and the rock cycle.

o There are two zones of the atmosphere that are essential for life on Earth:
   - The troposphere contains greenhouse gases (GHGs) that regulate the temperature of Earth and is where all weather takes place.
   - The stratosphere contains the ozone layer, which protects the Earth from harmful ultraviolet radiation.

o Climate is dictated by a region's average temperature and precipitation, which is influenced by air circulation from convection cells and Earth's tilt, daily rotation around its axis, and annual revolution around the sun.

o Unique local weather events include:
   - monsoons
   - lake effect
   - rain shadow effect
   - El Niño

o Hydrosphere: Freshwater biomes and ocean zones are stratified by temperature, light, oxygen content, and nutrient levels. Salinity and seasonal changes in temperature influence water density and thus vertical movement between the layers.

o The movement of ocean water around the globe is due to differences in density between layers, prevailing winds, regional air temperatures, and Earth's rotation. The resulting currents in turn influence weather patterns and thus terrestrial environments.

o Human water demands are influenced by activities such as:
  • irrigation for agriculture
  • domestic and public use
  • livestock, aquaculture
  • thermoelectric supply
  • industry
  • mining
  • hydraulic fracturing or "fracking," which uses pressurized liquid to fracture rock and open reservoirs of oil or natural gas

o Not all regions of the world have equal water supplies. Therefore, the possible activities and their proportional demand on the water resources will differ by region.

o The soil that makes up part of the lithosphere is the combination of rock and organic matter formed by:
  • physical weathering
  • chemical weathering
  • biological weathering
  • decomposition of organic matter

o As rock weathers into soil, the soil develops horizons or stacked layers, each with its own properties of texture, color, and mineral/chemical properties. These horizons take hundreds to thousands of years or longer to develop. Here are the soil-forming factors that interact to make this happen:
  • climate
  • organisms (biological activity)
  • relief (topographical relief, that is)
  • parent material
  • time
  • human influence

o Soil arability has been compromised by human activities, including:
  • agricultural techniques
  • erosion
  • deforestation
  • irrigation techniques

Chapter 5
The Inhabitants of Planet Earth and Their Relationships

As you probably learned in your biology class, the nonliving components of Earth are known as its **abiotic** components. These include the atmosphere, hydrosphere, and lithosphere—the things we studied in the last chapter. In this chapter, we'll talk about the **biotic**, or living components, of Earth. We'll start by discussing elements that bridge the gap between the nonliving and the living—water, nitrogen, carbon, and phosphorus—and how they cycle through the environment. We'll then move on to a discussion of what types of ecosystems exist and how they're structured. We'll continue our review by discussing how energy flows through ecosystems, and we'll end the chapter with a review of how ecosystems change. Let's begin!

CYCLES IN NATURE

As you may have learned in your biology class, nutrients such as carbon, oxygen, nitrogen, phosphorus, sulfur, and water all move through the environment in complex cycles known as **biogeochemical cycles**. Well, you'll need to know a bit about these cycles for the exam, so we'll go through each of them here.

As you can probably tell from the collective name of these natural cycles, living organisms, geologic formations, and chemical substances are all involved in these cycles. Keep in mind that when we describe the movement of these inorganic compounds, it's important to understand both the destinations of the compounds and how they move toward their destinations. In other words, you'll need to know that water moves from the atmosphere to Earth's surface through precipitation, either in the form of snow or rainfall.

But let's talk about a few things that all of these cycles have in common before we go into each one in detail. First of all, the term **reservoir** is used to describe a place where a large quantity of a nutrient sits for a long period of time (in the water cycle, the ocean is an example of a reservoir). The opposite of a reservoir is an **exchange pool,** which is a site where a nutrient sits for only a short period of time (in the water cycle, a cloud is an example of an exchange pool). The amount of time a nutrient spends in a reservoir or an exchange pool is called its **residency time.** In the water cycle, water might exist in the form of a cloud for a few days, but it might exist as part of the ocean for a thousand years! Perhaps surprisingly, living organisms can also serve as exchange pools and reservoirs for certain nutrients; we'll delve into more about this later.

The energy that drives these biogeochemical cycles in the biosphere comes primarily from two sources: the sun, and the heat energy from the mantle and core of the Earth. The movements of nutrients in all of these cycles may occur via abiotic mechanisms, such as wind, or may occur through biotic mechanisms, such as through living organisms (as we mentioned earlier).

Another important fact to note is that while the **Law of Conservation of Matter** states that matter can neither be created nor destroyed, nutrients can be rendered unavailable for cycling through certain processes—for example, in some cycles, nutrients may be transported to deep ocean sediments where they are locked away interminably.

Though we won't get into a discussion of trace elements here, you should also know that certain trace elements such as zinc, copper, and iron are necessary in small amounts for living organisms. Trace elements can cycle in conjunction with the major nutrients, but there's still much to be discovered about these elements and their biogeochemical cycles. For this exam, just know that there are certain trace elements required by living things that cycle, along with the major elements, through the biosphere.

Let's start with perhaps the best-known biogeochemical cycle: the water cycle.

The Water Cycle

As you might imagine, the water that exists in the atmosphere is in a gaseous state, and when it condenses from the gaseous state to form a liquid or solid, it becomes dense enough to fall to the Earth because of the pull of gravity. This process is formally known as **precipitation.** When precipitation falls onto the Earth, it may infiltrate the surface and percolate through soil and rock until it reaches the water table to become **groundwater,** or it may travel across the land's surface as **runoff** and enter a drainage system, such as a stream or river, which will eventually deposit it into a body of water such as a lake or an ocean. Lakes and oceans are reservoirs for water. In certain cold regions of Earth, water may also be trapped on Earth's surface as snow or ice; in these areas, the blocks of snow or ice are reservoirs.

Water is also cycled through living systems. For example, plants consume water (and carbon dioxide) in the process of photosynthesis, in which they produce carbohydrates and oxygen. Because all living organisms are primarily made up of water, they act as exchange pools for water.

Water is returned to the atmosphere from both Earth's surface and from living organisms in a process called **evaporation.** Specifically, animals respire and release water vapor and additional gases to the atmosphere. In plants, the process of **transpiration** releases large amounts of water into the air. Finally, other major contributors to atmospheric water are the vast number of lakes and oceans on Earth's surface. Incredibly large amounts of water continually evaporate from their surfaces.

Take a look at the following graphic, which shows all of the forms that water takes in the biosphere and atmosphere.

The Water Cycle

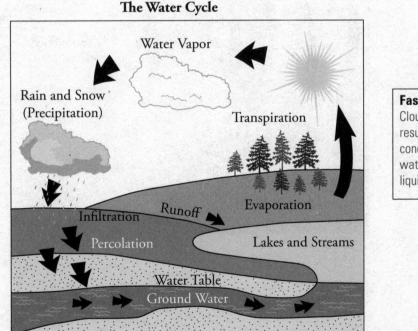

Fast Fact
Cloud formation results from condensation when water vapor becomes liquid or solid.

The Carbon Cycle

Now, let's talk about carbon. The key events in the carbon cycle are **respiration,** in which animals (and plants!) breathe in oxygen and give off carbon dioxide, and **photosynthesis,** in which plants take in carbon dioxide, water, and energy from the sun to produce carbohydrates. In other words, living things act as exchange pools for carbon.

When plants are eaten by animal consumers, the carbon locked in the plant carbohydrates passes to other organisms and continues through the food chain (more on this later in the chapter). In turn, when organisms—both plants and animals—die, their bodies are decomposed through the actions of bacteria and fungi in the soil; this releases CO_2 back into the atmosphere.

One aspect of the carbon cycle that you should definitely be familiar with for the exam is this: when the bodies of once-living organisms are buried deep and subjected to conditions of extreme heat and extreme pressure, this organic matter eventually becomes oil, coal, and gas. Oil, coal, and natural gas are collectively known as fossil fuels, and when fossil fuels are burned, or **combusted,** carbon is released into the atmosphere. Finally, carbon is also released into the atmosphere through volcanic action.

There are three major reservoirs of carbon: the first is the world's oceans, because CO_2 is very soluble in water. The second large reservoir of CO_2 is Earth's rocks. Many types of rocks—called carbonate rocks—contain carbon, in the form of calcium carbonate. Finally, fossil fuels are a huge reservoir of carbon, as well.

The Carbon Cycle

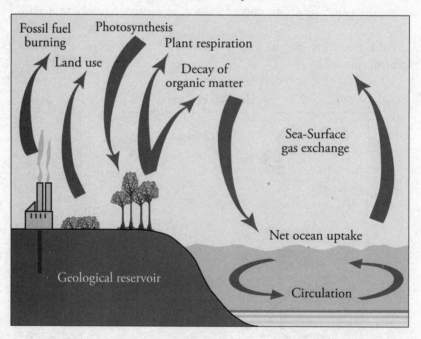

The Nitrogen Cycle

Earth's atmosphere is made up of approximately 78 percent nitrogen and 21 percent oxygen. (The other components of the atmosphere are trace elements and the greenhouse gases.) Nitrogen is the most abundant element in the atmosphere. For this reason, it might not seem like living organisms would find it difficult to get the nitrogen they need in order to live. But it is! This is because atmospheric N_2 is not in a form that can be used directly by most organisms. In order to keep this rather complicated cycle straight, let's look at it in steps.

Step 1: Nitrogen fixation—In order to be used by most living organisms, nitrogen must be present in the form of ammonia (NH_3) or nitrates (NO_3^-). Atmospheric nitrogen can be converted into these forms, or "fixed," by atmospheric effects such as lightning storms, but most nitrogen fixation is the result of the actions of certain soil bacteria. Fixing is the process that allows nitrogen to be made biologically available, much as photosynthesis makes carbon biologically available. One important soil bacteria that participates in nitrogen fixation is *Rhizobium*. These nitrogen-fixing bacteria are often associated with the roots of legumes such as beans or clover. In the future, we may be able to insert the genes for nitrogen fixation into crop plants, such as corn, and reduce the amount of fertilizer that is used.

Step 2: Nitrification—In this process, soil bacteria converts ammonia (NH_3) or ammonium (NH_4^+) into nitrites (NO_2) and then to one of the forms that can be used by plants—nitrate (NO_3^-).

Step 3: Assimilation—In assimilation, plants absorb ammonium (NH_3), ammonia ions (NH_4^+), and nitrate ions (NO_3^-) through their roots. Heterotrophs, or organisms that receive energy by consuming other organisms, then obtain nitrogen when they consume plants' proteins and nucleic acids.

Step 4: Ammonification—In this process, decomposing bacteria convert dead organisms and other waste to ammonia (NH_3) or ammonium ions (NH_4^+), which can be reused by plants or volatilized (released into the atmosphere).

Step 5: Denitrification—In denitrification, specialized bacteria (mostly anaerobic bacteria) convert ammonia back into nitrites and nitrates, and then into nitrogen gas (N_2) and nitrous oxide gas (N_2O). These gases then rise to the atmosphere.

The Nitrogen Cycle

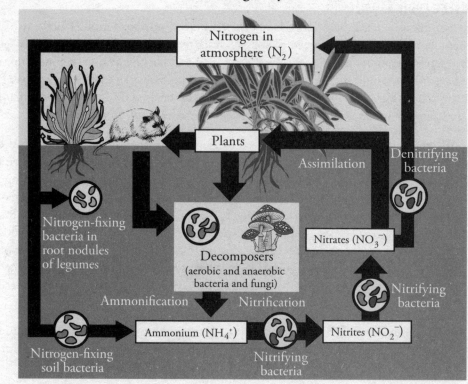

The Phosphorus Cycle

The **phosphorus cycle** is perhaps the simplest biogeochemical cycle, mostly because phosphorus does not exist in the atmosphere outside of dust particles. Phosphorus is necessary for living organisms because it's a major component of nucleic acids, ATP (cellular energy), cell membranes, and other important biological molecules. One important idea for you to remember about the phosphorus cycle is that phosphorus cycles are more local than those of the other important biological compounds.

For the most part, phosphorus is found in soil, rock, and sediments; it's released from these rock forms through the process of chemical weathering. Phosphorus is usually released in the form of phosphate (PO_4^{3-}), which is soluble and can be absorbed from the soil by plants. Symbiotic relationships that form between fungi and plants are known as *mycrrohizae*. In these relationships, *mychorrhizal* fungi colonize the root system of a host plant, which increases the water and nutrient absorption capabilities of the plant, while the plant provides the fungi with carbohydrates formed from photosynthesis. You should know that phosphorus is often a **limiting factor** (any factor that controls a population's growth—food, space, water) for plant growth, so plants that have little phosphorus are stunted.

Phosphates that enter the water table and travel to the oceans can eventually be incorporated into rocks in the ocean floor. Through geologic processes, ocean mixing, and upwelling, these rocks from the seafloor may rise up so that their components once again enter the **terrestrial cycle.** Take a look at the phosphorus cycle shown in the diagram.

Humans have affected the phosphorus cycle by mining phosphorus-rich rocks in order to produce fertilizers. The fertilizers placed on fields can easily leach into the groundwater and find their way into aquatic ecosystems where they can cause eutrophication. **Eutrophication** occurs when a body of water receives excess nutrients. The abundance of nutrients can cause an overgrowth of algae and deplete the water of oxygen.

The Phosphorous Cycle

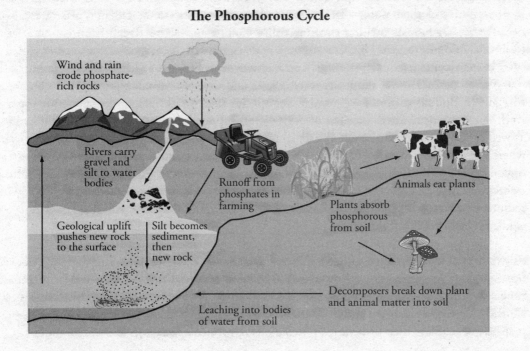

We are almost done with the chemistry. But we need to discuss one more element before we move on to discuss the biosphere, and that's sulfur.

Sulfur

The last biogeochemical cycle we'll talk about is the **sulfur cycle.** Sulfur is one of the components that make up proteins and vitamins, so plants and animals both need sulfur in their diets. Plants absorb sulfur when it is dissolved in water, so they can take it up through their roots when it's dissolved in groundwater. Animals obtain sulfur by consuming plants.

Most of the Earth's sulfur is tied up in rocks and salts or buried deep in the ocean in oceanic sediments, but some sulfur can be found in the atmosphere. The natural ways that sulfur enters the atmosphere are through volcanic eruptions, certain bacterial functions, decomposition in estuaries, and the decay of once-living organisms. When sulfur enters the atmosphere through human activity, it's mainly via industrial processes that produce sulfur dioxide (SO_2) and hydrogen sulfide (H_2S) gases. We'll talk more about sulfur and how it contributes to air pollution in Chapter 9.

All right! It's time to move on to our discussion of the biotic components of Earth. Let's start with a review of how energy moves through ecosystems.

THE WORLD'S ECOSYSTEMS

Because different geographic areas on Earth differ so much in their abiotic and biotic components, we can easily place them in broad categories. The two largest categories are broken down in this way: ecosystems that are based on land are called **biomes,** while those in aqueous environments are known as **aquatic life zones.** Aquatic ecosystems are categorized primarily by the salinity of their water—freshwater and saltwater ecosystems fall into separate categories. Freshwater biomes include streams, rivers, ponds, and lakes. Recall from chapter 4 that freshwater biomes (aquatic life zones) are divided into layers that vary in terms of temperature, oxygen content, and nutrients, or into zones describing what life they can support. The shallows, or littoral zone, support rooted plants and diverse animals that thrive on abundant sunlight (including turtles, frogs, and other species that travel back and forth from water to land. The limnetic zone (the surface of open water) supports photosynthesizing phytoplankton, as well as zooplankton, insects, and fish. The depths, or profundal zone, are the levels of water where light can't reach, so photosynthesizing plants and plankton don't live there. Instead, organisms adapted to little light, colder temperatures, and less oxygen thrive there. The bottom, or benthic zone, of a body of freshwater is inhabited by organisms that live on, in, or below the sediment surface, including bottom-feeders, scavengers, and decomposers (including microorganisms like bacteria and fungi).

Saltwater biomes include oceans, coral reefs, marshland, and estuaries. The divisions between ocean aquatic life zones also follow changes in light, temperature, and oxygen. In the coastal zone, life thrives due to abundant sunlight and oxygen and the proximity of the sediment surface, allowing for varied niches. In addition, costal zones border and extend into estuaries, beaches, and marshes, which have their own varied populations of organisms adapted to their conditions.

The euphotic zone, much like the limnetic zone in freshwater biomes, supports algae as well as fish. Algae in marine biomes supply a large portion of the Earth's oxygen, and also take in carbon dioxide from the atmosphere. Since most red light is absorbed in the top 1 meter of water, and blue light doesn't usually penetrate deeper than 100 meters, photosynthsizers have adapted mechanisms to address the lack of visible light. How deep the euphotic zone extends depends on the turbidity of the water in a given area.

The middle region, or bathyal zone, is colder and darker and does not receive enough light to support photosynthesis, so the density of organisms that live there is less. It's difficult for many fish to live there because of the lack of nutrients, and those that do often lack eyes since there's so little sunlight. It is populated by organisms like sponges and sea stars, as well as larger predators like squid, octopus, sharks, and large whales.

The deepest zone, called the abyssal zone, is extremely cold and dark and has very low oxygen levels, but the level of nutrients is high because of the sinking of decaying plant and animal matter from the higher levels. Without plants, the base level of the food chain in this aquatic zone is decomposers. Many of the creatures adapted to live here produce bioluminescence in order to attract prey or mates, and most are adapted to the cold, low oxygen, and intense pressure, using slower metabolisms and the ability to eat more when food is available to help them survive.

Land environments are separated into biomes based on factors such as climate, geology, soils, topography, hydrology, and vegetation. Although it might seem that each biome listed in the table on the following page is very distinct, in reality, biomes blend into each other; they do not have distinct boundaries. The transitional area where two ecosystems meet actually has a name—these areas are called **ecotones.** Another important term that you should be familiar with for the exam is **ecozones:** ecozones (also called **ecoregions**) are smaller regions within ecosystems that share similar physical features.

Types of Ecosystems			
Biome	**Annual Rainfall, Soil Type**	**Major Vegetation**	**World Location**
Deciduous forest (temperate and tropical)	75–250 cm, rich soil with high organic content	Hardwood trees	North America, Europe, Australia, and Eastern Asia
Tropical rainforest	200–400 cm, poor quality soil	Tall trees with few lower limbs, vines, epiphytes, plants adapted to low light intensity	South America, West Africa, and Southeast Asia
Grasslands	10–60 cm, rich soil	Sod-forming grasses	North American plains, prairie, and savanna; Russian steppes; South African velds; Argentinean pampas
Coniferous forest (Taiga)	20–60 cm—mostly in summer, soil is acidic due to vegetation	Coniferous trees	Northern North America, Northern Eurasia
Tundra	Less than 25 cm, soil is permafrost	Herbaceous plants	The northern latitudes of North America, Europe, and Russia
Chaparral (scrub forest or shrubland)	50–75 cm—mostly in winter, soil is shallow and infertile	Small trees with large, hard evergreen leaves, spiny shrubs	Western North America, the Mediterranean region
Deserts (cold and hot)	Less than 25 cm, soil has a coarse texture (i.e., sandy)	Cactus, other low-water adapted plants	30 degrees north and south of the equator
Temperate rainforest	over 140 cm, soil richer than that in tropical rainforests	Coniferous and broadleaf trees, epiphytes, mosses, ferns, and shrubs	North America, South America, South Africa, Europe, Russia, Northeast Asia, Australia, New Zealand
Savanna	10–30 cm almost all in rainy season, soil is porous and has only a thin layer of humus	Grasses with more widely spaced trees	Australia, South America, India, and half of Africa

Not surprisingly, each biome has specific characteristics that determine the types of organisms that are capable of living in it. Some of these characteristics are the type and availability of nutrients, the ecosystems' temperature, the availability of water, and how much sunlight the region receives. One important law to be familiar with for this test is the **Law of Tolerance.** The Law of Tolerance describes the degree to which living organisms are capable of tolerating changes in their environment. Living organisms exhibit a range of tolerance, and even individuals within a population tolerate changes to their environment differently. This concept is the basis for natural selection, which drives evolution (more on this later in the chapter).

Another important law for you to know is the **Law of the Minimum,** which states that living organisms will continue to live, consuming available materials until the supply of these materials is exhausted.

Ecosystem Diversity

The term *biodiversity* is used to describe the number and variety of organisms found within a specified geographic region, or ecosystem. It also refers to the variability among living organisms, including the variability within and between species and within and between ecosystems. Therefore, when we talk about the biodiversity of an area, we must specify the aspect of biodiversity that we're describing, or else the term is too vague to be comprehensible. **Species richness** refers to the number of different species found in an ecosystem. In general, however, biodiversity in an ecosystem is a good thing. The more biodiversity in a certain species within an ecosystem, the larger and more diverse the species' gene pool, and the greater its chance of adaptation and thus survival.

FOOD CHAINS AND FOOD WEBS

You might recall that the nonliving components of the environment are known as the abiotic components. These include the atmosphere, hydrosphere, and lithosphere. Remember them from Chapter 4? Well, it's time to begin our study of the living, biotic components of Earth. Together, all of the living things on Earth constitute the biosphere.

All living things can be classified by how they obtain food. You might recall that plants and some cyanobacteria are capable of making their own food through photosynthesis, and that some animals (for example, mice) eat plants. Some animals (for example, humans) eat both plants and animals, and some animals (for example, wolves) eat only other animals. There are actually two fancy terms that are normally used to describe these broad categories of organisms: **Autotrophs** are those organisms that can produce their own organic compounds from inorganic chemicals, while **heterotrophs** obtain food energy by consuming other organisms or products created by other organisms.

Finally, as unpleasant as it might be to think about, some animals feed only on the remains of other plants and animals! All of these different types of living things fall into specific categories—and you will definitely need to memorize all of these terms before the test, if you don't already know them!

Producers

Producers are organisms that are capable of converting radiant energy, or chemical energy, into carbohydrates. The group of producers includes plants and algae, both of which can carry out photosynthesis. The overall reaction of photosynthesis is shown below.

$$12H_2O + 6CO_2 + \text{solar energy} \longrightarrow C_6H_{12}O_6 + 6O_2 + 6H_2O$$

While most producers make food through photosynthesis, a few autotrophs make food from inorganic chemicals in **anaerobic** (without oxygen) environments, through the process of chemosynthesis. Chemosynthesis is only carried out by a few specialized bacteria, called **chemotrophs,** some of which are found in hydrothermal vents deep in the ocean. This unbalanced reaction is shown below.

$$CO_2 + 4H_2S + O_2 \longrightarrow CH_2O + 4S + 3H_2O$$

At this point, let's discuss a few other environmental science terms that you'll be required to know for the exam. The **Net Primary Productivity** (**NPP**) is the amount of energy that plants pass on to the community of herbivores in an ecosystem. It is calculated by taking the **Gross Primary Productivity,** which is the amount of sugar that the plants produce in photosynthesis, and subtracting from it the amount of energy the plants need for growth, maintenance, repair, and reproduction. NPP is measured in kilocalories per square meter per year ($kcal/m^2/y$). In other words, the Gross Primary Productivity of an ecosystem is the rate at which the producers are converting solar energy to chemical energy (or, in a hydrothermal ecosystem, the rate of productivity of the chemotrophs). Perhaps not surprisingly, the net productivity of an ecosystem is a limiting factor for its number of consumers. A limiting factor is a factor that controls a population's growth. It can be many things: space, available food, water, nutrients, and as we just mentioned, the net productivity of an ecosystem.

Consumers

Consumers are organisms that must obtain food energy from secondary sources, for example, by eating plant or animal matter. There are a number of different types of consumers, all of which you should commit to memory!

- **Primary consumers:** This category includes the herbivores, which consume only producers (plants and algae).

- **Secondary consumers:** An organism that consumes a primary consumer is a secondary consumer.

- **Tertiary consumers:** An organism that consumes a secondary consumer is a tertiary consumer.

- **Detritivores:** The organisms in this group derive energy from consuming nonliving organic matter such as dead animals or fallen leaves. They include termites, earthworms, and crabs.

- **Decomposers:** These are organisms that consume dead plant and animal material. The process of decomposition returns nutrients to the environment.

- **Saprotrophs:** These are decomposers that use enzymes to break down dead organisms and absorb the nutrients; they include bacteria and fungi.

Note that one organism may occupy multiple levels of a food chain. When eating a hamburger with toppings, you are a primary consumer because you are eating tomatoes and lettuce, and a secondary consumer by eating the beef.

Let's move on and talk about how energy flows through all of these different types of organisms in ecosystems.

Food Chains

As you probably recall, energy flows in one direction through ecosystems: from the sun to producers, to primary consumers, to secondary consumers, to tertiary consumers. In an ecosystem, each of these feeding levels is referred to as a **trophic level**. With each successive trophic level, the amount of energy that's available to the next level decreases. In fact, only about 10 percent of the energy from one trophic level is passed to the next; most is lost as heat, and some is used for metabolism and anabolism. Interestingly enough, this is why food chains rarely have more than four trophic levels.

Food chains are usually represented as a series of steps, in which the bottom step is the producer and the top step is a secondary or tertiary consumer. In food chains, the arrows depict the transfer of energy through the levels, and in fancier food chains, the relative biomass (the dry weight of the group of organisms) of each trophic level will often be represented.

Here's a simple food chain.

Food Chain

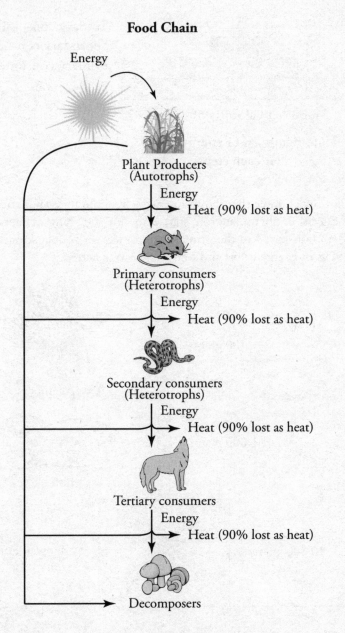

Energy

Plant Producers
(Autotrophs)

Energy

→ Heat (90% lost as heat)

Primary consumers
(Heterotrophs)

Energy

→ Heat (90% lost as heat)

Secondary consumers
(Heterotrophs)

Energy

→ Heat (90% lost as heat)

Tertiary consumers

Energy

→ Heat (90% lost as heat)

Decomposers

What we're showing here is a typical terrestrial food chain, but keep in mind that there are aquatic food chains as well, with algae and different types of fish.

One final note about food chains: in a food chain, only about 10% of the energy is transferred from one level to the next. The other 90% is used for things like respiration, digestion, running away from predators—that is, it's used to power the organism doing the eating! This is known as the **10% Rule**. In other words, the producers have the most energy in an ecosystem; the primary consumers have less energy than producers; the secondary consumers have less energy than the primary consumers; and the tertiary consumers will have the least energy of all. The amount of energy (in kilocalories) available at each trophic level organized from greatest to least is an **energy pyramid**.

Energy Pyramid

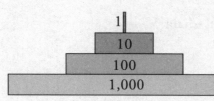

1	Tertiary consumers
10	Secondary consumers
100	Primary consumers
1,000	Producers

Hypothetical units of energy

Note that 90% of energy is lost
at each step

Consider the following example that shows an energy pyramid for an aquatic food chain. In this example, the producers, or phytoplankton, start with 100,000 grams of energy, but only 10,000 grams of energy are transferred to the primary consumers, or zooplankton. With each step up the food chain, 90% of energy is lost and only 10% is transferred.

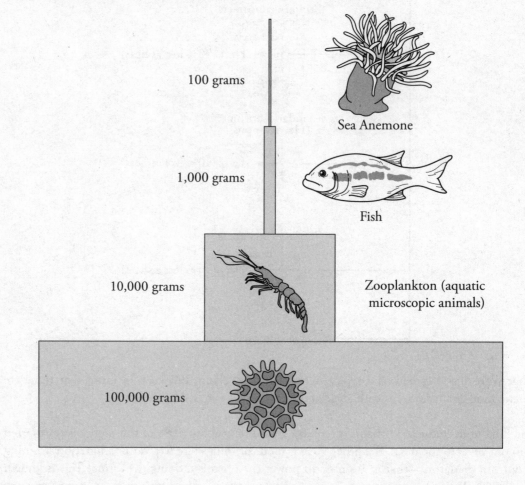

100 grams — Sea Anemone

1,000 grams — Fish

10,000 grams — Zooplankton (aquatic microscopic animals)

100,000 grams

Biomagnification

Food chains represent the flow of energy in an ecosystem, but other things can flow through food chains, too—including environmental toxins. While the amount of useable energy decreases at every level of the food chain, the concentration of certain toxins increases at each successive level, since most toxins cannot be broken down by organisms. The term **bioaccumulation** is used to describe the accumulation of a substance, such as a toxic chemical, in the tissues of a living organism, such as a producer or a primary consumer.

Biomagnification is the term used to describe the increasing concentration of these toxin molecules at successively higher trophic levels in a food chain. Keep in mind that although really any type of molecule could be described using the terms bioaccumulation and biomagnification, generally these terms are used to describe toxins and heavy metals. And certainly, this is how they'll be used on the test.

> There are also pyramid models for biomass and numbers at each trophic level. They demonstrate the decreasing number of organisms at each level, which is why tertiary consumers are the first to be affected by environmental problems.

PCB Bioaccumulation

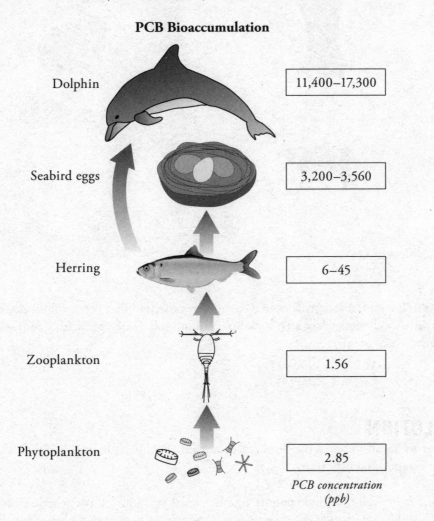

Dolphin	11,400–17,300
Seabird eggs	3,200–3,560
Herring	6–45
Zooplankton	1.56
Phytoplankton	2.85

PCB concentration (ppb)

Food Webs—Tangled Food Chains

As you're probably already aware, food chains are an oversimplified way of demonstrating the myriad feeding relationships that exist in ecosystems. Because there are so many different types of species of plants and animals in ecosystems, their relationships in real-world ecosystems are much more complicated than can be depicted in a single food chain. Therefore, we use a **food web** in order to represent feeding relationships in ecosystems more realistically.

Food Web

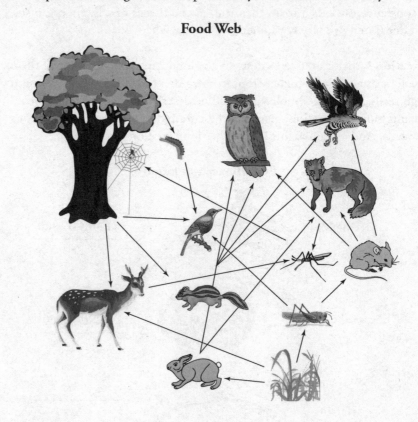

Again, this is a typical terrestrial food web, but keep in mind that very complicated aquatic food webs exist, as well! Let's take a step back for a minute and discuss the setting for food chains and food webs—ecosystems.

EVOLUTION

Biodiversity in all forms is the result of **evolution.** Evolution is the change in a population's genetic composition over time.

We use a figure called a **phylogenetic tree** to model evolution. Phylogenetic trees can be very broad, like the one on the next page, which encompasses many types of species, or they can be very specific and describe the evolutionary relationships that exist between two species (or even the genomes of one species!).

While you won't need to know much about evolution for this exam, you will need to have a rough idea of how and why it takes place, so we'll run through that now. Without trying to recreate the evolution of all living organisms, we will limit our discussion to a description of how new species are formed. This process is called **speciation**.

Strictly speaking, a **species** is defined as a group of organisms that are capable of breeding with one another—and incapable of breeding with other species. As you may recall, individual organisms that are better adapted for their environment will live and reproduce, ensuring that their genes are part of their population's next generation. This is what Charles Darwin meant by **evolutionary fitness.**

Phylogenetic Tree

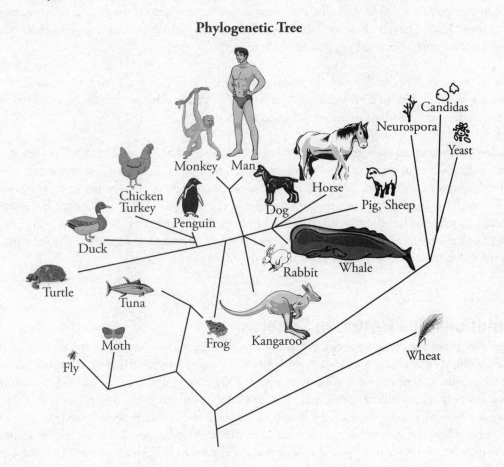

How Evolution Works

When a habitat (an organism's physical surroundings) selects certain organisms to live and reproduce and others to die, that population is said to be undergoing **natural selection**. In natural selection, beneficial characteristics that can be inherited are passed down to the next generation, and unfavorable characteristics that can be inherited become less common in the population. It is important to remember that natural selection acts upon a whole population, not on an individual organism during its lifetime. What changes during evolution is the total genetic makeup of the population, or **gene pool,** and natural selection is one of the mechanisms by which evolution operates.

Fast Fact
The term *survival of the fittest* does not necessarily refer to the fastest or the strongest. It refers to those organisms that produce offspring that will go on to also produce offspring.

The other way evolution operates is **genetic drift**. Genetic drift is the accumulation of changes in the frequency of alleles (versions of a gene) over time due to sampling errors—changes that occur as a result of random chance. For example, in a population of owls there may be an equal chance of a newly born owlet having long talons or short talons, but due to random breeding variances, a slightly larger number of long-taloned owlets are born. Over many generations, this slight variance can develop into a larger trend, until the majority of owls in that population have long talons. These breeding variances could be a result of a chance event—such as an earthquake that drastically reduces the size of the nesting population one year. Small populations are more sensitive to the effects of genetic drift than large, diverse populations.

When a population displays small-scale changes over a relatively short period of time, **micro-evolution** has occurred. **Macroevolution** refers to large-scale patterns of evolution within biological organisms over a long period of time.

Just as new species are formed by natural selection and genetic drift, other species may become extinct. **Extinction** occurs when a species cannot adapt quickly enough to environmental change and all members of the species die.

Biological extinction is the true extermination of a species. There are no individuals of this species left on the planet (for example, the dodo bird or passenger pigeon). **Ecological extinction** is when there are so few individuals of a species that this species can no longer perform its ecological function (for example, alligators in the Everglades in the 1960s or wolves in Yellowstone before re-introduction in the last decade). **Commercial or economic extinction** is where a few individuals exist but the effort needed to locate and harvest them is not worth the expense (for example, the groundfish population of the Grand Banks off the Maritimes of Canada).

Relationships Between Species

Let's talk more about how species get along together in ecosystems. You probably recall from your biology class that a group of organisms of the same species is called a **population,** and when populations of different species occupy the same geographic area, they form a **community.** Every species within a community has an ecological niche. A species' **niche** is described as the total sum of a species' use of the biotic and abiotic resources in its environment. The niche describes where the species lives, what it eats, and all of the other resources the species utilizes in an ecosystem. Another term you should know for the exam is habitat—a habitat is the area or environment where an organism or ecological community normally lives or occurs.

Species can be generalist or specialist. A specialist species is one that has a narrow niche and can only live in a certain habitat. A generalist species is one that has a broad niche, is highly adaptable, and can live in varied habitats. Specialist species tend to have an advantage when their environments are relatively unchanging, while generalist species have the advantage in habitats that undergo frequent change.

Some species interact quite a bit with other members of their population; for example, some animals form herds, while other species are loners—like bears. The reasons for these different levels of sociability are largely competition, predation, and a general need to exploit the resources in the environment.

Competition arises when two individuals—of the same species or of different species—are competing for resources in the environment. When the two individuals that are competing are of the same species, this is called **intraspecific competition,** and when they are of different species, it's called **interspecific competition.** The resources that are competed for can be food, air, shelter, sunlight, and various other factors necessary for life; individuals may be competing to live in a fallen tree, to catch a running rabbit, or to mate with the most desirable female in the population. The competitor who is "most fit" eventually wins and obtains the resource. That's right—the others are eliminated by competition.

One more thing about competition: when two different species in a region compete and the better adapted species wins, this phenomenon is called **competitive exclusion. Gause's principle** states that no two species can occupy the same niche at the same time and that the species that is less fit to live in the environment will relocate, die out, or occupy a smaller niche. When a species occupies a smaller niche than it would in the absence of competition, this compromised niche is called its **realized niche.** (The niche it would have if there were no competition is known as its **fundamental niche.**) Direct competition can also be avoided in the cause of **resource partitioning**. This occurs when different species use slightly different parts of the habitat, but rely on the same resource. For example, there are five species of warblers that can all live in the same pine tree. They can coexist because each species feeds in a different part of the tree: the trunk, at the ends of the branches, and at other sites. Keep in mind that many types of species can engage in both short- and long-term migration, for reasons including food and water availability, temperature changes, mating opportunities, and safety from predation. This means that a given species might be part of several different communities at different times, and might fill a given niche in each of those communities only some of the time. All right, moving on!

Know the three types of relationships between species: competition, predation, and symbiotic relationships.

Although it's relatively easy to observe competition between animals, competition between plants is much more subtle and occurs much more slowly. However, if you have a few years to kill, spend some time in your backyard watching the trees and other plants grow. You'll see that they compete for sunlight and for ground space; they even produce chemicals that inhibit other plants' growth!

The second important type of interspecies interaction is predation. **Predation** occurs when one species (a **predator**) feeds on another (**prey**), and it drives changes in population size. For example, in a year in which rainfall is relatively high in some regions, rabbits have plenty of food; this enables them to reproduce very successfully, and the number of rabbits in a population will increase dramatically. In turn, if the coyote is a predator of the rabbit, coyotes will have plenty of food, and their population will also boom. However, if the following year the rainfall is below average, there will be less grass. Then the population of rabbits will decline, and this will result in a decline in the population of coyotes. As a final note about predation: while it's tempting to think of predation existing only between animals, remember that herbivores prey on plants and zooplankton on phytoplankton!

A third type of relationship that exists between organisms is the symbiotic relationship. **Symbiotic relationships** are close, prolonged associations between two or more different organisms of different species that may, but do not necessarily, benefit each member. There are three types of symbiotic relationships, and you should be familiar with all of these for exam day. In mutualistic symbiotic relationships (**mutualism**), both species benefit; for example, this type of relationship exists between sea anemone and clown fish. The clown fish protects the sea anemone from some of its predators, while the stinging cells of the anemone protect the clown fish; the fish also eats some of the detritus left behind when the anemone feeds. In commensalistic symbiotic relationships (**commensalism**), one organism benefits while the other is neither helped nor hurt. One example of this type of relationship exists between trees and epiphytes (bromeliads and some orchids). The trees are not affected by the epiphytes growing in them, and the epiphytes benefit by collecting water running down the bark and get better access to light than they would on the ground. Finally, **parasitism** is a relationship in which one species is harmed and the other benefits; for example, the relationship that exists between fleas and dogs.

ECOSYSTEM SERVICES

Ecosystem services are benefits that humans receive from the ecosystems in nature when they function properly. There are four categories: provisioning services: providing humans with water, food, medicinal resources, raw materials, energy, and ornaments; regulating services: waste decomposition and detoxification, purification of water and air, pest and disease control and regulation of prey populations through predation, and carbon sequestration; cultural services: use of nature for science and education, therapeutic and recreational uses, and spiritual and cultural uses; and supporting services (the ones that make other services possible): primary production, nutrient recycling, soil formation, and pollination.

Human disruptions to ecosystem services can detrimentally affect our ability to benefit from them, resulting in ecological and economic consequences for us.

HOW ECOSYSTEMS CHANGE

When something changes in an ecosystem, the effects of that change can quickly spread, partly because food webs link species together. Food webs contain positive and negative feedback loops, so that when one species is added or removed, the rest of the food web is affected, sometimes drastically.

The more genetically diverse a given population is, the better it can cope with an environmental disturbance.

Believe it or not, oftentimes the biotic balance in a community is maintained by a single species, known as the **keystone species.** The name *keystone* comes from the last stone placed in an arch bridge, which is the key stone. A keystone species is a species whose very presence contributes to an ecosystem's diversity and whose extinction would consequently lead to the extinction of other forms of life. For example, fig trees are the keystone species in a tropical forest; likewise, wolves were introduced back into Yellowstone Park because without wolves to control the number of herbivores, the ecosystem had drastically changed. As a general rule, if the keystone species is removed from an ecosystem, then the ecosystem completely changes.

Indicator species are species that are used as a standard to evaluate the health of an ecosystem. They are more sensitive to biological changes within their ecosystems than are other species, so they can be used as an early warning system to detect dangerous changes to a community. Trout are a common indicator species, because they are particularly sensitive to pollutants in water. The disappearance of trout from a particular habitat is a warning that that habitat is becoming polluted.

Indigenous species are those that originate and live or occur naturally in an area or environment. With increasing frequency, however, new species are being introduced into ecosystems by chance, by accident, or with intention. While some introduced species cannot find a niche and die out, many others are quite happy in their new environment, and compete successfully with the indigenous species. One example of this is grey squirrels, which were introduced to England in 1876. The grey squirrel competed with England's native species of squirrel, the red squirrel, and today there are fewer than 30,000 red squirrels alive in England. Another example of the harm that introduced species can do was seen when, in 1904, a fungus was introduced accidentally into the deciduous forests of the eastern United States. This fungus caused a blight that killed nearly all of the chestnut trees by the early 1950s.

Although some people don't like to use the term *invasive species* because they feel that it's derogatory, it is often used to describe introduced species. Two other examples of invasive species are zebra mussels, which were introduced into the Great Lakes when ships dumped ballast water into the lakes, and the quickly growing vine kudzu, which was originally introduced in the southeastern United States in order to control the problem of erosion.

Ecological Succession

Communities are not static; they are constantly changing. Species of plants and animals are continually coming and going, evolving and dying out. Some of the changes that take place in a geographic area are predictable ones that can be described as **ecological succession.**

If ecological succession begins in a virtually lifeless area, such as the area below a retreating glacier, it is called **primary succession. Secondary succession** is ecological succession that takes place where an existing community has been cleared (by disturbance events such as fire, tornado, or human impact), but the soil has been left intact. The organisms in the first stages of either type of succession are referred to as **pioneer species**, and typically have wide ranges of environmental tolerance. These pioneers, over time, usually adapt to the particular conditions of the habitat. This may result in the origin of new species. The communities in each stage of succession facilitate the environmental changes that will allow the next stage to take over. The final stage of succession, in which there is a dynamic balance between the abiotic and biotic components of the community, is referred to as the **climax community.**

Ecological Succession

Bare rock

↓

Lichen, Algae, Mosses, Bacteria
(Break down rock and leave organic debris
which together form soil)

↓

Grasses
(Add organic matter to soil and anchor it in place)

↓

Small herbaceous plants
(Continue to add organic matter to soil)

↓

Small bushes
(Add shelter and shade for other plants)

↓

Conifers
(Create additional habitats)

↓

Short-lived hardwoods such as
dogwood and red maple
(Can tolerate shade of conifers but
are short-lived and vulnerable to damage)

↓

Long-lived hardwoods
(More specialized, hardier hardwoods
such as oak and hickory)

How does a new habitat full of bare rocks eventually turn into a forest? The first stage of the job usually falls to a community of lichens. Lichens are hardy organisms. They can invade an area, land on bare rocks and erode the rock surface, and over time turn them into soil. Lichens are pioneer organisms. Once lichens have made an area more habitable, other organisms can settle in. Lichens are replaced (out-competed!) by mosses and ferns, which in turn are replaced by tough grasses, then low shrubs, then conifers, then short-lived hardwood trees such as dogwood and red maple trees, and finally long-lived hardwood trees. Refer to the flowchart of ecological succession for a deciduous forest. Note that the stages are classified by the major new plant group, but remember that with the introduction of each new plant species comes an array of different animal species that exploit it.

What happens when an ecosystem is threatened or habitat is lost? This habitat loss tends to lead to a loss of specialist species, which then can lead to a loss of generalist species. Additionally, species that require large territories tend to suffer, and their numbers are reduced. When a large proportion of a population is lost, this leads to a bottleneck, which can reduce genetic diversity within the species. Succession in a disturbed ecosystem will affect the total biomass, species richness, and net productivity over time. Ecosystems with larger numbers of species (species richness) tend to recover more easily from disruptions.

We're almost done with this chapter, but you need to know a few more terms and concepts before you can move on to the population chapter. The following material will almost certainly be asked about on the test.

When the size of an organism's natural habitat is reduced, or when, for example, development occurs that isolates the habitat, this process is called **habitat fragmentation.** Habitat fragmentation can be quite damaging. As you know, ecosystems are not isolated; they abut each other and meet at wide and overlapping boundaries, called **ecotones.** At these boundaries, there is greater species diversity and biological density than there is in the heart of ecological communities, and this is called the **edge effect.** Some species can only live on the edge of certain habitats, and if the boundaries of a habitat are changed, a new edge is created, damaging both the edge and interior habitats.

The **theory of island biogeography** is a field that studies species richness and diversification in isolated communities: oceanic islands, and also other isolated ecosystems, such as mountain peaks, oases, seamounts, and fragments of habitat separated by human development. The number of species found on an island or in an isolated area is determined by two factors: immigration and extinction. Since islands are often colonized by new species arriving from elsewhere, immigration is a main factor. Once established on an island or in an isolated ecosystem, many species evolve to become specialists as an adaptation to the limited resources available. Immigration becomes a factor again when invasive species arrive, since the typically-generalist invasives may outcompete the native specialists and threaten their long-term survival.

When the changes taking place in a geographic area result from less-predictable events and have more drastic consequences for an ecosystem, they may be considered disruptions. Human-made disruptions such as pollution, habitat destruction, and depletion of natural resources will a major focus in the coming chapters. But what about natural disruptions? Some natural disruptions to ecosystems have environmental consequences that can exceed those caused by humans. Earth's climate has changed, over the course of geological time, many times and for varied reasons, including internal causes (changes in the type and distribution of species and the effects they produce on climate and changes in ocean-atmosphere circulations) and external ones (changes in Earth's orbit, solar output, volcanism, plate tectonics and the resulting size and configuration of continents, and asteroid impacts). Some of these factors cause periodic or episodic change, while others cause change at random times, and the timescales involved vary greatly.

When global climate change occurs, the effects are far-reaching. For example, changes in the amount of glacial ice on Earth have caused sea levels to vary quite significantly over geological history, which in turn changes the size and shapes of landmasses. When big changes occur in the environment, habitats can change on enormous scales, which in turn can cause extinctions, bottlenecks, and short- and long-term migrations among the species inhabiting the affected ecosystems.

Whew. You're done with this chapter! Before moving on to the next chapter (Population Ecology), answer all of the questions in the following drill—and don't forget to use the techniques you learned in Part III.

CHAPTER 5 KEY TERMS

Make sure you know these words and how to use them in your essays.

Cycles in Nature
biotic
abiotic
biogeochemical cycles
reservoir
exchange pool
residency time
Law of Conservation of Matter
precipitation
groundwater
runoff
evaporation
transpiration
respiration
photosynthesis
combusted
nitrogen fixation
nitrification
assimilation
ammonification
denitrification
phosphorus cycle
terrestrial cycle
eutrophication
sulfur cycle

Food Chains and Food Webs
10% Rule
autotroph
heterotroph
chemotroph
producer
anaerobic
Net Primary Productivity and Gross Primary
    Productivity
consumers: primary, secondary, tertiary
detritivore
saprotroph
decomposer
trophic level
food chain
energy pyramid
limiting factor
bioaccumulation
biomagnification

food web

Ecosystems
biomes
aquatic life zones
ecotone
ecozone
deciduous forests
coniferous forests
tropical rainforest
grasslands
tundra
chaparral
deserts
Law of Tolerance
Law of the Minimum
Species richness

Evolution
evolution
phylogenetic tree
speciation
competitive exclusion
resource partitioning
microevolution
macroevolution
population
community
niche
habitat
competition: intraspecific, interspecific
Gause's principle
realized niche, fundamental niche
predation
predator
prey
symbiotic relationship
mutualism
commensalism
parasitism
species
natural selection
gene pool
genetic drift
extinction

biological extinction
ecological extinction
commercial/economic extinction
evolutionary fitness

How Ecosystems Change
keystone species
indicator species
indigenous species
invasive species

ecological succession
primary succession
secondary succession
pioneer species
climax community
habitat fragmentation
ecotones
edge effect
island biography theory

Chapter 5 Drill

Directions: Each of the questions or incomplete statements below is followed by four suggested answers or completions. Select the one that is best in each case. For answers and explanations, see Chapter 13.

1. The relationship between a tick and a bird is best described as which of the following?

    (A) Commensalism
    (B) Mutualism
    (C) Parasitism
    (D) Neutralism

2. When two species live in the same habitat and use exactly the same resources, which of the following will probably occur?

    (A) The two species can live together indefinitely.
    (B) One of the species will eventually go extinct.
    (C) One species will evolve into a parasite.
    (D) The two species do not interact.

3. Organisms use different resources in the same habitat, and in this way avoid competition. This is referred to as

    (A) the Law of Tolerance
    (B) hunting and gathering
    (C) a predator-prey relationship
    (D) resource partitioning

4. Which of the following is true about the roles of both parasites and predators in ecosystems?

    (A) Predators and parasites can act as environmental resistance and allow the host population to grow.
    (B) Predators are generally smaller and parasites support many predators.
    (C) Predators generally have specialized means to capture prey.
    (D) Parasites and predators eliminate the weak and sick, leaving the strongest to reproduce.

5. All of the following are true concerning the characteristics of a climax community EXCEPT

    (A) the adult plants are small in size
    (B) there are many different species of plants
    (C) there is a mixture of decomposers, producers, and consumers
    (D) most of the organisms are specialists in their niche requirements

6. Which of the following describes the direction of the flow of energy in a food chain?

    (A) From parasite to host
    (B) From predator to prey
    (C) From prey to predator
    (D) From one mutual to another

7. Which of the following element's cycles includes long-term storage in rocks and a short storage time in the atmosphere?

    (A) Sulfur
    (B) Carbon
    (C) Nitrogen
    (D) Calcium

8. The current trend where some species of bacteria have become resistant to antibiotics is best described as

    (A) genetic diversity
    (B) speciation
    (C) extinction
    (D) microevolution

9. Large herds of grazing mammals are most likely to be located in a

    (A) rain forest
    (B) estuary
    (C) coniferous forest
    (D) grassland

10. Which of the following organisms is not likely to be found as a member of the detritus food web?

    (A) Ants
    (B) Earthworms
    (C) Fungi
    (D) Deer

Free-Response Question

1. Students from a local high school participated in a study of Hillside Pond. After safely taking samples of some small fish, a fish-eating hawk, some pond water, some zooplankton, and a fish that preys on the small fish, they determined the average concentration of compound "X" in each sample. The table below summarizes their data.

Organism	Compound "X" concentration
Small fish	0.1 ppm
Hawk	3.0 ppm
Pond water	0.1 ppb
Zooplankton	0.2 ppb
Predatory fish	1.0 ppm

(a) **Describe** ONE process that would cause compound "X" to contaminate the pond's water.

(b) **Draw** a food chain that illustrates the correct trophic order in the pond. Include the concentrations of compound "X" for each part of the chain.

(c) **Describe** a process that would explain the different concentrations of compound "X" in each organism.

(d) **Describe** ONE real-life example of a substance that behaves like compound "X" in the oceans. Give ONE negative effect that the substance might have on humans.

Summary

o   The five major biogeochemical cycles you should know are:
   • water
   • carbon
   • nitrogen
   • phosphorus
   • sulfur

o   Each cycle has both a natural cycle and anthropogenic influence.

o   Food chains and webs are made up of the following different categories of species:
   • autotrophs
   • producers
   • detritivores
   • heterotrophs
   • consumers
   • decomposers

o   All food chains move energy through the various trophic levels that make up the system.
   • All energy originates from the sun.
   • The 10% Rule describes the movement of energy between the levels.
   • Bioaccumulation organisms at lower trophic levels accumulate persistent toxins.
   • Biomagnification describes the movement of substances, such as toxins, between the trophic levels.

o   There are several major biomes on Earth (deciduous forest, tropical rainforest, grasslands, taiga, tundra, chaparral, and deserts) and aquatic life zones that make up the biosphere. Terrestrial biomes are defined by their average rainfall and annual temperature, while aquatic zones are categorized by the salinity. The species that survive in each zone have a specific range of tolerance. Ecotones are the regions where different biomes overlap.

o   The biodiversity on Earth is the result of speciation and natural selection of traits. Extinction allows for niches to become available and new traits or species to evolve to use those niches.

o   Organization of living things can be thought of as follows:

Biosphere ⟶ Ecosystems ⟶ Communities ⟶ Populations (of species with unique niches)

o   Species relationships within a community include the following:
   • competition
   • predation
   • parasitism
   • resource partitioning
   • commensalism
   • mutualism

o   Communities can have species with very specific roles, such as:
   • keystone species
   • indicator species
   • indigenous species

o   Ecological succession describes how ecosystems recover after a disturbance in terms of a disturbance event (primary or secondary succession) and the stage of succession (pioneer species, mid-succession species, climax communities).

Chapter 6
Population
Ecology

Topics in this chapter can be found in the "Populations" unit in the new Course and Exam Description.

In this chapter, we'll start by discussing some important characteristics of populations and then lead you through a short section on how and why populations grow. Next, we'll get to the heart of the topic in a section specifically devoted to human population growth. Remember to use the techniques you learned in Part III as you complete the drills—the more practice you have using those techniques, the better prepared you'll be on test day.

SOME IMPORTANT TERMS USED TO DESCRIBE POPULATIONS

A **population** is defined as a group of organisms of the same species that inhabits a defined geographic area at the same time. Individuals in a population generally breed with one another, rely on the same resources to live, and are influenced by the same factors in their environment.

Two important characteristics of populations are the density of the population and how the population is dispersed. **Population density** refers to the number of individuals of a population that inhabit a certain unit of land or water area. An example of population density would be the number of squirrels that inhabit a particular forest. **Population dispersion** is a little more complicated; this term refers to how individuals of a population are spaced within a region. There are three main ways in which populations of species can be dispersed, and you should know all of them for the test.

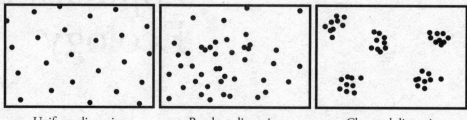

| Uniform dispersion | Random dispersion | Clumped dispersion |

- **Uniform:** The members of the population are uniformly spaced through-out their geographic region. This is seen in forests, in which trees are uniformly distributed so that each receives adequate light and water. Uniform dispersion is often the result of competition for resources in an ecosystem.

- **Random:** The position of each individual is not determined or influenced by the other members of the population. An example is seen in species of plants that are interspersed in fields or forests—the location of their growth is random and relative to other species, not their population. This type of dispersion is relatively uncommon.

- **Clumped:** The most common dispersion pattern for populations. In this type of dispersion, individuals "flock together." This makes sense for many species—many species of plants tend to grow together in a location or habitat that is near their parents and suits their requirements for life; fish swim in schools to avoid predation; and birds and many other animals migrate in groups.

POPULATION GROWTH

So, we know what populations are and how they're dispersed, but how do populations grow? What determines if they will or will not grow? When populations do grow, what are the trends? These are all questions that you'll need to be able to answer on test day. Let's review the basics of population size and growth before we get into a more specific discussion of how human population growth occurs.

The **biotic potential** of a population is the amount that the population would grow if there were unlimited resources in its environment. This is not a practical model for population growth simply because in reality the amount of resources in the environments of populations is limited.

As we reviewed in the last chapter, in every ecosystem, members of a population compete for space, light, air, water, and food. The **carrying capacity** (**K**) of a particular species in a particular environment is defined as the maximum population size for the species that can sustainably be supported by the available resources in that environment. As you might expect, a given geographic region will have different carrying capacities for populations of different species—because different species have different requirements for life. For example, within a certain area, you would expect a population of bacteria to be quite a bit larger—in terms of the number of individuals—than a population of zebras. This is because individual bacteria are much smaller than individual zebras; thus, each bacterium requires fewer resources to live than each zebra. These differences in population size may be driven not just by the different sizes of individual organisms of each species, but by each species' resource requirements and the particular array of resources available in the area.

Population Growth Graphs

If we looked at the growth of a population of bacteria in a petri dish with plenty of food, the curve produced by plotting the increase in their number over time would be in the shape of a J, because the bacteria would grow exponentially. The exponential growth curve is shown below.

Exponential (Unrestricted) Growth

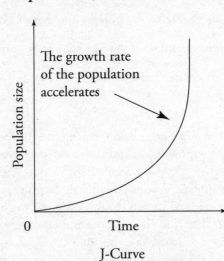

J-Curve

Now, we said that this **exponential population growth** rate is seen where resources are unlimited, but in nature, such ideal conditions are rare and fleeting. In a more realistic model for population growth, after the initial burst in population, the growth rate generally drops, and the curve ultimately resembles a flattened S.

This type of growth, which is a much better model for what exists in natural settings, is called **logistic population growth**. The logistic growth model basically says that when populations are well below the size dictated by the carrying capacity of the region they live in, they will grow exponentially, but as they approach the carrying capacity, their growth rate will decrease and the size of the population will eventually become stable. This logistic growth is shown below.

Logistic (Restricted) Growth

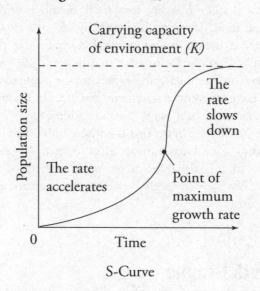

S-Curve

We can predict long-term population growth rates using a model called the Rule of 70. The **Rule of 70** says that the time it takes for a population to double can be approximated by dividing 70 by the current growth rate of the population. For example, if the growth rate of a population is 5 percent, then the population will double in 14 years ($\frac{70}{5}$ percent = 14 years). The Rule of 70 can be used to estimate the number of years for any variable to double, the doubling time.

Not surprisingly, the rate of growth of a population depends on the species. Species can be divided into two groups based on their reproductive strategies: the *r*-selected pattern or the *K*-selected pattern. ***r*-selected organisms** have populations below the carrying capacity of their environment, which means that population growth is constrained only by the species' own biological limits. Competition for resources in *r*-selected species' habitats is usually relatively low. These organisms tend to be small and have short lifespans; they mature and reproduce early in life and have many offspring at once—they may have so many that they only reproduce once in a lifetime. Thus they have a high capacity for reproductive growth. Some examples of *r*-selected species are bacteria, algae, and protozoa. In these species, little or no care is given to the offspring, but due to the sheer numbers of offspring in the population, enough of the offspring will survive to enable the population to continue. On the other hand, **K-selected organisms** have populations whose growth is limited by the carrying capacity of the environment; they live in stable environments where competition for resources is relatively high. These organisms tend to be large and have longer life spans. They mature and reproduce later in life after years of parental care, produce fewer offspring per reproduction event (though they tend to reproduce more than once in their lifetimes), and devote significant time and energy to the nurture of offspring. For these species, it is important to preserve as many members of the offspring as possible because they produce so few: parents have a tremendous investment in each individual offspring. Some examples of *K*-selected species are humans, lions, and cows. Invasive species tend to be *r*-selected, while the species most adversely affected by invasives tend to be *K*-selected. Many species lie on the continuum between these two strategies, and some can change strategies in different conditions or at different times, but the groups are useful for broad comparisons.

Survivorship curves represent the number of individuals in a population born at a given time (called a **cohort**) that remain alive as time goes on. Different species have different survivorship curves depending on their life cycles and life strategies, including *r*- and *K*-selection.

Survivorship Curve

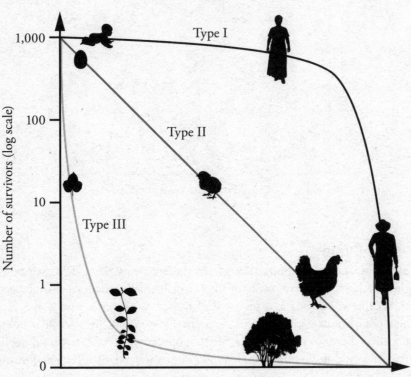

Survivorship indicates the probability that a given organism will live to a certain age.

Population Cycles

When we observe populations in their natural habitats, there are two distinct patterns that occur. These are the boom-and-bust cycle and the predator-prey cycle. Let's look at both of them, as they are important for the exam.

Boom-and-Bust Cycle

The **boom-and-bust cycle** is very common among *r*-strategists. In this type of cycle, there is a rapid increase in the population and then an equally rapid drop off. These rapid changes may be linked to predictable cycles in the environment (temperature or nutrient availability, for example). These cycles may reflect regular changes in rainfall, temperature, or nutrient availability over the course of the year. Or they may reflect longer and less regular cycles. When the conditions are good for growth, the population increases rapidly. When the conditions for that population worsen, its numbers rapidly decline. You might say that their strategy is "get it while the getting's good." Study the graph below so you can see this type of cycle in action.

Boom-Bust Cycle

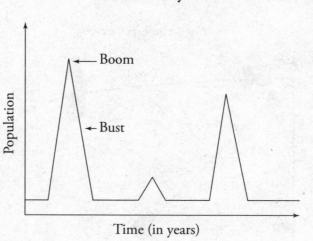

Predator-Prey Cycle

Remember the rabbit and coyote populations from the last chapter? We discussed how in a year of relatively high rainfall, rabbits have plenty of food, which enables them to reproduce very successfully. In turn, because the coyote is a predator of the rabbit, coyotes would also have plenty of food, and their populations would also rise rapidly. However, if the rainfall is below average a few years later, then there would be less grass, the population of rabbits would decline, and the coyote population would decline in turn. The graph of the predator-prey relationship looks like the following.

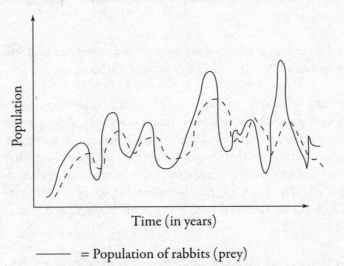

Predator-Prey Cycle

—————— = Population of rabbits (prey)

- - - - - = Population of coyotes (predator)

Something important to notice in this graph is that the coyote population does not change at exactly the same time as the rabbit population. The coyote population actually rises *after* the rabbit population does. That is because the rabbit population has to have time to build up to fairly high levels before the coyotes can find enough to eat. When there is enough food, the coyote mothers have enough energy to give birth to and feed their pups. Only then can the coyote population increase.

The predator-prey cycle also plays a role in understanding why many endangered species are large carnivores. Large predator populations can suffer directly if humans alter their natural habitats, but they can also suffer indirectly if humans kill off their prey. If the prey population falls so low that the predator cannot find food, then the predator population will decline, sometimes to the point of extinction.

Factors Influencing Population Growth

There are population-limiting factors that are purely the result of the size of the population itself. For example, in many populations of species in nature, birth and death rates are influenced by the density of the population. Other **density-dependent** factors that influence population size are increased predation (which occurs because there are more members of the population to attract predators); competition for food or living space; disease (which can spread more rapidly in overcrowded populations); and the buildup of toxic materials.

Some population-limiting factors operate independently of the population size. These **density-independent** factors will change the population's size regardless of whether the population is large or small. Independent factors include fire, storms, earthquakes, and other catastrophic events.

> **Remember Your Bio Definitions!**
> A carnivore is an animal that consumes only other animals, a herbivore consumes only plants, and an omnivore consumes both plants and animals.

Type I (*K*-selected)—The convex shape of this curve indicates that most individuals in the population survive into adulthood, with a sharp increase in mortality as the population approaches the species' maximum age.

Type II—Mortality and survival rates are fairly constant throughout life. Many bird species, mice, and some species of lizards exhibit this straight-line pattern.

Type III (*r*-selected)—The convex shape indicates that most offspring die young, but if they live to a certain age, they will live a longer life. Species with this curve produce high numbers of offspring that encounter bottlenecks to survival that wipe out most young, and parents provide little or no nurture to their young. Examples include plants that produce millions of seeds throughout their lifetimes, and most marine invertebrates. A clam, for example, produces millions of eggs, but the larvae are highly vulnerable to dying off from ocean currents and predators. The individuals that live long enough to develop their shell, however, will live to advanced age.

Most actual populations exhibit some combination of these patterns. For example, at different points in human history or in different societies, infant mortality has been unusually high, resulting in a sharp dip in the survivorship curve before it flattens out to the typical convex shape. Crustaceans like crabs and lobsters are most vulnerable while molting (replacing the hard shell). Since these species molt regularly throughout their lives, their survivorship curves show a stair-step pattern.

Now that you have a basic understanding of how and why populations change in size, let's move on to discuss human populations more specifically.

HUMAN POPULATIONS

You might have heard something about human population growth as you read the news or studied biology and earth science in school. But do you know how many humans are on the planet now? Do you know how fast the human population is growing? You'll need to know for test day.

How Many People Are There in the World?

According to the Central Intelligence Agency's World Fact Book, the world population as of January 2020 is approximately 7.6 billion. The birth rate has actually fallen in the United States and most developed (industrialized) countries worldwide. But that only means that, in prosperous countries, the population is increasing more slowly, and overall the world's population is still increasing. Take a look at the following table, which lists the world's most populated countries as projected for July 2020.

Top 10 Most Populous Countries	
Country	**Estimated Population (July 2020 projected)**
China	1,394,015,977
India	1,326,093,247
United States	332,639,102
Indonesia	267,026,366
Pakistan	233,500,636
Nigeria	214,028,302
Brazil	211,715,973
Bangladesh	162,650,853
Russia	141,722,205
Mexico	128,649,565

Source: Central Intelligence Agency, The World Fact Book Database.

The following graph shows how the world's human population has increased since the 1950s, and how it is predicted to increase, though less rapidly, into the 2050s.

World Population, 1950–2050

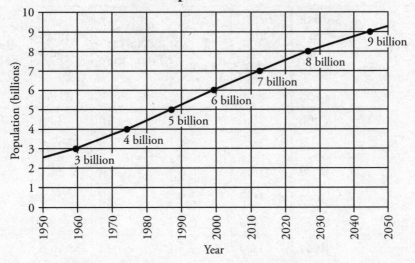

Projected global population growth 1950–2050. Source: U.S. Census Bureau, International Data Base, August 2016 update.

The population of many of the countries shown in the table is currently increasing in size. We can determine the rate of population change per year of a country by using the following formula.

$$\text{Population Change Over Time} = \frac{(\text{birth rate} + \text{immigration rate}) - (\text{death rate} + \text{emigration rate})}{10}$$

All rates in the formula are raw numbers per 1,000 people per year. For example, the birth rate (the **crude birth rate**) is equal to the number of live births per 1,000 members of the population in a year, and the death rate (or **crude death rate**) is equal to the number of deaths per 1,000 members of the population in a year. A simpler version of the above formula, called the **Actual Growth Rate**, considers only the birth and death rates, excluding immigration and emigration: (birth rate − death rate) / 10.

The table below gives the 2020 estimated growth rates for a selection of 11 countries, to show the range in this figure. Also shown are the crude birth and death rates as well as key demographic information on median age and population density per unit area. Note that some countries are experiencing negative growth. The statistics on median age are one indication of quality of life, while the information on population density suggests that not all densely populated countries are poor and not all sparsely populated countries are prosperous.

2020 Growth Rates Chart

Country	Growth Rate % Per Year	Birth Rate Per 1,000	Death Rate Per 1,000	Rank in World Population	Median Age	Population Density (per mile²)
Nigeria	+2.53	34.6	9.1	6	18.6	571
Mexico	+1.06	17.6	5.4	10	29.3	176
India	+1.1	18.2	7.3	2	28.7	1,192
Indonesia	+0.79	15.4	6.6	4	31.1	385
Canada	+0.81	10.2	7.9	38	41.8	11
United States	+0.72	12.4	8.3	3	38.5	93
United Kingdom	+0.49	11.9	9.5	22	40.6	717
China	+0.32	11.6	8.2	1	38.4	392
Russia	−0.16	10	13.4	9	40.3	23
Japan	−0.27	7.3	10.2	11	48.6	901
Latvia	+1.12	9.2	14.6	153	44.4	80

Source: Census Intelligence Agency, The World Fact Book Database. All data are 2020 estimates.

How Do Human Populations Change?

Populations can also change in number as a result of migration into and out of the population. Two important vocabulary words to describe human migration are **emigration,** which is the movement of people out of a population, and **immigration,** which is the movement of people into a population. In the Annual Growth Rate formula, immigration and emigration would also need to be expressed as rates per thousand in the population. Keep in mind that, in general, emigration and immigration are only small factors in the changes in size of human populations; however, the United States, unlike many other highly developed countries, has the third-largest population due to immigration.

Not surprisingly, a number of factors affect the total fertility rates in a population, and as a result, the population's birth rate. Among these are:

- the availability of birth control

- the demand for children in the labor force

- the base level of education for women

- the existence of public and/or private retirement systems

- the population's religious beliefs, culture, and traditions

The most significant additions to human populations are due to births, plain and simple. The term **total fertility rate (TFR)** is used to describe the number of children a woman will bear during her lifetime, and this information is based on an analysis of data from preceding years for the population in question.

Total fertility rates are predictions that provide a rough estimate, but they can't be depended on because they assume that the conditions of the past will be the conditions of the future.

The **replacement birth rate** of a human population refers to the number of children a couple must have in order to replace themselves in a population. While you might automatically think that the answer is always 2, in reality it is slightly higher to compensate for the deaths of children, the existence of non-child-bearing females in the population, and other factors. In developing countries, the replacement birth rate can be as high as 3.4 because of higher mortality rates! If the fertility rate is at replacement levels, a population is considered relatively stable.

The **infant mortality rate** is the number of deaths of children under 1 year old per 1,000 live births. Obviously, whether mothers have access to good healthcare and nutrition has the greatest effect on the infant mortality rate. Other factors are sanitation, clean drinking water, environmental conditions, and political infrastructure. Changes in these factors can lead to changes in the infant mortality rate over time. As we mentioned earlier, despite the relatively recent drop in total fertility rates worldwide, the world's population is still increasing. This is because many members of the human population are future parents, so even if they only reproduce at a replacement rate, there will be an overall increase in the total population. Now, let's look at a graph of how the overall world population growth rate has changed since 1750 and is projected to change through 2100.

World Projected Population Growth Rate, 1750–2100

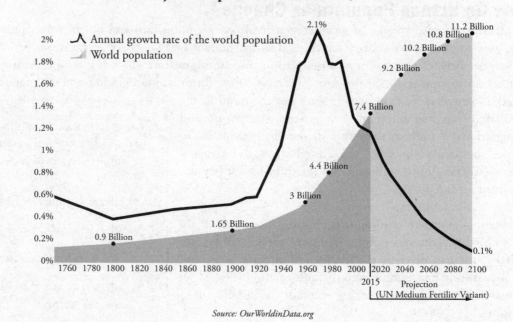

Source: OurWorldinData.org

Though not shown on this graph, the human population has actually been growing exponentially for more than three centuries. But what did we learn earlier in this chapter? That no population can grow exponentially indefinitely...we'll talk more about that in a bit. For now, let's discuss some factors that affect the growth rates of human populations.

Perhaps not surprisingly, there is a strong empirical correlation between the education level of women and the growth rate of populations. Additionally, the reason that religion and culture are predictors of birth rates is that in some countries, certain groups have a proclivity toward reproduction for religious reasons. The reason that the world's population has grown so considerably, especially in the past 100 years, is not because of an increased number of births, but because of the significant drop in the world death rate. People are living longer lives, and there are far fewer infant deaths today than there were 100 years ago. This is due, in large part, to the Industrial Revolution, which improved the standard of living for millions living in industrialized nations. Other causes of the extension of the human life span are the development of clean water sources and better sanitation, the creation of dependable food supplies, and better health care. In general, the overall health of a population can be estimated by examining the expected life span of individuals and the mortality rate of infants.

On the other hand, there are obviously factors that limit human population growth. We've already seen that the availability of birth control and education for women limit the total fertility rate; negative density-independent effects, such as major storms, fires, heat waves, or droughts can also be limiting factors. Additionally, density-dependent effects of a population nearing its carrying capacity, such as decreased access to clean water and air, lack of resources including food, disease outbreaks, and territory size also can limiting growth, showing that the carrying capacity of a given area, and of Earth as a whole, are ultimate factors that will have to be reckoned with.

Age-Structure Pyramids

Age-structure pyramids (also called **age-structure diagrams**) are useful for graphically representing populations. Some age-structure diagrams group humans into three categories by age: those who are **pre-reproductive** (0–14), those who are **reproductive** (15–44), and those who are **post-reproductive** (45 and older). Each of these groups at the same stage of life is also called a **cohort**. Age pyramids, such as the one shown below, group members of the population strictly by age, with each decade representing a different group. The *x*-axis contains the information relating to the percent or number of individuals in each of the age groups.

We can use age-structure pyramids to predict population trends; for example, when the majority of a population is in the post-reproductive category, the population size will decrease in the future because most of its members are incapable of reproducing. The opposite is true if the majority of a population is in the pre-reproductive category; these populations will increase in size as time goes on. Take a look at the shapes of the example age-structure diagrams below representing Nigeria and the United States. As you can see, Nigeria has a large number of pre-reproductive and reproductive members in its population, while the United States has a fairly even distribution. From this, we can see that the population of Nigeria should increase significantly over time—it has what is referred to as **population momentum**—while the population of the United States should grow more slowly.

You should be able to identify the growth rate of a country based upon its age-structure pyramid. Here are the four main types:

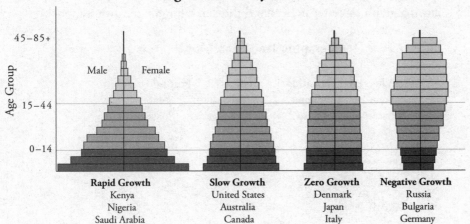

The *x*-axis contains the information relating to the percent or number of individuals in each of the age groups. (On the AP Exam, when a question asks for the percent or number of individuals in a particular age group, be sure to count both males and females.)

The Demographic Transition Model

The **demographic transition model** is used to predict population trends based on the birth and death rates of a population. In this model, a population can experience zero population growth via two different means: as a result of high birth rates and high death rates; or as a result of low birth rates and low death rates. When a population moves from the first state to the second state, the process is called **demographic transition.** The four states that exist during this transition are the following:

1. **Preindustrial state:** In this state, the population exhibits a slow rate of growth and has a high birth rate and high death rate because of harsh living conditions. Harsh living conditions can be considered environmental resistance, an umbrella term for conditions that slow a population's growth.

2. **Transitional state:** In this second state, birth rates are high, but due to better food, water, and health care, death rates are lower. This allows for rapid population growth. Birth rates remain high due to cultural or religious traditions and a lack of education for women.

3. **Industrial state:** In the third state, population growth is still fairly high, but the birth rate drops, becoming similar to the death rate. Many developing countries are currently in the industrial state.

4. **Postindustrial state:** In the final state, the population approaches and reaches a zero growth rate. Populations may also drop below the zero growth rate (as we saw for Japan, Latvia, and Russia in the table of growth rates).

Check out the following graph to better understand the demographic transition model.

Demographic Transition Model

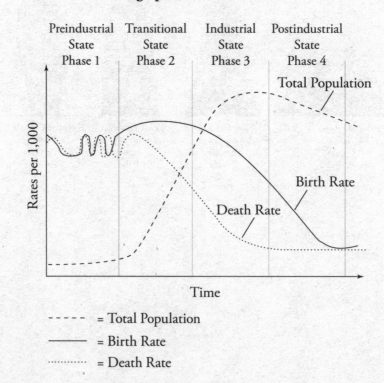

You will most likely see questions that will ask you specifically to describe the demographic transition model or apply it to hypothetical or real population states, so make sure you understand this material. If you don't, go back to your textbook for further explanation!

THE IMPACT OF HUMANS ON EARTH

As you're probably well aware, humans have the greatest impact on the environment of any living species on Earth, and the increase in our population over the last few centuries has seriously and dramatically changed the face of the Earth.

Four of the most significant factors that have contributed to the increase in the world human population are improved nutrition, the availability of clean water, newly implemented systems for sanitary waste disposal, and better medical care.

Another significant factor is the increase in food production. Almost half of Earth's land surface is currently devoted to various ways of producing food for humans; specifically, 12 percent of Earth's land is composed of farms, 11 percent is composed of forests planted by humans, and 26 percent is used for grazing livestock! As we mentioned earlier, this enormous amount of food production takes its toll on the land, and we now face excessive and harmful erosion, in addition to a variety of environmental problems that have resulted from the wide-scale use of irrigation. Finally, the widespread use of pesticides and fertilizers for increased crop yields leaves large amounts of harmful chemical residues in the soil and water.

In response to these problems, the agricultural industry is continuing to invent and promote soil conservation techniques, organic farming, more efficient irrigation methods, and genetically modified crops. However, these new techniques will need to be implemented in all countries in order to be effective globally. In many cases, new techniques introduce new problems— for example, crop sizes may increase, but pesticides must then be used in order to protect the larger crops.

The use of **genetically modified organisms (GMOs)** is also controversial. Inserting strands of DNA that code for resistance to pests or for larger crop size may lead to less genetic diversity. This in turn can lead to the likelihood of crops being susceptible to future pests and diseases. Also, there is no clear data regarding the effect these GMOs may have on human health in the future.

What Happens When There Aren't Enough Resources?

Our bodies need certain nutrients to keep them healthy and to help resist disease. Some nutrients, or **macronutrients,** are needed in large amounts. These include proteins, carbohydrates, and fats. Other nutrients are needed in smaller amounts; these are called **micronutrients**. Micronutrients include vitamins, iron, and minerals such as calcium. When people are deprived of food, one result is the onset of hunger. Technically speaking, **hunger** occurs when insufficient calories are taken in to replace those being expended. **Malnutrition** is poor nutrition that results from an insufficient or poorly balanced diet; those whose diets lack essential vitamins and other components often suffer from it. A third term used to describe those who aren't receiving sufficient

Poverty or Government Warfare?

According to Alex de Waal, author of *Mass Starvation: The History and Future of Famine*, the majority of famines in modern times are caused not by insufficient resources to produce or buy food, but are the result of deliberate warfare by certain governments against their own people.

resources is **undernourished.** Undernourished people have not been provided with sufficient quantity or quality of nourishment to sustain proper health and growth. According to the Food and Agriculture Organization, or FAO, 795 million people on Earth are undernourished. Some 780 million of these people are living in developing countries, but, perhaps surprisingly, the remaining 15 million are living in developed nations. On the other hand, 38 percent of adults in the United States are considered obese and, globally, 2.1 billion people are overweight. Why does this dichotomy exist?

While the reasons for hunger are many and complex, the simplest explanation for the problem is poverty. Our planet produces sufficient food to feed today's world population, but many people lack the money to buy food or the resources to produce it.

All over the world, human communities are trapped in a cycle of poverty, resource degradation, and high fertility. For example, in the first third of the twentieth century, Asia, Africa, and Latin America produced enough grain that it was not necessary for them to import it from other countries. However, because of their constantly increasing populations, all of these countries are now importing grain; this is an ominous sign of impending problems with hunger in these countries.

Encouragingly, China, Thailand, and Indonesia are working hard to implement government reforms that will increase the quality of life for their citizens. In China, for example, as a result of reform and development in rural areas, the number of people in the country without enough food and clothing has decreased from 250 million in 1978 to 23.65 million in 2006. Furthermore, new initiatives, such as the Zero-Hunger Initiative for West Africa, the Asia-Pacific Zero Hunger Challenge, and the Hunger-Free Latin America and the Caribbean Initiative aim to eradicate hunger and achieve food security by 2030.

Hunger in America

Despite its high obesity rate, there are hungry people even in the United States, one of the richest countries in the world—lots of hungry people. Thankfully, the number of hungry people in the United States is less now than it was when international leaders set hunger-cutting goals at the 1996 World Food Summit. At this summit, government leaders pledged to cut the number of Americans living in hunger from 30.4 million to 15.2 million by 2010, but this goal has not been met. According to the United States Department of Agriculture, 40.0 million Americans were considered "food insecure" in 2016. Additionally, more than 13 percent of the U.S. population relies on food stamps. That's 42.6 million Americans!

U.S. Households with Children by Food Security Status of Adults and Children 2018

Food-insecure households—13.9%

Food insecurity among adults only in households with children—6.8%

Food-insecure children—7.1%

Low food security among children—6.5%

Very low food security among children—0.6%

Food-secure households 86.1%

Source: USDA, Economic Research Service using data from U.S. Department of Commerce, U.S. Census Bureau, 2018 Current Population Survey Food Security Supplement.

But why are American people hungry when they live in a nation that's known as the world's breadbasket? Again, the main reason is poverty. Many neighborhoods in which the majority of the citizens who reside there are low-income are often called **food deserts**. This is because access to fresh, healthy food is difficult. The residents rely on low-quality processed foods for subsistence. Lack of education on healthy food choices also contributes to poor nutrition.

In the mid-1990s, a call for welfare reform resulted in the passing of the Personal Responsibility and Work Opportunity Reconciliation Act (PRWORA). The premise of this welfare reform, according to its proponents, was that people who are able to work should be encouraged to find employment, so that they will not remain dependent on government assistance. The act limited the number of people who qualified for food stamps and limited the duration that people could receive food stamps and public support.

While at the time it was generally agreed that welfare reform was necessary, many families are now reaching their deadline for public assistance. Since the implementation of this act, and in the future, it will be of crucial importance for state and local groups to find ways to support the truly needy.

Many grocery stores, restaurants, and individuals waste a lot of food. Food is often discarded because of the "sell by" date, when actually it is still fine to eat. Fruits and vegetables are discarded if they do not look perfect. The dumpsters behind delis and bagel stores are filled with bread at the end of the day, which could have gone to a shelter or food pantry.

In the United States, there are a number of charitable agencies that provide food at no or low cost to those in need. One example is **Feeding America,** which makes use of food that would otherwise go to waste. Feeding America receives food from food processors and distributors and redistributes it via food banks. The organization helps to feed more than 45 million Americans each year.

Global Hunger Policies—For Good or Ill

While social reform is a viable solution to the problem of hunger in the United States, often the only solution in developing countries is to enable communities to become self-sufficient in food procurement. These destitute communities either need monetary resources that will allow them to purchase necessary food supplies or the resources that would enable them to produce their own food.

Another issue that must be dealt with in the struggle against global hunger is that many countries just can't produce the food they need in order to feed their citizens. In these cases, the first viable solution mentioned above—that of providing monetary resources to people so they can purchase food—is rendered irrelevant.

The World Trade Organization (WTO) controls the policies of international trade. Unfortunately for smaller, poorer nations, the economically strong nations of the world have more influence over the creation of policies by the WTO. Oftentimes, trade policies undercut prices for developing nations, which makes it difficult for them to enter the world market. Also, many of the poorest countries have corrupt governments that prevent the distribution of food aid to their own citizens.

Another problem for developing countries is that a trade imbalance exists between them and developed nations. In many poorer nations, national resources have been degraded in an effort to reduce the national debt. The people in these poverty-stricken nations have nothing to export—except labor. Relatively recently, companies in the United States and other developed nations have begun outsourcing jobs to developing countries. Many Americans are appalled at the terrible conditions under which these overseas laborers work—their hours are abnormally long and they are paid almost nothing. Despite these conditions, the competition between poor communities to secure contracts with companies overseas is fierce because often the alternative is continued poverty.

So, Where Do All These People Live?

The majority of humans live in some type of community, and the largest percentage of the human population lives in relatively large communities and urban centers.

Since the development of ancient civilizations, humans have lived together in large centralized communities, or cities. A couple of ancient cities that you may be familiar with are Rome (in what is now Italy) and Athens (in Greece). However, never before have the urban centers of the world grown as quickly as they are growing now. If we traced the growth of urban areas in the United States, we would find that before the Civil War (around the 1850s), only about 15 percent of the population lived in a city. Around the time of World War I (1920), that number grew to encompass about 50 percent of the total population of the United States, and today it hovers around 75 percent.

Globally, almost half of the world's population today lives in an urban area. In the United States, this is partly due to the fact that our aging population has largely moved into the cities to have greater access to health services, employment opportunities, and cultural activities.

When considering those who live in urban areas, we also count those who reside in satellite communities, or **suburbs.** In recent times, with lower oil prices making it easier for people to afford gasoline to commute to and from their jobs in cars, people have moved out of city centers in order to have more living space. Interestingly, people who live in the suburbs, on average, occupy eleven times more space than do those who live in the city. One of the advantages of living in the suburbs is that people have their own land space—a backyard—which they need not share with others.

The term used to describe the emigration of people out of the city and into the suburbs is **urban sprawl.** In some areas of the United States, urban sprawl takes over vast tracts of land. In Colorado, for example, population growth has resulted in a number of new communities between Denver and Boulder; when traveling between the two cities, it is now difficult to determine where the Denver metro area ends and the city of Boulder begins.

When urban areas grow too large and become too dense, distributing water to all citizens becomes increasingly difficult. Coupled with this is the strain on the water supply—more people means more water use. In many of these newly crowded areas, water shortages have led to the implementation of restrictions on water usage.

Another problem that results from the increase in the populations of cities is what to do with all of the waste that's created. When you think about it, almost all human activities create waste—when you go to your local coffee shop and get a cup of coffee, you probably don't think much of it. However, if you get a cup of coffee every morning in a paper cup, five days a week for the 52 weeks of the year, then at the end of the year you've accumulated more than 250 cups! That's a significant pile of garbage.

There are also urban areas that contain abandoned factories or former residential sites. These are referred to as **brownfields**. Any type of redevelopment of these areas is hindered by the possibility that the soil and water are contaminated.

Transportation Alternatives

While many people find the suburbs a pleasant place to live, ecologists and city planners have recently come to realize that this urban sprawl may reduce quality of life for all urban dwellers. One major concern of policy-makers and citizens in metro areas is what to do about transportation. Ideally, people would be encouraged to use mass transit or participate in carpools rather than drive separately in personal vehicles. Having fewer cars on the road decreases air pollution from automobile emissions and makes for less congestion on roadways.

Other environmentally conscious or "green" modes of transportation include bicycles, motor scooters, and electric bikes. Larger cities often opt to build subway systems, but they are extremely expensive to develop and are only cost-effective when there are enough people who will pay to use them. However, city buses are an option for both large and small cities. Fleets of buses are less expensive than subways to create and maintain, and although they contribute to road congestion, they decrease congestion by accommodating more people per vehicle. In addition, many cities have buses that are hybrids to reduce harmful greenhouse gases.

Rapid-rail or light rail systems are more common in Japan and Western Europe than in the United States, but as of recently they're being considered as an option for cities that lack subways. Rapid rail systems work by magnetic levitation; suspended above a track, a train moves along as a result of strong attractive and repulsive magnetic forces.

Building Sustainable Cities

In order for cities to be sustainable, city planners and developers must build and manage cities to work with, and within, their natural settings, instead of merely placing buildings and structures in these settings.

There are certain cities in the United States and elsewhere in the world that are setting examples of progressive thinking in conservation and ecology. For example, the city of Boulder, Colorado, has long been recognized for its forward-thinking, green policies. Bicycle paths cover the city, allowing cyclists to move freely from one area of the community to another. Buses move around the city and in and out of Denver and the surrounding communities, which enables people to commute to work without using their cars and creating more emissions. For those who need to drive to work, the city encourages carpools and provides parking areas for those who carpool. Additionally, the city's strong recycling programs help reduce the amount of material that's added to landfills. The city is ringed by open spaces that can be used by the city population for recreation. These areas are also leased to local ranchers for grazing cattle. Boulder's citizens have voted for tax increases to purchase additional communal green space for the city!

Other cities that have been held up as models for city planning are Curitiba, Brazil, and Portland, Oregon. Curitiba has an excellent mass transit system as well as bicycle paths and pedestrian walkways. The city provides recycling programs, job training, health care, and environmental education for its citizens. Likewise, in the 1970s the state of Oregon became determined to head off urban sprawl, and the city of Portland established zoning policies and restrictive growth policies for urban areas that were adopted statewide. City developers were encouraged to invest in established neighborhoods rather than develop undisturbed areas. The city of Portland established Metro, a regional body that deals with land use, city planning, and the development of natural areas. Light rail systems were developed, and Metro began to encourage neighborhood self-sufficiency in order to keep the number of people who need to commute for food or other supplies to a minimum.

Now and in the future, it will be important for city planners to deal with new problems created as a result of urban sprawl. City planners and developers must take environmental concerns into consideration: providing green spaces and transportation alternatives and planning for the supply of water are all relatively new challenges for those involved in building cities. According to the Global Health Observatory (GHO), the global population went from 34% urban in 1960 to 54% urban in 2014, and the urban population continues to grow.

Big Cities in Less-Developed Countries

So far, we've made it sound like the cities of the world are dealing well with the boom in their population, but this is not true globally. Some cities, called **megacities**, have grown in excess of 10 million people very rapidly. In less-developed countries, this increase in the population size of major cities has many very negative effects. Among the worst effects is a deficiency of housing or habitable areas for the burgeoning population. As a result, people are homeless, become "squatters," or make their homes in areas that are completely undeveloped—areas that have no water, electricity, or stable, durable housing.

Some of the reasons people in less-developed countries are moving to cities are similar to those of people in developed countries: for example, cities have more opportunities for employment. However, these people often have other motivations that drive them out of the country, such as war, religious or cultural persecution, or the degradation of their environment.

Ecological Footprint

One concept that you should definitely be familiar with for this exam is that of the ecological footprint. An **ecological footprint** is used to describe the environmental impact of a population or individual person. It is defined as the amount of Earth's surface that's necessary to supply the needs of, and dispose of the waste of, a particular population or individual. Americans have one of the largest ecological footprints: we require about 9.7 hectares per capita (per person). One hectare is 10,000 square meters, or about 2.5 acres. America's amount is comparatively enormous—the ecological footprint of Indonesia is only 1.1 hectares per capita. In general, affluent populations have a much higher ecological footprint than non-affluent ones.

We can use a mathematical model to describe the impact that humans have on the environment. Nicknamed the **IPAT model**, it is written as

$$I = P \times A \times T$$

In the model, I = the total impact, P = population size, A = affluence, and T = level of technology.

While you probably won't be asked to use this model to calculate the impact of populations on the exam, it's a good idea to know this formula exists and that the variables of a population's size, affluence, and level of technology all influence its environmental impact.

Mass Extinction

During the past 500 million years, there have been five major extinction events, during which a significant percentage of all life on Earth, life inhabiting the oceans included, was exterminated by some catastrophic event. The most famous incident was the asteroid impact 65 million years ago near the present-day Yucatan Peninsula, which eliminated the dinosaurs and other large reptiles, vacating numerous niches that were quickly filled by mammals through radiative adaptation. Radiative adaptation is when, through evolutionary forces of natural selection, a species rapidly diversifies into numerous new species to take advantage of newly freed-up resources. The lesson to gain from this information is that life on Earth is very adaptive and capable of surviving great stress. Presently, humans are stressing the planet and may render Earth uninhabitable for our species. Life, however, will certainly continue.

Threatened and Endangered Species

Another way humans impact Earth is through their interaction with animals. Human activities have caused or contributed to the **extinction** of many species. The International Union for Conservation of Nature (IUCN) evaluates the conservation status of plant and animal species. A species is designated as **critically endangered** if the species is under a very high risk of extinction; **endangered** if the species is likely to become extinct; and **vulnerable** if the species is likely to become endangered if no action is taken. Species assigned to any one of these three categorizations are considered **threatened species.** The IUCN maintains a Red List of Threatened Species that is updated regularly. In 2000, the number of threatened species was just over 10,000. As of 2019, this number hit 30,000. Refer to the figure below.

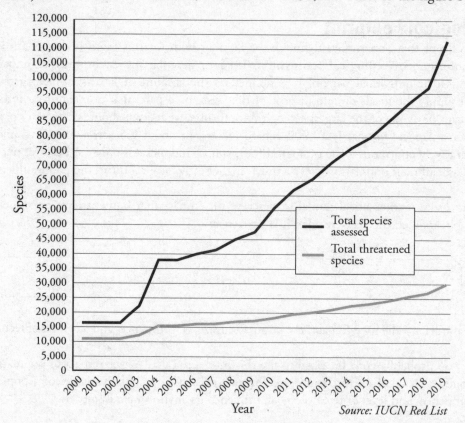

Source: IUCN Red List

Extinctions have happened throughout Earth's history. This natural rate of extinction—occurring apart from widespread events—is called the **background extinction rate**. Knowledgeable scientists estimate that the current extinction rate is between 50 and 500 times higher than in the past, likely due to human influence. Extinctions can happen anywhere in the world, but the rates are particularly high in the tropics (mostly on mountains and islands, which are home to isolated small populations that are especially vulnerable to both natural and human-caused changes to their environments).

The species that are most endangered have several factors in common: they require large ranges of habitat to survive, have low reproductive rates, have specialized feeding habits, have some sort of value to humans (medicinal or food), and have low population numbers.

Humans play a major role in the extinction of species because of our destruction of animal and plant **habitats**. Poverty and rapid population growth cause people to use destructive practices, such as slash-and-burn farming, that destroy species' habitats. When we build roads or cities, habitats are lost or **fragmented** (broken into smaller pieces); this fragmentation may prevent the free movement of a species to find mates or escape danger, or it may reduce the area a species has for all the activities of its life cycle below a critical threshold. Finally, we cause habitat **degradation** by adding pollutants to the environment. Other factors that can contribute to extinction are invasive species and the direct hunting or **overexploitation** of a species or desired products. Dr. Norman Myers coined the term **biodiversity hot spot** to describe a highly diverse region that faces severe threats and has already lost 70 percent of its original natural habitat by area.

There are things we can do to reduce the rate of extinction. Living sustainably and conserving resources helps lower the demand that destroys habitats. Making it illegal to trade in specific organisms means that those organisms will not be hunted or collected. We can also help organisms on a species-by-species approach. Zoos and other institutions have captive breeding programs in which endangered species are bred under human control until their populations are high enough to be reintroduced into the wild. We can conserve habitats by requiring that large tracts of land be set aside and protected from human activity. In these protected habitats, organisms can find their niche and survive without risk of human interference. National parks and animal sanctuaries are two examples of protected habitats.

There are many United States laws that have been passed to reduce the rates of extinctions and protect specific organisms. The following are relevant examples:

Date	Name of Legislation	What It Did
1972	**Marine Mammal Protection Act**	This act protected marine mammals from falling below their optimum sustainable population levels.
1973	**Endangered Species Act** Program for the protection of threatened plants and animals and their habitats	The act prohibited the commerce of those species considered to be endangered or threatened.
1973	Convention on International Trade in Endangered Species of Wild Flora and Fauna (**CITES**)	This international treaty bans the capture, exportation, or sale of endangered and threatened species.

> **Endangered Species Act** and **CITES** are part of the required environmental legislations to know for the exam, so make sure to memorize those two!

Use the acronym **HIPPCO** to memorize the human factors that can cause extinction and decrease biodiversity.

- **H**abitat destruction/fragmentation
- **I**nvasive species
- **P**opulation growth
- **P**ollution
- **C**limate change
- **O**verharvesting/overexploitation

Now you're ready for the key terms review and the following drills. Remember to use our techniques as you go through them.

CHAPTER 6 KEY TERMS

Know these terms backward and forward so that you can spout them in your sleep.

Population Growth

population density

population dispersion: random, clumping, uniform

biotic potential

exponential growth

carrying capacity

logistic population growth

Rule of 70

r-selected, *K*-selected

survivorship curve

cohort

boom-and-bust cycle

overshoot

collapse

perdator-prey cycle

density-dependent and -independent factors

Human Populations

formulas: Population Change over Time, Annual Growth Rate, crude birth rate, crude death rate

emigration

immigration

total fertility rate (TFR)

replacement birth rate

age-structure pyramids: pre-reproductive, reproductive, post-reproductive

population momentum

demographic transition states: preindustrial, transitional, industrial, postindustrial

environmental resistance

genetically modified organisms (GMOs)

macronutrient, micronutrient

hunger, malnutrition

undernourished

food desert

Feeding America

suburbs

urban sprawl

brownfields

megacity

ecological footprint

IPAT model

Threatened and Endangered Species

mass extinction

extinction

threatened

endangered

critically endangered

vulnerable

background extinction rate

fragmented habitat

habitat degradation

overexploitation

biodiversity hot spot

Marine Mammal Protection Act

Endangered Species Act

CITES

HIPPCO

Chapter 6 Drill

Directions: Each of the questions or incomplete statements below is followed by four suggested answers or completions. Select the one that is best in each case. For answers and explanations, see Chapter 13.

1. Populations have all the following characteristics EXCEPT

   (A) density
   (B) dispersion
   (C) habitat
   (D) gene pool

2. Which of the following describes individuals leaving a population?

   (A) Birth rate
   (B) Carrying capacity
   (C) Immigration
   (D) Emigration

3. A population has a growth rate of 2 percent per year. How long will it take for this population to double?

   (A) 70 years
   (B) 40 years
   (C) 35 years
   (D) 15 years

4. An age-structure pyramid is used to

   (A) study the immigration rates in a population
   (B) calculate the doubling time of a population
   (C) study the carrying capacity of a habitat
   (D) study the number and ages of people in a country

5. Which of the following are exhibited by *K*-select organisms?

   I. Slow maturation
   II. Many small offspring
   III. Reproduction occurs late in life

   (A) I only
   (B) II only
   (C) III only
   (D) I and III only

6. A population cycle that is marked by regular increases and decreases in its numbers is correctly said to be

   (A) boom-and-bust
   (B) irruptive
   (C) stable
   (D) logistic

7. The demographic transition model is used to study the

   (A) effects of migration patterns
   (B) influence of industrialization on population growth or decline
   (C) location of large population centers
   (D) benefits of mass transportation projects

8. Which disease is having a severe negative impact on the population in sub-Saharan Africa today?

   (A) Lung cancer
   (B) Heart disease
   (C) HIV/AIDS
   (D) Alzheimer's

9. Which of the numbers below is closest to the population of India?

   (A) 1.3 billion
   (B) 1 billion
   (C) 700 million
   (D) 400 million

10. Which of the following is a density-independent population factor?

    (A) Number of parasites in the population
    (B) Number of predators in the population
    (C) Competition for resources
    (D) Habitat destruction

11. When a population encounters environmental resistance it is most likely to

    (A) continue its high growth rate
    (B) develop new mutations and continue growing
    (C) slow down its growth rate
    (D) move to a higher growth rate

12. A population's growth can best be calculated using which of the following?

    (A) (Births + immigration) – (deaths + emigration)
    (B) Immigration + emigration
    (C) Emigration + births
    (D) (Births + emigration) – (deaths + immigration)

13. Overexploitation of a species can happen by all of the following EXCEPT

    (A) excessive hunting
    (B) use of a species for food
    (C) use of species as a pet
    (D) habitat conservation

14. Poverty can affect population in all of the following ways EXCEPT

    (A) causing premature deaths
    (B) increasing the total fertility rate
    (C) decreasing the total fertility rate
    (D) forcing the use of resources in unsustainable ways

Free-Response Question

1. A habitat's carrying capacity imposes limits on the growth of populations and their consumption of resources.

    (a) **Define** the term *carrying capacity*. Give TWO examples of how carrying capacity can impose limits on a population.
    (b) **Explain** how a population's consumption of natural resources might be controlled. Give TWO examples of how nature slows down the consumption of natural resources by a population.
    (c) **Describe** TWO ways human activity can raise a habitat's carrying capacity for humans.

Summary

- Population patterns: The distribution of a population is influenced by the density a particular species can handle based on its resource use and interspecific competition.

- Population growth: The biotic potential of a population is the amount that the population would grow if there were unlimited resources.

- Limits in the ecosystem define the maximum population size, known as the carrying capacity (K).

- The rate of growth of the population depends on the species.
  - K-selected species
  - r-selected species

- Cyclic patterns in population size can be influenced either by the abiotic environmental changes or the population changes of their predators or prey (food sources).

- Human populations: There are almost 7.5 billion humans on Earth. Although the growth rate has decreased, the population is still increasing.

- Actual growth rate is determined by the difference in birth rate and death rate. Population change over time is also influenced by immigration and emigration rates.

- Total fertility rate (TFR) is greatly influenced by:
  - education and employment opportunities for women (most important)
  - availability of birth control
  - child labor
  - existence of a retirement system
  - cultures, traditions, and religious beliefs

- Replacement birth rates must remain just above 2 due to deaths before reaching reproductive age and females in the population who do not bear children; rates are almost always higher in developing countries due to higher infant and child mortality rates.

- Age-structure pyramids are an essential diagram to be able to read.
  - Large pre-reproductive cohorts imply population growth.
  - Large post-reproductive cohorts imply population decrease.
  - Pyramids which depict little difference between cohorts suggest a stable population with little to no population change.

o   Demographic transition
   • Populations typically achieve zero net population growth toward the end of the industrial state, moving from a period of high birth/high death rates to low birth/low death rates.
   • During the industrial state, however, there is often a large growth in population because the decreasing birth rate lags behind the decreasing death rate.

o   The increasing population of the human species has had major impacts on the planet and its inhabitants:
   • resource depletion
   • hunger and malnutrition
   • threatened and endangered species
   • extinctions
   • development of suburbs/urban sprawl and megacities

o   The ecological footprint of a person varies greatly from country to country. With the growing world population, rising expectations of prosperity, and increasing reliance on technology, our species' collective footprint is now exceeding the space available on Earth.

o   The main factor responsible for threatened and endangered species is habitat loss, but other human influences also play a role:
   • invasive species
   • human population growth
   • pollution
   • climate change
   • overharvesting/overexploitation

o   Calculators are allowed on the current exam; make sure you know how to use your calculator to figure out the following:
   • the Rule of 70 for population doubling
   • a country's Population Change over Time as influenced by four parameters: birth rate, death rate, immigration rate, and emigration

o   Required Environmental Legislation
   • Endangered Species Act of 1973
   • Convention on International Trade in Endangered Species of Wild Fauna and Flora (CITES), an international agreement drawn up in 1973 and enacted in 1975

Chapter 7
Resource
Utilization

Topics in this chapter can be found in the "Land and Water Use" unit in the new Course and Exam Description.

Let's take a step back and review the fundamental themes of this chapter. First of all, a **resource** is strictly defined as any substance, capability (such as work performed by humans or animals), or other asset that is available in a supply that can be accessed and drawn on as needed. Resources are utilized for the effective functioning of an organism or community, often for economic gain. Within the scope of Environmental Science, we typically understand resources to mean natural resources—resources that occur in nature. That is to say, they exist apart from humans, without human effort or intervention prior to our exploitation of them. What natural resources are we talking about? Well, all of Earth's substances and materials that humans rely on to live—namely the land and water and the things growing there or found there. Natural resources can also refer to properties of the natural world such as gravity, magnetic power, and other forces of potential use to humans. As we'll see in this chapter, humans use the land and water for countless reasons.

We will begin our discussion with a description of the resources of the world—including what happens if people don't get enough resources and who has too few. We'll then go through the resources gleaned from agriculture, forests, oceans, and mining. We'll end with a (short!) discussion of the economics behind our resource use. Let's begin!

SHARE AND SHARE ALIKE?

When people talk about managing common property resources such as air, water, and land, **"The Tragedy of the Commons"** often comes to mind. This is an important concept introduced by the English economist William Forster Lloyd in 1833 and later applied to the field of natural resource management by Garrett Hardin in a 1968 paper published in *Science* magazine. In his paper, Hardin referenced the example used by Lloyd, in which a piece of open land, a commons, was to be used collectively by the townspeople for grazing their cattle. Each townsperson who used the land continued to add one cow or ox at a time until the common was overgrazed. Hardin eloquently says, "Each [person] is locked into a system that compels him to increase his herd without limit—in a world that is limited. Ruin is the destination toward which all [people] rush, each pursuing his own best interest in a society that believes in the freedom of the commons. Freedom in a commons brings ruin to all."

The tragedy of the commons serves as a foundation for modern conservation. **Conservation** is the management or regulation of a resource so that its use does not exceed the capacity of the resource to regenerate itself. This is different from **preservation,** which is the maintenance of a species or ecosystem in order to ensure their perpetuation, with no concern as to their potential monetary value.

In this chapter, we'll continue to show how human economics influence how we interact with Earth's resources. Bear in mind that natural resources are drawn from the biotic and abiotic components of functioning ecosystems, and so our exploitation of those resources necessarily affects the functioning of those ecosystems. Human impact in turn affects the ability of those ecosystems to continue providing the resources. When we humans exploit a resource for the functioning of society or for economic gain, we are placing an economic value on it; therefore, natural resources are described in terms of their value as **ecosystem capital** or natural capital.

Some Terms Used to Describe Resources

Let's start by discussing the two main types of resources.

- **Renewable resources** are resources such as plants and animals. These resources can be regenerated quickly. Water is an abiotic substance that's renewable because it can be used over and over again and because sources of water are replenished naturally through the water cycle. Certain natural sources of energy—such as the sun, the wind, and the tides—are also considered renewable because their occurrence in nature is perpetual and not depleted with use. The time necessary for hardwood trees to mature (about 50 years) is widely considered the crossover point from renewable resources to nonrenewable resources. But, in purely practical terms, a resource is renewable if it can be replenished within the time it takes to draw down its supply. Bear in mind that even renewable resources must be carefully managed in order to conserve their sources and insure an ongoing supply.

- **Nonrenewable resources** are things like minerals and fossil fuels. Nonrenewable resources are typically formed by very slow geologic processes, so we consider them incapable of being regenerated within the realm of human existence.

There are a couple more terms you should know before we dive into our review of the major resources available to humans on Earth; these are consumption and production. The **consumption** of natural resources refers to the day-to-day use of environmental resources such as food, clothing, and housing. On the other hand, **production** refers to the use of environmental resources for profit. An example of this might be a fisherman who sells his fish in a market. Got those terms? Let's move on.

AGRICULTURE

How do resources relate to your dinner? Well, 77 percent of the world's food comes from croplands, 16 percent comes from grazing lands, and 7 percent comes from ocean resources. Despite the importance of our ever-increasing population, fewer people than ever in the history of the United States now farm the land. Why is this? The short answer is that it has a lot to do with increasing urbanization (as we mentioned in the last chapter) and industrialization. Now that machines are readily available to work the land and harvest crops, farms have become more like factories—currently only 2 percent of the United States population is directly employed in agriculture. Farms in the United States today are quite a bit larger than farms of the past. The average farm is 434 acres, or a little less than an American football field, while in the early 20th century the average farm size was about 100 acres.

The use of machinery in farming has allowed a farmer to work more land more efficiently; however, one of the drawbacks of the machinery is the amount of fossil fuel needed to power it. As the cost of fuel rises, the cost of food will also rise.

This rise in agricultural productivity can be tied to new pesticides and fertilizers, expanded irrigation, and the development of new high-yield seed types. However, it has also resulted in a significant decrease in the genetic variability of crop plants and led to huge problems in erosion.

Traditional Agriculture and the Green Revolution

Throughout most of history, agriculture all over the world was such that each family grew crops for itself, and families relied primarily on animal and human labor to plant and harvest crops. This process is called **traditional subsistence agriculture,** and it provides enough food for one family's survival. Traditional subsistence agriculture is currently practiced by about 42 percent of the world's population, predominantly in developing nations. Such intensive mixed farming allows people to settle permanently and subsist without having to migrate seasonally. Extensive subsistence agriculture results in low amounts of labor inputs per unit of land.

Do not confuse the Green Revolution, which is about farming, with the Green Movement, which is about conservation.

One form of traditional agriculture that's still practiced in many developing countries today is a method called **slash and burn,** a practice that dates back to early humankind and is especially common in the tropics. In slash and burn, an area of vegetation is cut down and burned before being planted with crops. Tropical soils are typically thin and poor, and whatever fertility they hold is rapidly depleted by the deforestation and subsequent farming. Therefore, the farmer must leave the area after a relatively short time and find another location to clear. Practiced indiscriminately on a broad scale, slash-and-burn agriculture has led to rapid deforestation of the tropical rainforest.

The **Green Revolution**, which occurred in the 1950s and 1960s, is generally thought of as the time after the Industrial Revolution when farming became mechanized and crop yields in industrialized nations boomed. Such innovations also allowed farmers in the Third World to increase crop production on small plots of land. There is a second green revolution, which promotes integrated pest management and organic methods, such as fertilizers that are not synthetic.

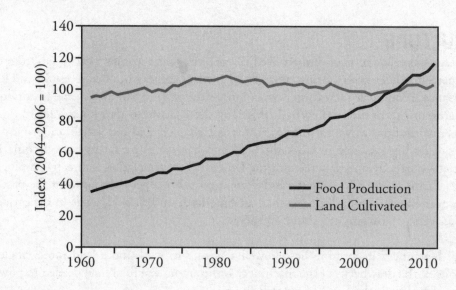

Fertilizers and Pesticides

One factor that contributed to the Green Revolution was an increase in the use of fertilizers and pesticides. Interestingly, when the non-native settlers (the first white settlers) planted their first corn crops, certain tribes of Native Americans taught them to plant fish leftovers (the inedible parts of the fish) along with the corn seed. The fish acted as a natural fertilizer for the crops.

As you can see, manures and other organic materials have been used as fertilizers by farmers for many years. However, the development of inorganic (chemical) fertilizers brought about the huge increases in farm production seen during the Green Revolution. It's estimated that if chemical fertilizers were suddenly no longer used, then the total output of food in the world would drop about 40 percent!

Of course, there are downsides to the widespread use of chemical fertilizers, including the reduction of organic matter and oxygen in soil; the large amounts of energy needed to produce, transport, and supply the fertilizers; and the fact that once the fertilizers are washed into watersheds, they are dangerous pollutants.

Likewise, the increased use of pesticides in the Green Revolution has significantly reduced the number of crops lost to insects, fungi, and other pests, but these chemicals have also had an effect on ecosystems in and surrounding farms. It's estimated that the average insect pesticide will only be useful for 5–10 years before its target pest evolves to become immune to its effects through natural selection; therefore, new pesticides must constantly be developed. However, even with this constant development, crop loss due to pests has not decreased since 1970, although the use of pesticides has tripled!

Because the use of pesticides is so prevalent in the United States, Congress passed the Federal Insecticide, Fungicide, and Rodenticide Act (FIFRA) in 1947 and amended it in 1972. This law requires the EPA to approve the use of all pesticides in the United States.

Irrigation

Another major contributor to the increased crop yields seen in the Green Revolution was advanced irrigation techniques, which allowed crops to be planted in areas that normally would not have enough precipitation to sustain them. However, repeated irrigation can cause serious problems, including a significant buildup of salts on the soil's surface, which makes the land unusable for crops. To combat this **salinization** of the land, farmers have begun flooding fields with massive amounts of water in order to move the salt deeper into the soil. The drawback to this, however, is that the large amounts of water can waterlog plant roots, which will kill the crops, and this process also causes the water table of the region to rise. Furthermore, the water for these irrigation farms comes from underground water tables called aquifers. These aquifers are being depleted at a rapid rate and large-scale, grain-producing countries such as India, China, and the United States are examples of those caught in this predicament.

Integrated Pest Management (IPM)

When dealing with pests, **integrated pest management** is a more environmentally sensitive approach than chemical pesticides that uses a combination of several methods. Rather than try to get rid of every single pest on the farm, IPM tries to keep the pest population down to an economically viable level. Some of the methods include introducing natural insect predators to the area, intercropping, using mulch to control weeds, diversifying crops, crop rotation, releasing pheromone or hormone interrupters, using traps, and constructing barriers. People using IPM consider using chemical pesticides only in the worst-case scenario.

Genetically Engineered Plants

The third and last significant contributor to the Green Revolution was the introduction of genetically engineered plants. In genetic engineering, scientists try to improve plants by adding genes from one species to another to encourage desirable characteristics, such as longer shelf life, disease/drought/pest resistance, faster growth, and higher crop yields. One beneficial example of this method was the development of golden rice, which contains vitamin A and iron. The introduction of this rice addresses two of the serious health problems that are seen in developing nations: vitamin A deficiency, which can result in blindness and other serious health problems; and iron deficiency, which leads to anemia.

However, there are many problems that arise from **genetically modified organisms** (GMO), as well. Therefore, GMOs have become a very controversial topic. Because this is a relatively new technology, scientists don't know exactly how GMOs will affect the planet ecologically. Genetically modified plants discourage biodiversity, which may harm beneficial insects and organisms, could pose new allergen risks, may increase antibiotic resistance, and could cause new pesticide-resistant pests. Many farmers and consumers are also concerned that cross pollination can contaminate other crops, including organic farms that choose not to use GMOs, or cause unwanted mutations with unknown results.

Monotonous Monoculture

Believe it or not, three grains provide more than half of the total calories that are consumed world-wide! These three crops are rice, wheat, and corn, and the phenomenal increase in the yield of these crops was a result of genetic engineering. Genetic engineers discovered a way to cause plants to divert more of their photosynthetic products (called **photosynthate**) to grain biomass rather than plant body biomass.

It's estimated that of the roughly 30,000 plant species that could possibly be used for food, only 10,000 have been used historically with any regularity. Today, 90 percent of the caloric intake worldwide is supplied by just fourteen plant species and eight terrestrial animal species! In other words, today's agriculture represents a major reduction in agricultural biodiversity.

Much of the farming that occurs today is characterized by **monoculture.** In a monoculture, just one type of plant is planted in a large area. Monocultures became common in the era of early political civilizations, when farms produced a staple crop in order to feed whole societies and armies. As we discussed earlier, this has proved to be an unwise practice for numerous reasons. **Plantation farming,** which is practiced mainly in tropical developing nations, is a type of industrialized agriculture in which a monoculture cash crop such as bananas, coffee, or vegetables, is grown and then exported to developed nations.

Soil Degradation

Have you ever read *The Grapes of Wrath* by John Steinbeck? Well, the story in this book took place in the 1930s, when droughts in the Great Plains reduced the area to a giant **Dust Bowl**. Although the drought was the major cause of the Dust Bowl, farming practices used at that time also contributed to the destruction of the land.

In an effort to address the Dust Bowl and other agricultural problems, the United States Soil and Conservation Service (today it's called the National Resources Conservation Service) was established, and it passed the Soil Conservation Act in 1935. Conservation districts were set up by the Service, and these franchises provided education to farmers.

Today, farmers can protect soil from degradation in numerous ways. The practice of **contour plowing,** in which rows of crops are plowed across the hillside, prevents the erosion that can occur when rows are cut up and down on a slope. **Terracing** also aids in preventing soil erosion on steep slopes. Terraces are flat platforms that are cut into the hillside to provide a level planting surface; this reduces the soil runoff from the slope. Additionally, **no-till methods** are quite beneficial; in no-till agriculture, farmers plant seeds without using a plow to turn the soil. Soil loses most of its carbon content during plowing. Plowing accelerates the decomposition of organic matter in the soil, decreasing soil fertility and releasing carbon dioxide gas into the atmosphere. (And as you know, increased levels of CO_2 in the atmosphere have been associated with global climate change!)

Finally, **crop rotation** can provide soils with nutrients when legumes are part of the cycle of crops in an area. An alternate to crop rotation is **intercropping** (also called **strip cropping**), which is the practice of planting bands of different crops in a field. This type of planting can also prevent some erosion by creating an extensive network of roots. As you might be aware, plant roots hold the soil in place and reduce or prevent soil erosion.

The Livestock Business

Perhaps not surprisingly, the introduction of all these new agricultural techniques has significantly affected the livestock business. Free-range grazing, which simply implies that animals are able to move about outdoors and eat the foods they are adapted to eat, is a new term for the traditional way livestock animals were fed: by grazing on the land. In contrast, new meat production industry techniques include feedlots, or concentrated animal feeding operations (CAFOs), in which animals are confined and concentrated into smaller spaces in order to keep costs down and quickly get livestock ready for slaughter. They tend to be crowded and are often fed grains or feed rather than grass. Feedlots often require the use of antibiotics to prevent the spread of disease among animals densely packed together, and create problems in disposing of animal waste, which can contaminate ground and surface water. Manure is not used as fertilizer due to difficulty with transport. It has instead become the most widespread source of water pollution in the United States. Free-range animals tend to be free from antibiotics and their waste can be used as fertilizer, but since this method requires large areas of land, the meat produced is more expensive for consumers. As long as the grazing area is sufficient for the number of animals, livestock grazing is a sustainable practice. If, however, grass is consumed by animals at a faster rate than it can re-grow, land is considered **overgrazed.** Overgrazing is harmful to the soil because it leads to erosion and soil compaction. Overgrazing can cause desertification: the degradation of low-precipitation regions toward being increasingly arid until they become deserts. One solution to the problem of overgrazing is similar to crop rotation—animals can be rotated from site to site and away from their source of water. Another solution involves the overall control of herd numbers.

Various tracts of public lands are available for use as rangeland, and cooperation between government agents, environmentalists, and ranchers can help avoid problems of overgrazing on these lands. The Bureau of Land Management is responsible for managing federal rangelands.

Grazing animals also consume 70 percent of the total grain crop consumed in the United States, making them expensive food stuff. Meat production is less efficient than agriculture: it takes approximately 20 times more land to produce the same number of calories from meat as from plants. One possible solution is for people to consume less meat overall: this could reduce emissions such as carbon dioxide, methane, and nitrogen oxides; conserve water and reduce water pollution; reduce the use of antibiotics and growth hormones; and improve topsoil.

FOREST RESOURCES

Many environmentalists are concerned about the deforestation that is taking place in North America. It is interesting to note that the number of trees growing in North America is approximately the same as 100 years ago, but only 5 percent of the original forests are left. The numbers are approximately the same because of the number of trees growing in national parks and tree plantations. What does this mean? It means that most of the trees in North America are young, and that most forests have been harvested and replanted, and have undergone significant succession.

Deforestation

Deforestation, or the removal of trees for agricultural purposes or purposes of exportation, is a major issue for conservationists and environmentalists. Worldwide, industrialized countries have a higher demand for wood and less deforestation, while developing countries exhibit a smaller demand for wood, but more deforestation. This can be partly explained by the fact that the deforestation that occurs in developing countries primarily takes place because land is being cleared for pastures and farms. Industrialized countries also import lumber from developing countries.

Nearly all of the deforestation that takes place in North America is done in order to create space for homes and agricultural plots. In sites where deforestation is occurring, the impact on resident ecosystems is significant. Take, for instance, Canada's Vancouver Island. On this island, whole mountainsides have been stripped bare of the centuries-old forests that once existed. While the lumber industry tries to offset this destruction by planting new trees, the saplings, which won't be harvestable for another 50 years, are no substitute for forests of 300-foot giant redwoods. Remember how we talked about ecological succession in Chapter 5? Where do you think all of the plants and animals that relied upon this old growth forest ecosystem (which was a climax community) went to live?

Despite the moral questionability of this habitat destruction, the lumber industry will not be asked to leave Vancouver Island. This is because it's the island's most important source of income. Fifty cents of every dollar the island earns comes from lumbering—this number easily beats the island's income from tourism, which is the runner-up. This type of deforestation, also called **clear-cutting,** has other consequences as well. The areas affected experience a great deal of runoff due to the loss of root structure, which leads to more erosion. The soil is washed into freshwater streams and rivers and makes the environment less suited for salmon. The loss of shade also leads to higher stream temperatures, which also affect aquatic organisms.

Another environmentally negative by-product of deforestation is seen in countries with tropical forests. In these forests, when trees are removed and farms are placed in the cleared land, the already-poor soil is further degraded, and the area can only support crops for a short time. Usually, once the soil will no longer support a crop, the land will be used for grazing, but the soil becomes more and more depleted over time until it has no use for humans.

Additionally, any forest is made up of trees, which absorb pollutants and store carbon dioxide. Cutting and burning trees releases carbon dioxide (along with preventing those trees from absorbing it in the future), so deforestation contributes to climate change. The negative repercussions of clearing tropical rainforests—the losses in biodiversity, and the erosion and depletion of nutrients in the soil, and the release of carbon dioxide—seem to outweigh the economic gains in many people's opinions. However, for those who would like to take a stand by refusing to purchase wood from tropical rainforests, it is often difficult to determine which wood products come from tropical rainforests cleared for slash-and-burn agriculture and which come from sustainable forests. Various organizations, such as the nonprofit group the Forest Stewardship Council, have developed certifying procedures based on standards that will encourage only the use of wood from sustainable forests.

How Can We Use Forests Sustainably?

There are three major types of forests, which are categorized based on the age and structure of their trees. An **old growth forest** is one that has never been cut; these forests have not been seriously disturbed for several hundred years. Not surprisingly, the controversies that revolve around the issue of deforestation are primarily centered on instances in which deforestation is occurring in old growth forests. As we mentioned in the last section, old growth forests contain incredible biodiversity, with myriad habitats and highly evolved, intricate niches for a multitude of organisms. **Second growth forests** are areas where cutting has occurred and a new, younger forest has arisen naturally. About 95 percent of the world's forests are naturally occurring, and the remaining forests are known as **plantations** or **tree farms.** Plantations are planted and managed tracts of trees of the same age (because they were planted by humans at the same time) that are harvested for commercial use.

It makes sense that those in the forestry business would be concerned about finding a way to promote sustainable forestry, because without forests they have no way of perpetuating their income. From an economic viewpoint, the forest must be managed to continually supply humans' need for wood. The management of forest plantations for the purpose of harvesting timber is called **silviculture.** This relatively modern field has a basic tenet to create a sustainable yield; to do this, humans must harvest only as many trees as they can replace through planting. There are two basic management plans that attempt to uphold this tenet.

- **Clear-cutting** is the removal of all of the trees in an area. This is typically done in areas that support fast growing trees, such as pines. Obviously, this is the most efficient way for humans to harvest the trees, but it has major impacts on the habitat, as in our previous example of Vancouver Island.

- **Selective cutting** is the removal of select trees in an area. This leaves the majority of the habitat in place and has less of an impact on the ecosystem. When selective cutting is used, it's quite difficult to remove these trees from the forest, though. This type of **uneven-aged management** is more common in areas with trees that take longer to grow or if the forester is only interested in one or more specific types of trees that grow in the area. Another type of uneven-aged management occurs in **shelter-wood cutting**. For shelter-wood cutting, mature trees are cut over a period of time (usually 10–20 years); this leaves some mature trees in place to reseed the forest.

In the case of **agroforestry**, trees and crops are planted together. This creates a mutualistic symbiotic relationship between the trees and crops—the trees create habitats for animals that prey upon the pests that harm crops, and their roots also stabilize and enrich the soil.

National Forest Policy

The federal government owns about 28 percent of all land in the United States. The need to preserve some of the land was recognized by President Lincoln when he set aside a park in Yosemite, California, as a land grant (the precursor to the National Park System). In 1916, the National Park System was created in part to manage and preserve forests and grasslands. Today, in addition to the National Park System, there are several ways the federal government controls forested land: Wilderness Preservation Areas are open only for recreational activities with no logging permitted. The National Forest System, Natural Resource Lands, and National Wildlife Refuges are the other groups of federally controlled lands that allow logging with a permit.

One more point about managing treed areas: recent times have seen an increase in the number of greenbelts, nationally. **Greenbelts** are open or forested areas built at the outer edge of a city. Since no one is permitted to build in them, they can increase the quality of life for people living nearby. They also border cities, putting limits on their growth. Sometimes, satellite towns are built outside the greenbelt and interconnected with the city by highways and mass transportation methods; in this way, we can add green spaces in urban areas.

Natural Events (That Create Problems for Humans) in Forests

Tree diseases (usually caused by fungal pathogens) and insect pests of trees are two natural problems in forested areas. These can create problems for humans (in addition to the trees) because, oftentimes, they affect the quality of the food and the number of trees that are available for use. Some of the most devastating pathogens and diseases are non-native invasive species, introduced—intentionally or not—through human travel and commerce. Humans manage these natural events in many different ways: by removing infected trees, by removing select trees or planting them sparsely to provide adequate spacing between them, by using chemical and natural pest controls, by carefully inspecting imported trees and tree products, and by developing pest- and disease-resistant species of trees through genetic engineering.

Forest fires are another natural occurrence. There are three major types of fires that occur in forests, and you should be familiar with them for the test.

- **Surface fires** typically burn only the forests' underbrush and do little damage to mature trees. These fires actually serve to protect the forest from more harmful fires by removing underbrush and dead materials that would burn quickly and at high temperatures if they accumulate, escalating more severe fires.

- **Crown fires** may start on the ground or in the canopies of forests that have not experienced recent surface fires. They spread quickly and are characterized by high temperatures because they consume underbrush and dead material on the forest floor. These fires are a huge threat to wildlife, human life, and property.

- **Ground fires** are smoldering fires that take place in bogs or swamps and can burn underground for days or weeks. Originating from surface fires, ground fires are difficult to detect and extinguish.

One final note about forest fires: most people believe that forest fires are a bad thing despite the fact that they are part of the natural life of a forest. Some trees and plants even need fire in order for their seeds to germinate. The U.S. Forest Service started an advertising campaign to warn people about the ravages of fires and soon adopted "Smokey the Bear" to help get the message out. This policy reduced the amount of fires, but it also created conditions for more destructive fires. Under natural conditions, fires burn every few years and consume the fuel (dry leaves, needles, and wood) on the forest floor. However, if there are fewer fires, the amount of fuel can build up to very high levels. When this large amount of fuel ignites, the fires are much hotter and the flames much larger, causing more damage than if the fuel supplies were low. One way to solve the fuel buildup issue is to implement controlled burns, also called prescribed fires. These are small fires started when the conditions are just right and which lower the amounts of fuel. Obviously, this practice must be implemented with great caution and can be quite controversial.

Have you got all that information about forests? Let's move on to another vast resource that exists on Earth—oceans.

OCEAN RESOURCES

The term **fishery** is used in several ways, but it is primarily defined as the industry or occupation devoted to the catching, processing, or selling of fish, shellfish, or other aquatic animals. In the economic sense, a fishery is the sum of all activities on a given marine resource.

Worldwide, about one billion people depend on fish as their main source of food, and about 35 million people are currently employed in the fishing industry. Incredibly, about 172 million metric tons of fish are harvested each year—approximately 75 percent of this total amount is consumed as food by humans, and the other 25 percent is used for other purposes.

For many years, nations were subject to what is known as the 12-mile limit—this limited each nation's territorial waters to just 12 miles from shore. However, in the late 1960s, the depletion of a number of offshore fisheries inspired the United Nations to host a series of international conferences to address the problems of fish scarcity. The result of this conference was that nations were authorized to extend their limits of jurisdiction to 200 miles from shore. The depletion of marine fisheries worldwide came to be seen as a further example of the tragedy of the commons but on an international scale. A new term was coined to recognize this shift, the **Tragedy of Free Access.**

Today, fishermen must go farther and farther out to sea to catch fish and need to rely on more sophisticated methods for finding them. Sonar mapping, thermal sensing, and satellite navigation are just a few of the advances that have aided fisherman as fish become scarcer and harder to locate.

By-Catch and Overfishing

Most of the fish that are harvested worldwide come from **capture fisheries;** they are caught in the wild and not raised in captivity for consumption. Some of the techniques that have been developed in order to improve fishing yields are creating problems that relate to overfishing. One of these problems is known as by-catch. **By-catch** refers to species of fish, mammals, and birds that are caught during fishing operations but are not the target fish. Some fishing methods that result in by-catch are **drift nets,** which float through the water and indiscriminately catch everything in their path; **long lining,** which is the use of long lines that have baited hooks and will be taken by numerous aquatic organisms; and **bottom trawling,** in which the ocean floor is literally scraped by heavy nets that scrape away or smash everything in their path, including corals and other delicate marine life on underwater mountains known as seamounts. Some advances that have been made in the fishing industry in an attempt to mitigate the problems of by-catch are restrictions on the use of drift nets, the installation of ribbons on bait hooks that scare away birds and prevent them from being caught, and bans on bottom trawling.

The National Oceanic and Atmospheric Administration worked with fisheries that used trawling for shrimp to design an apparatus called a Turtle Excluder Device (TED). It is located at the end of the trawling net and will eject large organisms such as sea turtles and sharks from the net while keeping most of the shrimp.

How Many Fish Are Left?

It has recently been reported that about 69 percent of the major fish stocks of the world are either overexploited or fully exploited. Close to another 20 percent of the stocks are moderately overexploited, and about 8 percent are either recovering from depletion or depleted; and this is mostly due to overfishing.

One partial solution to the problem of overfishing is **aquaculture,** which is the raising of fish and other aquatic species in captivity for harvest. In general, the fish that are raised in captivity are those with the highest economic value—for example, salmon and shrimp. Various different methods are used in aquaculture—some fish are raised totally in captivity and then harvested, while others (like salmon) are initially hatched in captivity, but then released into the wild and captured later. Some saltwater aquaculture is performed in shallow coastal areas, though this is generally for raising seaplants and mollusks.

While aquaculture, also known as **fish farming**, does help to meet worldwide demands for fish, it is not a panacea for all of our fishery problems. One concern about aquaculture is the possibility of the accidental release of farmed fish into the wild, which has the potential to introduce new diseases to ocean fish and contaminate the native gene pool. Carp released from fish farming in Southern states is currently causing issues in the Mississippi River and the Great Lakes. Another problem lies in the fact that many fish that are raised in captivity are carnivorous and are fed captured wild fish, which defeats the purpose of the attempt to kill fewer wild fish!

Most of the public outcry about the endangered animals of the sea has centered on two groups: dolphins and whales. Dolphins are a high-profile by-catch, and as you may have noticed, many cans of tuna now advertise as having been caught using "dolphin safe" nets. The slogan "Save the Dolphins" has been frequently employed by international marine conservation groups. However, that slogan is impossible to obey unless humans first work to save the natural *habitat* of these creatures.

The International Whaling Commission (1974) regulates whaling. Recent policies implemented by the IWC allow the capture of a certain number of whales annually—by Norway for human consumption and by Japan for scientific use. It has recently come to light, however, that Japan has been eating the whales it catches, and they have stated that their rationale is that whales eat too many fish that could instead be caught by humans. Another industry that has recently been criticized for damaging whales' ecosystems is the tourism industry—whale-watching tours are said to disrupt whale migration patterns and cause the whales undue stress.

Two Endangered Aquatic Ecosystems

As we've reviewed in earlier chapters, coral reefs are structures found in warm, shallow tropical waters that represent diverse and ecologically crucial ecosystems. Coral reefs are created by small marine animals (called **cnidarians**), which are involved in mutualistic relationships with photosynthetic algae called **zooxanthellae.** Reefs provide local populations with a great variety of seafood and are popular recreational areas for humans.

In many areas of the world, exploitation has led to severe and irreversible damages to these reefs. One example of such irreversible coral damage is **coral bleaching**. In coral bleaching, higher-than-usual water temperatures cause the death of the zooxanthellae, and this in turn causes the death of the coral reef. While some bleaching is normal, high water temperatures can be caused by weather fluctuations such as El Niño, and since this is a periodic event, coral bleaching is an ongoing concern.

Another threatened aquatic ecosystem is the mangrove swamp. Mangrove swamps are coastal wetlands (areas of land covered in fresh water, salt water, or a combination of both) found in tropical and subtropical regions, and they are threatened by activities such as shrimp aquaculture and the degradation of the Western coastlines. Mangroves are characterized by trees, shrubs, and other plants that can grow in brackish tidal waters and are often located in estuaries, which, as you learned earlier, are areas where freshwater meets salt water. In North America, mangrove swamps are found from the southern tip of Florida along the entire Gulf Coast to Texas; Florida's southwest coast supports one of the largest mangrove swamps in the world.

A huge diversity of animals is found in mangrove swamps. Because these estuarine swamps are constantly replenished with nutrients transported by freshwater runoff from the land, they support a bursting population of bacteria, other decomposers, and filter feeders. These ecosystems also sustain billions of worms, protozoa, barnacles, oysters, and other invertebrates, which in turn feed fish and shrimp, which support wading birds, pelicans, and, in the United States, the endangered crocodile.

The importance of mangrove swamps has been well established. They function as nurseries for shrimp and recreational fisheries, exporters of organic matter to adjacent coastal food chains, and enormous sources of nutrients valuable to plants, wildlife, and ecosystem function. Their physical stability also helps to prevent shoreline erosion, shielding inland areas from severe damage during hurricanes and tidal waves.

Along with the Whaling Commission, there are many laws and regulations pertaining to preserving ocean resources. Luckily, the College Board now only requires you to know the following:

Date	Name of Legislation	What It Did
1973	Endangered Species Act	Provided broad protection for species of fish, wildlife, and plants that are listed as threatened or endangered in the United States or elsewhere
1975	CITES (the Convention on International Trade in Endangered Species of Wild Fauna and Flora)	An international agreement between governments that ensured that international trade in species of wild animals and plants do not threaten their survival

We're done with our discussion of resources from the sea. Let's move on to talking about resources that come from underground.

MINING

Mining is the excavation of earth for the purpose of extracting ore or minerals. We can divide mineral resources into two main groups according to how they're used. **Metallic minerals** are mined for their metals (for example, zinc), which can be extracted through smelting and used for various purposes. **Nonmetallic minerals** are mined to be used in their natural state—nothing is extracted from them. Examples of nonmetallic minerals are salt and precious gems. Here are two more terms you should know for the exam, if you don't already: a **mineral deposit** is an area in which a particular mineral is concentrated. An **ore** is a rock or mineral from which a valuable substance can be extracted at a profit.

The cost of extracting minerals depends on numerous factors, including the location and size of the mineral deposit. Additionally, the impetus for mining certain deposits more than others is often purely based on the value of the mineral resource. Understandably, the higher the value of the resource, the more money and effort will be put into mining it.

Environmental concerns about mining do not center on the depletion of mineral resources from Earth's surface. Instead, they revolve around the damage that is done during the extraction process. The extraction of a mineral from Earth generally disrupts the ecosystem and scars the land. Sometimes the extraction leaves pollutants that result from the surface exposure of underground minerals, from transformation of these minerals during mining, from chemicals or other substances introduced during extraction, or even from the machinery used for extraction. One example of this is the deposition of iron pyrite and sulfur in the mining of coal. The acid forms as water seeps through mines and carries off sulfur-containing compounds. The chemical conversion of sulfur-bearing minerals occurs through a combination of biological (bacterial) and inorganic chemical reactions, and the result is the buildup of extremely acidic compounds in the soil surrounding the deposit. These compounds create acid mine drainage that can severely harm local stream ecosystems. In mining processes, waste material is called **gangue,** and piles of gangues are called **tailings.**

As the more accessible ores are mined to depletion, mining operations are forced to access lower grade ores. Accessing these ores requires increased use of resources that can cause increased waste and pollution.

Surface mining is the removal of large portions of soil and rock (this layer is called **over-burden**, and it is whatever material lies above an area of scientific interest) in order to access the ore underneath. An example is **strip mining**, which involves removal of the vegatation from an area, which makes the area more susceptible to erosion. This type of mining is only practical when the ore is relatively close to the surface, which is why it's used mainly for coal mining. This is the least expensive—and least dangerous—method of mining for coal. However, because strip mining requires removing massive amounts of top soil, it has a much greater impact on the surrounding environment than underground mining. The most extreme form of strip mining, mountaintop removal, transforms the summits of mountains and destroys ecosystems. This method is mostly associated with coal mining in the Appalachian Mountains. As coal reserves get smaller, due to a lack of easily accessible reserves, it becomes necessary to access coal through subsurface mining, which is very expensive. With **shaft mining**, vertical tunnels are built to access and then excavate minerals that are underground and otherwise unreachable.

Another environmental drawback to mining is that the refinement of these minerals often requires extensive energy input. For example, it takes approximately 15.7 kW of electricity to produce one kilogram of pure aluminum from its ore. On the other hand, recycling aluminum requires only 5 percent of the energy that's required to smelt it, and generates only 5 percent of the greenhouse gases. Recycle those soda cans!

After minerals have been extracted from their ore, they may be used in their rough form or further processed. Aluminum, for example, must be further refined after it is mined. Coal is an exception. After mining, it is transported to a power plant and burned in its original state. Sometimes two metals are combined to form a product; this is the case with stainless steel, which is a combination of iron and either nickel or chromium, and regular steel, which is 95.5 percent iron and 0.5 percent carbon. Because of the energy expended in mining and extraction, the steel industry is responsible for much of the air pollution that exists today!

Fortunately, air, land, and water harmed by mining can be reclaimed through **mine resto-ration** projects. In 1977, Congress passed the Surface Mining Control and Reclamation Act (SMCRA), which created one program to help coal mines manage pollutants and another to guide the reclamation of abandoned mines.

Mineral Production

The following table shows you the production (in thousands of metric tons) of some non-fuel mineral resources. While you will not have to memorize the amounts for the exam, you can appreciate how much is produced. Remember that these high production rates lead to the eventual depletion of these resources. Also, be ready to describe the impact of mining and mineral production on the environment.

Mineral	2019 Production (metric tons)
Copper	20,000
Phosphate rock (for fertilizer)	240,000
Bauxite (aluminum ore)	370,000
Iron ore (usable)	2,500,000
Zinc	13,000

As you can see from the chart, demand for mineral resources is very high. The increased demand for manufactured goods means that we need to extract more and more raw materials from the earth. The chart below, from the U.S. Geological Service, shows the size of the reserve for five selected mineral ores. For the exam, you should be aware of the need to use mineral resources in a sustainable manner.

Mineral	2019 Reserves (metric tons)
Copper	870,000
Phosphate rock (for fertilizer)	69,000,000
Bauxite (aluminum ore)	30,000,000
Iron ore (crude ore)	170,000
Zinc	250,000

Source: Mineral Commodity Summaries 2020 from the U.S. Department of Interior Geological Survey.

Finally, while there are several laws that govern mining in the United States, the following are required to know for the exam:

Date	Name of Legislation	What It Did
1980	Comprehensive Environmental Response, Compensation, and Liability Act (Superfund)	Regulated damage done by mining
1976	Resource Conservation and Recovery Acts (RCRA)	Regulated some mineral processing wastes

ECONOMICS AND RESOURCE UTILIZATION

The study of how people use limited resources to satisfy their wants and needs is called economics. As you can imagine, some of those needs are tangible, having a physical value (food, shelter, and clean air are examples), while others are intangible (such as recreational opportunity and the spiritual value of a forest's beauty). A resource can have both **tangible** and **intangible** properties.

A forest has value for supplying jobs and wood and for removing CO_2 from the atmosphere (tangible), as well as for its ability to nurture the human spirit and inspire artistic expression (intangible). When private citizens, governments, and corporations make a decision on how to use the forest, they must weigh the benefits (more jobs or lumber) against the costs—tangible and intangible—of cutting down the trees (less recreation space, the loss of biodiversity, decrease in CO_2 removal). This process is called **cost-benefit analysis**. While it may be easy to assign a monetary value to many tangible assets (like the amount of lumber that can be harvested from a forest), others are harder to quantify (such as ecosystem services like clean air, clean water, and biodiversity). It's even harder to assign a monetary value to most intangible assets (like the beauty of a forest), although impact on other intangibles (like recreational and tourism value) can be quantified. While cost-benefit analysis helps make decisions on how to use resources, you can see that the process is very difficult, and it can lead to different estimates by different groups.

Economists also want to figure out the cost of each step in a process. From our forest example, what is the cost to the economy of removing one more acre from the forest; or what is the benefit to us if we add one more acre to the forest? The additional costs are termed **marginal costs**; the added benefits are called **marginal benefits**. It is important to remember that resources are not free and unlimited. Some resources must be expended in order for us to use them. While we may benefit from more acres to hike in, the lumber company will suffer from not having as many trees to cut. In other words, marginal benefits and costs help us understand tradeoffs. By preserving a forest, we trade more hiking space with less profit for local economies.

As we use resources, there are often unwanted or unanticipated consequences of our using those resources, or **externalities**. These can be positive, when the result is good, and negative when the result is bad for the environment. Consider buying an air conditioner, for example. When you buy the A.C., there are costs—you pay for the labor, raw materials, and electricity to run it. After you buy the A.C., the dealer uses some of that money to pay employees to clean up litter on a highway. That cleanup benefits everyone (positive externalities), even those who do not buy the air conditioner. On the other hand, there are also negative externalities. If you run the A.C. a lot (perhaps during the day), you use a lot of electricity (and you can see just how much more when your electric bill spikes in the summertime). That electricity is generated by burning coal, and that causes acid rain. The damage done by the acid rain harms everyone—a negative externality.

One final thing to remember: the use of economics (cost-benefit analysis, marginal costs and benefits, and externalities) to make choices about dealing with environmental issues is morally neutral. These economic factors do not say anything about the ethics or fairness of those choices. There are situations in which we make decisions not based on the best balance between marginal costs and benefits, but on what is best for everyone. Take water pollution, for example. If a toxic chemical "X" is in a stream, there is a cost to clean it up. We make the decision to clean up most of chemical "X," even if the marginal costs exceed the marginal benefits, because removing chemical "X" will keep all of the people healthy.

We're done discussing resources. As we've alluded to many times, as populations increase, more pressure is placed on Earth's natural resources, and along with this comes the need for humans to find ways to develop and maintain those natural resources for direct human use. With this in mind, let's move on to the next chapter and review energy resources and consumption.

CHAPTER 7 KEY TERMS

Here are your key terms for Chapter 7. Learn them, love them, and commit them to memory.

Resources
resource
conservation
preservation
natural resources
ecosystem (or natural) capital
renewable resources
nonrenewable resources
consumption
production
The Tragedy of the Commons
The Tragedy of Free Access

Agriculture
subsistence agriculture
slash-and-burn agriculture
Dust Bowl
Green Revolution
salinization
integrated pest management
genetically modified organisms
monoculture
photosynthate
plantation farming
contour plowing
terracing
no-till methods
crop rotation
intercropping
strip cropping
overgrazing

Forestry
deforestation
old growth forest
second growth forest
plantations (or tree farms)
silviculture
clear-cutting

selective cutting
uneven-aged management
shelter-wood cutting
greenbelts
agroforestry
surface, crown, ground fires
controlled (prescribed) burns

Oceans
fishery
capture fisheries
by-catch
drift nets
long lining
bottom trawling
aquaculture
fish farming
cnidarians
zooxanthellae
coral bleaching

Mining
ore
metallic and nonmetallic minerals
mineral deposit
gangue
tailings
strip mining
overburden
mountaintop removal
shaft mining
mine restoration

Economics
cost-benefit analysis
tangible and intangible properties
marginal costs and benefits
externalities

Chapter 7 Drill

Directions: Each of the questions or incomplete statements below is followed by four suggested answers or completions. Select the one that is best in each case. For answers and explanations, see Chapter 13.

1. Which of the following correctly describes the process of smelting?

   (A) Separating the desired metal from other elements in the ore
   (B) Cleaning up drainage from mines
   (C) Detoxifying harmful chemicals
   (D) Removing ore from underground mines

2. In a very polluted river, it costs $3 per kilogram to remove the first 80% of the pollution. It costs $25 per kilogram to remove the last 20% of the pollutant. This phenomenon is correctly referred to as

   (A) cost-benefit analysis
   (B) external costs
   (C) marginal costs
   (D) marginal benefit

3. Which of the following correctly describes the process of clear-cutting?

   (A) Some mature trees are left to provide shade for younger trees.
   (B) Only trees with commercial value are cut down.
   (C) A few mature trees are left to reseed the land after cutting.
   (D) All the commercially usable trees in an area are cut down.

4. Moderate irrigation with groundwater over a long period of time can cause

   (A) salinization
   (B) waterlogging
   (C) desertification
   (D) succession

5. All of the following are problems created by the deforestation of rainforests EXCEPT

   (A) increased erosion
   (B) loss of biodiversity in the area
   (C) changes in local rainfall levels
   (D) an increase in the availability of grazing land

6. Greenbelts are useful to

   (A) slow the process of urban growth
   (B) get more crops out of farmland
   (C) maintain borders around a person's home property
   (D) prevent erosion

7. Which of the following government agencies is responsible for the management of federal rangeland?

   (A) The U.S. Park Service
   (B) The U.S. Bureau of Mines
   (C) The Bureau of Land Management
   (D) The Environmental Protection Agency

8. Which of the following is NOT a renewable resource?

   (A) Air
   (B) Soil
   (C) Copper ore
   (D) Water

9. Nations have overfished international waters and have depleted many commercially important fish species. This is a good example of which of the following?

   (A) International agreements
   (B) The Tragedy of the Commons
   (C) The Rule of 70
   (D) Trade barriers

10. Which of the following best describes industrialized agriculture?

    (A) Consumes large amounts of fossil fuels, pesticides, and water
    (B) Uses human labor and draft animals to grow crops
    (C) Rows of crop plants are interspersed with rows of trees
    (D) Uses little water or fossil fuels; relies on human labor

11. The international trade in endangered species is regulated by which of the following?

(A) The Endangered Species Act
(B) Marine Mammal Protection Act
(C) The National Environmental Policy Act
(D) CITES

12. Which of the following are problems that have emerged with the overuse of pesticides?

I. Better crop yield
II. Pesticide-resistant pests
III. Improved human health

(A) I only
(B) II only
(C) III only
(D) I and III only

13. Which of the following is true concerning the use of national parks?

(A) They can be used for cutting timber as well as recreation.
(B) They can be used for mining as well as recreation.
(C) They can be used only for camping, fishing, and boating.
(D) They can be used for the conservation of a natural habitat as well as livestock grazing.

14. The acid most commonly found in mine drainage is

(A) carbonic acid
(B) sulfuric acid
(C) hydrochloric acid
(D) acetic acid

15. The World Trade Organization strives to

(A) protect endangered species on land
(B) regulate the global fishing industry
(C) move toward the globalization of all the nations
(D) establish rules for the free flow of economic goods and services between countries

Free-Response Question

1. The irrigation of farmland is vital to the production of the world's food supply. In China, 87 percent of the water withdrawn is used for irrigation. In the United States, this figure approaches 41 percent. Most of the water is applied to the land in a process called gravity irrigation, in which the water is simply allowed to flow, via the force of gravity, into the fields.

    (a) **Describe** ONE positive and ONE negative aspect of gravity irrigation.
    (b) **Describe** ONE alternative to gravity irrigation. Give ONE positive and ONE negative effect of that practice.
    (c) Massive irrigation programs can also impact underground water supplies. **Describe** ONE negative impact that irrigation might have on those supplies.
    (d) Dams are often used to create irrigation water reservoirs. **Describe** TWO positive and TWO negative effects that a large dam would have on the immediate area around it.

Summary

- o "The Tragedy of the Commons" is essential to understanding the management of shared resources. Resources have both an innate value to the function of their ecosystem and an economic value to the humans who use them, known as ecosystem capital.

- o The rate of resource replenishing along with the human consumption and production rates:
  - Renewable resources are perpetual or regenerate quickly, less than the average human lifetime (≈50 years), or before it can be depleted by human consumption.
  - Nonrenewable resources renew at an insufficient rate for human use.

Agriculture

- o Agriculture has been divided into two main methods: traditional subsistence agriculture and industrial agriculture.

- o Traditional subsistence agriculture:
  - This type of agriculture is most common in developing nations.
  - It provides enough food for one's family.
  - The slash-and-burn method is often used, especially in undeveloped countries of the tropics.

- o Industrial agriculture or Green Revolution:
  - Industrial agriculture is most common in developed nations due to initial cost.
  - Chemical fertilizers and pesticides may increase food production but also require lots of energy to produce, can contaminate water supplies, and organisms can develop immunity to them.
  - Technologies for irrigation have allowed for food to be produced where it otherwise could not, but has developed issues like salinization and water wars.
  - Monocrop and plantation farming are easier to grow due to mechanized planting, but reduce native biodiversity.
  - GMOs can increase economic benefits but discourage biodiversity and have unknown health results.
  - Increased tilling and livestock overgrazing can cause major soil erosion.

Forest Management

- Deforestation is the largest concern in forestry management, primarily in developing countries for agricultural space and lumber exports. Minimal deforestation takes place in industrial countries, primarily for development.

- Even if replanted, the age of the forest is impacted and can change the native biome.

- Several national forest policies and programs have been established to protect the age and diversity of forests with different policies on use and management.
  - National Park System
  - Wilderness Preservation Areas
  - National Forest System
  - Natural Resource Lands
  - National Wildlife Refuges

- Forest fires can be essential to plant regeneration and forest health but can also burn uncontrolled and cause great economic loss.

Ocean Resources

- Improved technologies, fishing strategies, and human population growth have all contributed to harvesting fish faster than the populations replenish themselves.
  - Unintentional catches known as by-catch also increase the number of fish caught, yet they are often wasted instead of used.
  - Aquaculture is a partial solution, though has potential health impacts to both humans and the ecosystem.

- Coral reefs and mangroves are the two most susceptible marine biomes due to their proximity to humans, narrow ranges of tolerance, and huge biodiversity.

Mining

o   Both naturally used and metallic minerals have great value and multiple purposes in human lives, but their extraction and use also have a significant environmental impact:
   - extraction pollution
   - soil erosion and/or habitat loss
   - extraction and refinement require extensive energy input

Economics

o   The cost-benefit of a resource is complicated to assess because of the combination of its tangible and intangible values to humans and its intrinsic value to the ecosystem. Often, the price of a resource is not representative of its true cost because some costs of the product or unanticipated consequences are not included. These are called externalities.

Required Environmental Legislation

o   Endangered Species Act of 1973

o   CITES (the Convention on International Trade in Endangered Species), 1975

Chapter 8
Energy

Unlike the essential elements we discussed in earlier chapters, energy flows on a one-way path through the atmosphere, hydrosphere, and biosphere, and is essential for living organisms in many of its forms. At the most fundamental level, **energy** is defined as the capacity to do work. There are three types of energy: **potential energy** is energy at rest—it's stored energy—while **kinetic energy** is energy in motion. You might recall from your physics class that potential energy can be converted to kinetic energy. The third type of energy is in the form of **radiant energy,** or sunlight, and it is the only form of energy that can travel through empty space. Two terms that describe the movement of energy around Earth are **convection,** which is the transfer of heat by the movement of the heated matter, and **conduction,** which is the transfer of energy through matter from particle to particle. Keep in mind that different energy sources are capable of storing types of energy that differ in quality. For example, both wood and coal will burn to produce heat, but coal produces more heat because it contains higher **energy quality**. **Net energy yield** refers to the comparison between the energy cost of extraction, processing, and transportation, and the amount of useful energy derived from the fuel.

> It is important to note that energy cannot be created or destroyed, but it can be transferred to another form. Some energy is lost as heat, which increases entropy.

As you saw in our chapter about weather, convection and conduction are very important processes that drive the movement of water in the hydrosphere of Earth. This is just one way in which the flow of energy around the Earth affects every process—geological or biological—that takes place.

In this chapter, we'll begin with a discussion of the basic units of energy and review the two laws of thermodynamics that you'll need to know for the test. After that, we'll begin our discussion of Earth's energy resources—the Earth provides humans with resources of energy, just as it does the physical, natural resources that we learned about in the last chapter. Let's begin!

UNITS OF ENERGY

For this exam, you'll be expected to recognize the following units of energy and power:

> Topics in this chapter can be found in the "Energy Resources and Consumption" unit in the new Course and Exam Description.

- **Energy Units:** joule (J), calorie (cal), British thermal unit (BTU), and kilowatt hour (kWh), which is a measure of watts × time
- **Power Units:** watt (W) and horsepower (hp)

Remember, a watt is equal to volts × amperage. You should also be intimately familiar with the First and Second Laws of Thermodynamics, so let's review those before we move on—and make sure you memorize these before test day!

Laws of Thermodynamics

1. The **First Law of Thermodynamics** says that energy can neither be created nor destroyed; it can only be transferred and transformed. One example of such a transformation occurs in photosynthesis. In photosynthesis, radiant energy from the sun is converted to chemical energy in the form of the bonds that hold together atoms in carbohydrates.

2. The **Second Law of Thermodynamics** says that the entropy (disorder) of the universe is increasing. One corollary of this Second Law of Thermodynamics is the concept that, in most energy transformations, a significant fraction of energy is lost to the universe as heat; for example, as we reviewed in Chapter 5, in food chains only about 10 percent of the energy from one trophic level is available for the next higher energy level upon consumption.

Okay, those are the basics about energy that you'll need to know for the test. Now let's begin our review of the energy resources that exist on Earth.

NONRENEWABLE ENERGY

Perhaps surprisingly, one of our biggest uses of energy is in the production of electricity. In other words, we use tons of energy each year to produce electricity—another form of energy!

In general, electricity is produced in the following way: an energy source provides the power that heats up water, transforming it into steam, which then turns a turbine. Hence, the turbine converts kinetic energy (from the steam) into mechanical energy (the spinning of the turbine). Now here's where the generator comes in. The generator consists of copper wire coils and magnets, one of which is stationary (stator) and the other of which rotates (rotor). As the turbine spins, it causes the magnets in the generator to pass over the wire coils (or vice versa), generating a flow of electrons through the copper wire and thus producing an alternating current that passes into electrical transmission lines. In lieu of steam, flowing water or wind can also provide the power needed to turn the turbine and produce electricity.

So, where do we get the energy that we use to heat up that water in the first step of the creation of electricity? Well, the three main sources for electricity production in the United States are:

- fossil fuels (provide 65 percent of the world's electricity)

- nuclear energy (provides 20 percent of the world's electricity)

- renewable energy sources (provide 15 percent of the world's electricity)

Let's go through each of the types of energy above and see where they came from, what effect their use has on Earth, and how sustainable they are.

Fast Fact
Electricity production from oil sources has declined rapidly over the past 55 years, now only accounting for 4% of U.S. electricity production!

Quick Quiz!
Q: Which of the following energy sources contributes the least to global warming?

(A) Coal
(B) Natural Gas
(C) Oil
(D) Solar

Turn the page for the answer.

Fossil Fuels

During the Industrial Revolution (in the early 18th century), steam was produced almost exclusively through the burning of firewood and coal—and this, in turn, provided the energy for most mechanical processes. Today, oil is our primary power source. About 33 percent of total global energy production comes from oil products; the runner-up to oil is coal, and the runner-up to coal is natural gas. Together, these three fossil fuels provide 81 percent of the world's energy.

Fossil fuels, as the name indicates, are formed from the fossilized remains of once-living organisms. Over vast tracts of time, this organic matter was exposed to intense heat and pressure. Eventually, these forces broke down the organic molecules into oil, coal, and natural gas.

Oil, or petroleum, is made of long chains of hydrocarbons; and coal contains a mixture of carbon, hydrogen, oxygen, and other atoms. Natural gas is made mostly of methane gas (CH_4) with a mixture of other gases.

Generally, oil and natural gas are formed in the same areas. These materials are found deep in the Earth under both land and ocean floor, where they are stored in the pores (spaces) between rocks. Coal is found in long continuous deposits, called **seams,** at various depths underground. The seams represent areas where large amounts of plant remains were buried and eventually transformed into coal. We will cover the process of coal mining on the next page.

Certain types of geologists locate fossil fuel reserves. They plan and supervise the extraction of these fuels from the Earth. Using knowledge of geology and rock formations, these scientists make predictions about which sites are most likely to have fossil fuel deposits. They use **exploratory wells** to drill and sample a particular area. If an exploratory well hits a fossil fuel reserve, it can provide an estimate of the amount of fuel that can be obtained from that area; this is called the **proven reserve**. It is important to know that although exploratory wells can provide a fairly precise estimate of the size of a reserve, these numbers are just educated guesses (not so proven after all!). The amount of a resource that can be extracted from a reserve is dependent on the technologies available and the cost of extraction. If extraction costs are too high, it is not economically feasible to extract the resource. If a coal seam is buried very deeply, for instance, it may cost more money and fuel to extract it than the value of the seam.

What About Oil?

When oil is pumped up fresh from a reserve, it is called **crude oil**. Crude oil varies greatly from reserve to reserve. It can range from thin to viscous (thick); from high sulfur to low sulfur; it can even vary in color and odor.

There are three different methods of extracting oil. In primary extraction, the oil can be easily pumped to the surface. When some oil wells are tapped for the first time, there is a large release of oil and gas, a *gusher,* due to the pressure in the reserve. When the oil is harder to extract, people rely on pressure extraction, which uses mud, saltwater, and even CO_2 to push out the oil from the reserve. The final method utilizes steam, hot water, or hot gases to partially melt very thick crude oil and make it easier to extract. Oil reserves can also be found in rock (shale oil) and surface sands (tar sands). Tar sands are the dirtiest of all the oils extracted. Its method of extraction and processing are detrimental to the environment and use a great deal of energy. There was much controversy about the Keystone Pipeline, which ran from Alberta, Canada, south to the refineries along the Gulf Coast. Those in favor of the pipeline said it would bring much-needed jobs to the areas around the pipeline. Those who disapproved of it list the pollution it could bring, including contamination of groundwater.

> In 2010, the United States suffered one of its worst environmental disasters when an explosion on the Deep Water Horizon drilling rig caused oil to spill from the well into the Gulf of Mexico for three months. This was the largest marine oil spill in the history of the oil industry. Eleven men were killed during the explosion, and the spill caused a great deal of damage to both marine and wildlife habitats and local economies along the coast.

Drilling for oil is only moderately damaging to the environment because little land is needed to drill. However, since oil is transported thousands of miles by tankers, pipelines, and trucks, a lot of environmental damage can occur during transportation.

What About Coal?

Let's spend some time reviewing coal. The qualities of different types of coal are ranked by the number of BTUs that they produce upon burning. The purest coal is called **anthracite,** which is almost pure carbon. The second purest coal is **bituminous,** followed by **subbituminous,** and finally **lignite**—the least pure coal. Coal mining occurs through one of two processes—strip mining or underground mining—both of which can be hazardous and have serious environmental impacts. **Underground mining** involves sinking shafts to reach underground deposits. In this type of mining, networks of tunnels are dug or blasted, and humans enter these tunnels to manually retrieve the coal. After production stops at these mines, cave-ins can occur, causing massive slumping or **subsidence. Strip mining** involves the removal of the Earth's surface, all the way down to the level of the coal seam. The coal is then removed, the **overburden** (the Earth that was removed) is replaced and topped with soil, and the area is contoured and re-vegetated. Most states require strip-mine owners and operators to completely reclaim areas that are mined by taking all of the steps outlined above. However, the process of mining and removing the coal from the Earth leaves hazardous slag heaps containing sulfur that can leach out and enter the water table.

Fast Fact
Crude oil is not only used for the production of fuel products but also many items used every day in the form of plastic and petroleum jelly (Vaseline).

Coal Power Process

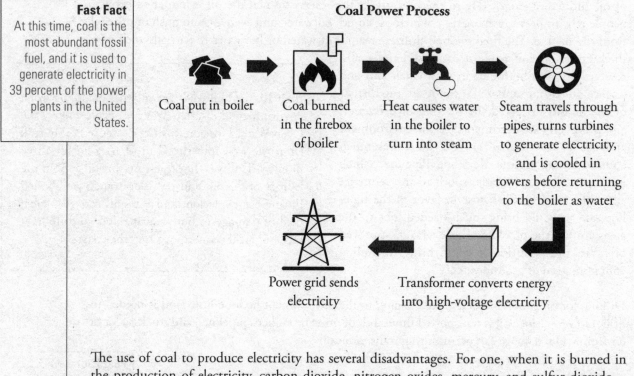

Coal put in boiler

Coal burned in the firebox of boiler

Heat causes water in the boiler to turn into steam

Steam travels through pipes, turns turbines to generate electricity, and is cooled in towers before returning to the boiler as water

Power grid sends electricity

Transformer converts energy into high-voltage electricity

The use of coal to produce electricity has several disadvantages. For one, when it is burned in the production of electricity, carbon dioxide, nitrogen oxides, mercury, and sulfur dioxide—all of which contribute to air pollution—are released as by-products. However, some of these by-products can be removed through the actions of **scrubbers,** which contain alkaline substances that precipitate out much of the sulfur dioxide. The neutral compound formed in the scrubber (calcium sulfate) is eliminated in waste sludge. Two other waste products produced by the burning of coal are **fly ash** and **boiler residue**—you should be familiar with both of these terms for the exam.

Another problem with coal is that it often contains a significant amount of the element sulfur, both in the form of iron sulfide (pyrite) and as organic sulfur. Sulfur is another contributor to air pollution. While iron sulfide can be removed by grinding the coal into small lumps and washing it, organic sulfur is only released during the **combustion** (burning) of coal. However, scrubbers can remove organic sulfur from the flue gases after the coal is burned. Another solution to this problem is to burn the coal with limestone—the liberated sulfur combines with the calcium in limestone to form calcium sulfate, which prevents it from being released through the flue. There are different types of smokestack scrubbing: wet scrubbing and baghouse, or cyclo, scrubbing. Wet scrubbing uses a fine mist of water to transform sulfur oxides (SO_x) from an air pollution issue to either a water pollution issue or to a commercial product, sulfuric acid. This is similar to how a rain storm will cleanse the air of pollen and dust that causes allergies in some people. Baghouse, or cyclo, scrubbers are very similar to large vacuum cleaners that either filter (baghouse) or spin (cyclo) particulates out of the effluent gases. Electrostatic filters use an electric charge to attract dust particulates to metal surfaces where they can be gathered and disposed of as solid waste. This is similar to your television screen being the dustiest surface in your living room due to electric properties of the device.

You've probably seen lots of news reports lately about the harmful effects—particularly to babies and very young children—of ingesting seafood that's contaminated with mercury. An EPA study found that one in six women of childbearing age in the Unites States may have blood mercury levels that could be harmful to a developing fetus. Coal-fired power plants are the major source of mercury pollution in the environment. Airborne mercury can travel hundreds of miles from its source before being deposited in the ground and lakes, streams, and other bodies of water, both directly from rainfall and as a result of runoff. Mercury in water can bioaccumulate in fish, which are then eaten by people. (Biomagnification is what we call this increase in the concentration of a substance at higher and higher levels of the food chain.) Abandoned metal and coal mines frequently produce **acid mine drainage**, highly acidic water which flows to surrounding areas. In 2012, the EPA issued the Mercury and Air Toxics Standards (MATS) rule that provides emission limits for mercury, particulate matter, hydrogen chloride, and hydrogen fluoride for approximately 600 coal and oil power plants. However, the Trump administration's desire to stop the "war on coal" recently led the EPA to state that political appointees are reviewing findings to determine whether the MATS rule should be reconsidered.

What About Natural Gas?

The third fossil fuel you need to know about is natural gas. Natural gas is made mostly of methane (CH_4) as well as pentane, butane, and several other gases in small quantities. As you learned earlier, natural gas is produced by the actions of heat and pressure over long periods of time. It is also produced by living organisms (mostly by anaerobic bacteria). Methane-producing bacteria can be found in landfills, swamps, and the intestines of various animals. Here's an interesting fact: while the largest source of methane is wetlands, the second largest source is our flatulent livestock.

Currently, natural gas is used for heating homes and cooking. It can also be burned to generate electricity. Some power plants are designed to switch between oil and natural gas fuels depending on the cost. The engines of cars and trucks can be modified to burn natural gas instead of gasoline. There is a landfill operator in the state of New Jersey who tested a process of trapping methane from a landfill, liquefying it, and then using the liquid methane to power the trucks that bring garbage to the landfill.

Because of its simple molecular structure, natural gas produces only carbon dioxide and water when it burns. It does not produce the oxides of nitrogen and sulfur associated with burning coal or oil. Before you get really excited about natural gas, you should be aware of its dangers. In an uncontrolled release (like a leak), it can cause violent explosions. It is also more difficult to transport than coal or oil. Because a tank can hold a small amount of gas, producers liquefy it by putting the gas under high pressure (**L**iquefied **N**atural **G**as). This process requires energy. Natural gas can also be transported by pipes. However, pipes carry the risk of leaks and explosions, and some habitats are damaged during the building of the pipe system.

Furthermore, methane is a potent greenhouse gas. Its chemical structure makes it 30 times more powerful than carbon dioxide at trapping heat in the atmosphere. And, as climate change continues to warm the Earth, the biochemical reactions by which certain bacteria produce methane will accelerate. This is a particular danger in habitats such as swamps and other wetlands as well as freshwater sediments. This

Current Day Application

Fracking, also called "hydraulic fracturing," is a process by which natural gas and oil are extracted from rock that lies deep underground. A deep well is drilled and then millions of gallons of toxic fracking fluid—a mix of water, sand, and harsh chemicals—are injected at a high enough pressure to fracture the rock and release the oil or gas. It's a highly controversial practice that has been linked to earthquakes in the states of Arkansas, Ohio, and Pennsylvania. It requires a large amount of water, which has to be safely stored after use due to chemical contamination.

situation where increasing concentrations of a greenhouse gas in the atmosphere further accelerate the production of greenhouse gases is just one of many feedback loops that makes the climate change unleashed by human activity so scary and difficult to quantify!

How Much Fossil Fuel Is Left?

In order to understand how long our accessible fossil fuel supplies will last, you should know how quickly we are using up those fuels. Let's take oil, the most widely used fuel, as an example. The table below shows the amount of oil (including crude oil, other petroleum liquids, and biofuels) that selected countries use each day. This data is from 2018, the last year for which the U.S. Energy Information Administration has this information.

Country	2018 Oil Consumption (millions of barrels per day)
United States	17.94
Saudi Arabia	12.42
Russia	11.40
Canada	5.38
China	4.81

> **Fast Fact**
> The increase in hydraulic fracturing and deep water drilling has been the result of new technology that allows for horizontal drilling. Several areas currently being surveyed for future drilling are major fisheries, such as the Grand Banks.

As you can see, the United States is by far the largest consumer of oil. A quick bit of addition shows that these five countries alone consume almost 51 billion barrels of oil each day! As you can imagine, that leads some scientists to ask questions about how long our supplies of oil (and the other fossil fuels) will last. One well-known authority on the future of oil production, the late M. King Hubbert, stated that the end of oil as a cheap and easily available form of energy is in the near future and that we must begin to develop alternative fuel sources. This theory is commonly known as **Hubbert peak** or "peak oil."

> **Fast Fact**
> In 1956, Hubbert predicted that the lower 48 states would reach peak oil production by 1970 and then decline. In fact, advances in technology have increased the amount of oil available in the United States today.

Nuclear Energy

Nuclear energy is the world's primary non-fossil fuel, nonrenewable energy source. In the United States, 20 percent of electrical energy is provided by nuclear power plants. Worldwide, more than 400 nuclear power plants produce approximately 13 percent of the world's electrical energy. The United States and France lead the world in the creation of new nuclear facilities.

Let's talk a little bit about what exactly these nuclear power plants do. The key reaction in the production of nuclear energy is **fission,** a nuclear reaction in which the nucleus of an atom is struck by a neutron and then splits into two smaller, lighter nuclei. The current nuclear plant technology involves the use of uranium-238, which is enriched with 3 percent uranium-235. Nuclear fission releases a large amount of heat, which is used to generate steam, which powers a turbine and generates electricity. **Breeder reactors** generate new fissionable material faster than they consume such material. Furthermore, they can use a more abundant form of uranium, uranium-238, or an alternative called thorium-232.

However, the products of nuclear fission are much more **radioactive** than the fuel used; in other words, they have unstable nuclei which emit energy over time as particles or photons. Radioactivity can be measured in curies. Radioactive materials differ in terms of rate of decay—this is generally thought of in terms of **half-life,** which is the time it takes for half of a given radioactive sample to degrade on average. A radioactive element's half-life can be used to calculate a variety of things, including the rate of decay and the radioactivity level at specific points in time. For example, if a given sample of a radioactive substance has an activity level of 10 curies and a half-life of 5 days, then after 5 days it will have an activity level of 5 curies; after 10 days it will have an activity level of 2.5 curies; after 15 days it will have an activity level of 1.25 curies; and so on.

Uranium-235 remains radioactive for a long time—its half-life is 703.8 million years—which leads to problems with the disposal of nuclear waste. So as it stands, nuclear power is considered a cleaner energy source than fossil fuels because it does not produce air pollutants, but it does release thermal pollution and hazardous solid waste.

However, the future of nuclear power will probably involve **nuclear fusion,** which is the process of fusing two nuclei—likely from the hydrogen isotopes tritium and deuterium, which have two and one protons, respectively (the common form of hydrogen has no protons). Fusion would have several advantages over fission as a source of nuclear power—less radioactivity and nuclear waste, more easy-to-obtain fuel supplies, and increased safety—but while research into this process has been ongoing since the 1940s, it hasn't produced a viable way to harness nuclear fusion into a useable power source yet.

In the United States, there are two types of nuclear reactors. They are known as boiling water reactors and pressurized water reactors.

- **Boiling Water Reactors**—These reactors use the heat of the reactor core to boil water into steam. This steam is piped directly to the turbines. The steam spins the turbines that generate the electricity. The water is cooled back to a liquid (by a heat exchanger), then pumped back to the core to be turned into steam again. This reactor uses two water circulation systems; one system makes steam and carries it to the turbine and the other cools the water from the core so it can be turned back into steam.

Boiling Water Reactor

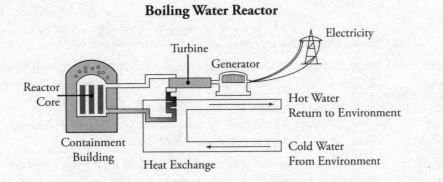

- **Pressurized Water Reactors**—These reactors also produce electricity by generating steam but contain three water circulation systems. Similar to boiling water reactors, the heat produced from the reactor core is used to heat the first water supply. However, the first water supply is kept under high pressure to prevent the water from boiling. It is then passed through the reactor heat exchanger, where heat from the first water supply is transferred to the second water supply. The second water supply is not kept under high pressure and forms steam to spin the turbines. A third water supply cools the steam from the turbines to regenerate the second water supply.

Pressurized Water Reactor

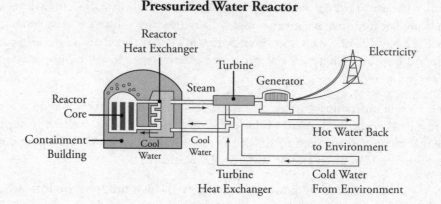

Some arguments in favor of the use of nuclear power include the fact that the production of nuclear energy produces no sulfur dioxide or nitrogen oxide and less carbon dioxide than does the production and combustion of fossil fuels.

Since the first testing of atomic weapons, people have been concerned about safety issues surrounding nuclear energy. The nuclear reactor accident that occurred at the Three Mile Island facility in Pennsylvania in 1979, the devastating explosion that occurred at the Chernobyl facility in the Ukraine in 1986, and the catastrophic failure at the Fukushima nuclear power plant in Japan in 2011 brought some major safety concerns to the public's attention.

The chart below shows a number of these issues.

Safety Issue	Description
Meltdown	Reactor loses coolant water, and thus the very hot core melts through the containment building. The radioactive materials could then get into the groundwater.
Explosion	Gases generated by an uncontrolled core burst the containment vessel and spread radioactive materials into the environment.
Nuclear weapons	Some of the by-products of the fission reaction can be remade into fission bombs, or "dirty bombs," that spread damaging radioactive isotopes.
Highly radioactive waste	No longer usable cores, piping, and spent **fuel rods** need to be stored for many centuries. The "spent" fuel can contain radioactive elements like plutonium-239, which has a half-life of 2.13×10^6 years.
Thermal pollution	The water used to cool turbines is returned to local bodies of water at a much higher temperature than when it was removed unless first cooled.
Radioactive elements	Gamma rays produced by radioactive decay can damage cells and DNA, which can cause breast, thyroid, stomach, and leukemia cancer. Damage to the immune system can also result.
Concern for one's safety	People suffer from mental stress, anxiety, and depression caused by concerns for their safety.

At this point, the cost for building a new nuclear power plant in the United States is prohibitive due to changing regulations; and worldwide, more than 100 nuclear power plants have been decommissioned. The main reason for this is the problem of the nuclear waste that's created. The United States' spent fuel is currently stored on site at nuclear power plants, but the plants are running out of space. While Yucca Mountain (in Nevada) was previously proposed as a final destination storage site in the past, it has not been finalized due to push back from Nevada residents. This has prompted the government to consider other storage options like Salado Salt in New Mexico, or deep holes in the earth's crust.

RENEWABLE ENERGY

Obviously, the advantage to discovering or developing renewable energy sources lies in the fact that these types of energy sources are bottomless. However, globally only about 15 percent of our energy needs are currently met using renewable resources.

Biomass is one of the most consistently and widely used renewable energy sources today. Biomass includes wood, charcoal (wood that has been baked to remove water and impurities), and animal waste products. Although biomass is renewable, as we discussed in Chapter 7, it is only renewable when it is used at a pace that allows time for replacement.

Using biomass as an energy source is beneficial because it is low-cost, readily available, and better for the environment than fossil fuels. Biomass can also help reduce the amount of waste in landfills, as it relies on the burning of organic matter. The major downsides of biomass include the initial cost and large land mass required for installation of a biomass boiler.

Biomass Power Process

Sunlight, carbon dioxide, and water → Photosynthesis causes plant growth → Organic matter burned; energy produced

One interesting fuel that's recently been developed is called gasohol, a type of synfuel. **Gasohol** is a gasoline extender made from a mixture of 90 percent gasoline and 10 percent ethanol, which is often obtained by fermenting agricultural crops or crop wastes. So, as you can see, it is partly derived from organic substrate. Gasohol has higher octane than gasoline and burns more slowly, coolly, and completely, thus resulting in reduced emissions of some pollutants. Despite those advantages, it also vaporizes more readily than gasoline, and has the potential to aggravate ozone pollution in warm weather. Ethanol and, similarly, methanol from organic sources are surface carbons and therefore add no net carbon to the atmosphere. However, because ethanol is carbon-based, carbon dioxide is released during its combustion. Ethanol-based gasohol is also expensive and energy-intensive to produce—one bushel of corn produces only two and a half gallons of ethanol.

This is clearly an inefficient way to use biomass. However, researchers are developing new and better ways to use biomass as a fuel. Biodiesel is a great example. While it is made primarily from virgin oils (such as soybean oil), it can also be made from waste vegetable oil. More and more restaurants are beginning to use their waste oil byproducts to produce biodiesel. In addition, it is possible to use algae to produce biodiesel. The major benefit of algae is that they can be grown in sewage water, allowing the production of biodiesel without using land normally used for food production. It is important to note, though, that the cost of producing is currently greater than that of their nonrenewable energy sources. Without government subsidies, it is tough to make biodiesel prices competitive with those of fossil fuels.

Hydroelectric Energy

Hydroelectric power works via the systematic placement of river dams that spin turbines to produce electricity. The process involved in creating electricity from hydroelectric power is not unlike that of geothermal energy; instead of having water pumped from beneath the Earth, however, water in a dam's reservoirs is responsible for spinning the turbines and generating electricity. One advantage of hydroelectric power is that its production releases no pollutants. However, hydroelectric power does produce thermal pollution. In addition, it requires that rivers be dammed, and this can change the rates at which rivers flow and also lead to the destruction of habitats. On the other hand, as water is held behind dams, new habitats, in the form of wetlands, are created.

Hydroelectric Power Process

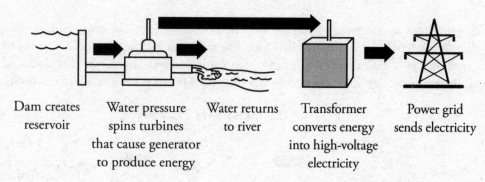

| Dam creates reservoir | Water pressure spins turbines that cause generator to produce energy | Water returns to river | Transformer converts energy into high-voltage electricity | Power grid sends electricity |

Fast Facts
The reservoir formed behind the dam leads to the destruction of habitats. Vegetation is drowned and will decompose, therefore still supplying CO_2 as a waste product.

Generally, a river has many dams along its length. The addition of each new dam leads to less and less water downstream and a change in the natural course of the river. In California, there is continued unrest in the rationing of water to agriculture of fish.

There are other problems associated with dams that you need to know about. One of them is silting. As water sits behind the dam, the normal sediments it carries have time to sink to the bottom. This puts additional weight on the structure and means that dams have to be built strong enough to hold back the many tons of sediment. This also means that the sediment is not passed farther down the river. The sediment that used to fertilize the flood plains of the river is now trapped behind the dam. In addition, the reservoir usually has a greater surface area than the preceding lake or river; this increased surface area can actually result in a higher rate of evaporation and water loss than before.

Another problem is that fish that spawn in the normally silty river no longer have a place to do so. Salmon and other anadromous fish breed in the streams where they hatched from eggs. Dams prevent the salmon from returning to their hatching streams. While **fish ladders** do let some fish return upriver, the number of fish that get through is so limited that the populations will still decline.

One more important note about hydroelectric power: the development of this as an alternative, renewable energy source is limited—simply because there are a limited number of rivers of sufficient flow and drop in the world that can be used for these purposes.

Solar Energy

While it is important to remember that we already obtain through the sun the energy we need to live—producers capture the sun's energy and convert it into chemical energy—solar energy also has the potential to supply many of our external energy needs, as well.

The use of solar energy actually dates back to Roman times; the Romans developed window glass, which allowed sunlight to come in and trapped solar heat indoors. Another interesting historical note is that the Swiss scientist Horace Bénédict de Saussure built a solar reflector in 1767 that could heat water and cook food.

Passive solar energy collection is the use of building materials, building placement, and design to passively collect solar energy (such as through windows) that can be used to keep a building warm or cool. On the other hand, **active collection** is the use of devices, such as solar panels, that collect, focus, transport, or store solar energy. Solar panels absorb solar energy and pass on the energy to tubes in which water is circulating; this heated water can be stored for later use. Direct collection of solar energy via **photovoltaic cells** (PV cells) produces electricity, which is then stored in batteries. When sunlight hits the PV cells, electrons are energized and can flow freely, producing an electric current. After this, one of two things can happen. If the home or building is connected to a regional electric grid, the energy produced is fed into the grid; this results in the electricity meter on the building actually spinning backward! Homeowners who have installed solar panels receive a credit against further charges from their electricity providers when the energy that they've fed into the grid exceeds the amount of energy the household uses. If the home or building is not connected to the local electric grid, the energy stored can be stored in batteries to be utilized later.

Solar Power Process

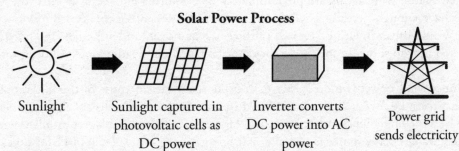

| Sunlight | Sunlight captured in photovoltaic cells as DC power | Inverter converts DC power into AC power | Power grid sends electricity |

While the use of solar energy produces no air pollutants, the production of photovoltaic cells does require the use of fossil fuels. The advantages of solar panels are that photovoltaic cells use no moving parts, require little maintenance, and are silent. However, not every location receives enough sunlight to make solar panels worthwhile. Also, the initial financial outlay for solar power is significant, although money is saved when the home is disconnected from the regional grid. Additionally, some states (such as New Jersey) give homeowners financial assistance for the installation of solar systems in their homes. Eventually, new technology should significantly lower the cost of solar systems.

Wind Energy

People have been using wind to produce energy for centuries. As early as the 17th century, windmills were so abundant in Schermerhorn (which is northwest of Amsterdam) that their turning paddles could be heard as far as 20 miles away! Windmills work in this way: wind turns the blades, or paddles, of the windmill and this drives a shaft that's connected to several cogs. The cogs then turn wheels that can perform mechanical work, such as grinding grain or pumping water. Although the Dutch windmill is a picturesque symbol of wind power, the modern wind **turbine** looks more like an airplane propeller. The wind that blows into the wind turbine spins the blades, and this, in turn, causes the machinery inside the base of the windmill to rotate. The base of the windmill is called the **nacelle,** and it houses a gearbox and generator as well as machinery that controls the turbine. Wind turbines can be designed to utilize the energy from wind at all speeds, or to function only when the wind is at a certain velocity.

Wind Power Process

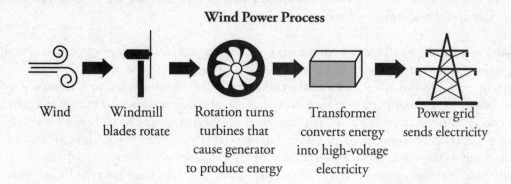

| Wind | Windmill blades rotate | Rotation turns turbines that cause generator to produce energy | Transformer converts energy into high-voltage electricity | Power grid sends electricity |

Wind energy is the fastest growing alternative energy source, and modern wind turbines are usually placed in groups called **wind farms** or parks. In the United States, the largest of these wind farms is located in Altamont Pass, California; this farm has several thousand wind turbines. Wind-generated power has been increasing at a rate of more than 30 percent per year and is projected to supply a full 20 percent of the world's energy needs by 2030. Although, in the United States, wind farms are predominately in California and Texas at this time, many locations have enough prevailing winds to make production of electricity from wind power feasible. Wind farms can also be located offshore in the ocean, and although they're currently only located near to shore, in the future they may be placed on floating docks in deep water.

At this time, wind power is more costly than using fossil fuels because of the initial outlay of capital that must be invested in order to build the windmills; windmills are also considered by many to be annoyingly loud and unattractive. However, perhaps the biggest problem with this type of renewable energy source is that alternate energy sources must be in place for times when there is no wind. In the 1990s, one other public concern about the use of wind turbines was that birds would be cut up and killed by the blades, but now we know that as long as wind farms are not located in the middle of migration routes, only one or two birds per turbine per year are killed—and this is far fewer than the number of birds killed by other types of towers. Finally, one tremendous advantage of using wind energy is that it produces no harmful emissions. Let's move on and talk about another type of renewable energy source—geothermal energy.

Geothermal Energy

Geothermal energy is a form of energy that's obtained from within the Earth; it's energy that's produced by harnessing Earth's internal heat. The interior of the Earth is still warm due to radioactive decay. Therefore, geothermal energy indirectly gains its energy from nuclear power. The greatly elevated temperatures within the Earth result in a buildup of pressure; some of this heat escapes through fissures and cracks to the surface. Some common examples of these fissures and cracks that you may have heard of are geysers, hydrothermal vents, and hot springs.

More specifically, in the process of geothermal energy production, the naturally heated water and steam from the Earth's interior turn turbines, and this creates electricity. Although surface water from geysers could be used, wells are typically drilled down into the Earth as far as thousands of meters to water that is 300–700 degrees Fahrenheit and then brought to the surface and converted to steam, which powers a turbine. Geothermal energy can also be used directly; in this process, the heated water is piped directly though buildings to heat them—this is a common method for heating homes in Iceland. In a sense, geothermal energy is renewable; however, if the groundwater is used at a faster rate than it is replaced, then this energy source is limited.

Geothermal Power Process

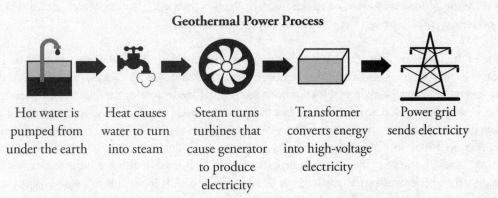

Hot water is pumped from under the earth → Heat causes water to turn into steam → Steam turns turbines that cause generator to produce electricity → Transformer converts energy into high-voltage electricity → Power grid sends electricity

Unfortunately, geothermal energy hasn't gained popularity, is expensive to install, and can potentially release toxic gas during the drilling process. The use of geothermal energy is also limited because only a few areas have geothermal sources to tap. Another problem with this renewable energy source is that the salts that are dissolved in the water corrode machinery parts. Additionally, some gases (such as methane, carbon dioxide, hydrogen sulfide, and ammonia) that are trapped in the water may be released as the water is utilized.

Other Sources of Energy

There are two other less widespread renewable energy sources: energy that can be harnessed from tidal movement in the ocean and hydrogen cells. You should be somewhat familiar with both of these energy sources for the exam.

Ocean Tides

The tidal movements of ocean water can be tapped and used as a source of energy. To harvest tidal energy, dams are erected across outlets of tidal basins. Incoming tides are sluiced through the dam, and the outgoing tides pass through the dam, turning turbines and generating electricity. Recently, ocean dams have been developed that allow energy to be harnessed from both the outgoing and incoming tides.

At this time, there is a tidal power plant installed in the East River in New York, NY. It harnesses enough power to power 10,000 homes. There are many different designs of ocean tide power plants in the idea stages. One of these involves having waves push into a chamber of air; the compressed air is then forced through a small hole at the turbine, and turns the turbine as it is released. An experimental prototype of this design has been installed off the coast of Scotland and is nicknamed the LIMPET (Land-Installed-Marine-Powered Energy Transformer).

Hydrogen Cells

Hydrogen is obtained from fossil fuels by a process called reforming. Hydrogen is very difficult to store and not very energy dense, but hydrogen fuel cells are considered by many to be the best, cleanest, and safest fuel source. Free hydrogen is not found on Earth, but it can be released through the process of electrolysis, in which hydrogen atoms are stripped from water, leaving the oxygen atom. Hydrogen can also be obtained from organic molecules, but the use of organic sources can release pollutants—as can the process of electrolysis if a fossil fuel, such as natural gas or coal, is used to drive the process. However, once the free hydrogen is released, it can be stored and then used to generate electricity through the reverse reaction of electrolysis.

One of the major benefits of the use of hydrogen fuel cells is that the only waste from the fuel cell is steam—water vapor. This technology has been used for decades in spacecrafts, but the high cost of the fuel cell and lack of hydrogen fuel stations has limited the technology to just a few test programs. The United States Department of Energy estimates that hydrogen fuel cells large enough to power light trucks and cars in the United States will require the production of 150 megatons of hydrogen per year (in 2004, only nine megatons of hydrogen were produced). Co-generation is also an energy source in which the waste heat from electricity generation is used in another industrial or residential process, thereby increasing the efficiency of the use of that fuel. It is possible to use the combustion of a fuel for more than one purpose simultaneously. Often a generator is driven by burning natural gas near the turbine. The gases are still very hot after they pass by the turbine. This formerly "waste" heat can then be used to transform water to steam to heat buildings or to run a chiller to cool buildings through a deep application of thermodynamic principles. Either way, the efficiency of the system is raised considerably by the dual use of the fuel combusted.

In order for hydrogen to become a truly viable option as a renewable energy source, an inexpensive and efficient way to produce hydrogen from nonfossil fuel sources must be developed. One of the most promising techniques for this involves the use of photovoltaic cells to harvest sunlight and then power the splitting of the water molecule.

ENERGY CONSERVATION: A FINAL NOTE

When we discuss **energy conservation**, we are basically referring to the practice of reducing our use of fossil fuels and reducing the impact we have on the environment as we produce and use energy.

One important form of energy conservation is the use of alternative fuel cars. They are gaining in popularity and acceptance. Hybrid vehicles are built with two motors: one electric and one gasoline-powered. The electric motor powers the car from 0 to about 35 miles per hour. Above 35 mph, the gasoline engine starts and helps to power the car. At highway speeds, both the electric and gas motors operate. The cars are designed so that when the brakes are applied, some of the energy is transferred from the brakes to recharge the electric motor's battery. Not only do these cars have good gas mileage, but they also produce far less CO_2 pollution. Several carmakers make models that use propane or natural gas as fuels, although these are not as common as hybrids. These generate only CO_2 and water as emissions, and they get good gas mileage. A problem is the lack of refueling stations, although devices are available that allow refueling from home. Cars can also be retrofitted with natural gas fuel tanks, so the driver can choose between gasoline or methane fuel.

Another type of alternative fuel is used cooking oils. The oils used in deep-fat fryers can be filtered and then burned in diesel-fueled cars, trucks, and buses. After starting the engine on pure diesel fuel, the driver switches to the **biofuel** to drive. At the end of the trip, the driver runs on pure diesel fuel again for a few minutes before shutting off the engine.

It has been argued that finding new fossil fuel sources would serve the same purpose as would reducing our current use of fossil fuels. However, this is not true—this statement does not take into consideration the fact that our use of fossil fuels has numerous negative effects on the environment. Additionally, in the long term it will not help us much to conserve fossil fuel resources—simply because these are not renewable energy sources—so they will eventually be depleted. Therefore, if we are to have dependable, long-term, renewable sources of energy, we must continue to develop, implement, and improve upon current renewable technology and methods.

On the legislative front, the United States has adapted the **CAFE,** or **Corporate Average Fuel Economy**, standards. These standards set mile-per-gallon standards for a fleet of cars. The goal of these standards is to reduce energy consumption by increasing the fuel economy of cars and light trucks. Review the CAFE standards described on page 250.

Finally, do not forget the role of mass transit in reducing pollution. Buses and trains can move many more people than cars. When the amount of pollution made by the vehicle is divided by all the passengers it is carrying, the bus or train generates far less pollution per person than a car.

In order to determine whether or not a resource should be used, citizens, governments, and businesses engage in a process called **cost-benefit analysis**. Costs and benefits can be either tangible (measurable), or intangible (immeasurable). Consider a corporation interested in clearing a forest for wood. The benefits are jobs and lumber (both tangible), while the costs are the loss of beauty and recreation opportunity (intangible), decreased biodiversity, and carbon dioxide removal (tangible).

In the next chapter, we will further discuss pollution—the effects it has on Earth and its inhabitants, the types of wastes that currently exist, and how we can manage them.

CHAPTER 8 KEY TERMS

Use some of your renewable brain energy to study these terms.

Energy
potential energy
kinetic energy
radiant energy
conduction
convection
energy quality
net energy yield
energy units
power units
First and Second Laws of Thermodynamics

Nonrenewable Energy
fossil fuels
seam
exploratory well
proven reserve
crude oil
tar sands
petroleum
anthracite
bituminous coal
subbituminous coal
lignite
underground mining
subsidence
fly ash
boiler residue
strip mining
overburden
scrubbers
acid mine drainage
Hubbert peak (peak oil)

Nuclear Energy
fission
breeder reactor
radioactive
half-life
nuclear fusion
fuel rod
boiling water reactor
pressurized water reactor

Renewable Energy
gasohol
biodiesel
hydroelectric power
fish ladder
passive solar energy collection
active collection
photovoltaic cells
nacelle
turbine
wind farm
geothermal energy
hydrogen cell
ocean tides
energy conservation
biofuel
CAFE standards
cost-benefit analysis

Chapter 8 Drill

Directions: Each of the questions or incomplete statements below is followed by four suggested answers or completions. Select the one that is best in each case. For answers and explanations, see Chapter 13.

1. A fuel's net energy yield is correctly defined as

   (A) how much of the fuel is left in the world
   (B) how much time it takes to extract and transport
   (C) a comparison between the amount of pollution the fuel generates and the amount of useful energy produced
   (D) a comparison between the costs of mining, processing, and transporting a fuel and the amount of useful energy the fuel generates

2. A regular light bulb has an efficiency rating of 3 percent. For every 1.00 joule of energy that bulb uses, the amount of useful energy produced is

   (A) 1.03 joules of light
   (B) 1.03 joules of heat
   (C) 0.97 joules of light
   (D) 0.03 joules of light

3. Methane gas and ethanol are two examples of biogases that are produced in which of the following processes?

   (A) The distillation of oil
   (B) The pressurization of natural gas
   (C) The anaerobic digestion of biomass
   (D) The catalytic reaction of coal and limestone

4. Hybrid car engines have which of the following types of motors?

   (A) Gasoline powered only
   (B) Natural gas powered only
   (C) Electric powered only
   (D) Gasoline- and electric-powered engines

5. All of the following are ways to increase energy efficiency EXCEPT

   (A) using low volume shower spray heads
   (B) insulating your home thoroughly
   (C) switching incandescent light bulbs to fluorescent bulbs
   (D) leaving room lights on

6. A typical coal-burning power plant uses 4,500 tons of coal per day. Each pound of coal produces 5,000 BTUs of electrical energy. How many BTUs are produced each day from this plant?

   (A) 4.5×10^{10}
   (B) 0.45×10^{10}
   (C) 11.5×10^{3}
   (D) 4.5×10^{8}

7. Which of the following produces the least amount of carbon dioxide while generating electricity?

   (A) Oil
   (B) Coal
   (C) Wind turbines
   (D) Wood

8. How much energy, in kWh, is used by a 100-watt computer running for 5 hours?

   (A) 500 kWh
   (B) 200 kWh
   (C) 50 kWh
   (D) 0.5 kWh

9. Photovoltaic cells produce electricity by

   (A) a system of mirrors that focuses sunlight onto a heat collection device
   (B) using the sun's energy to create a flow of electrons in a material such as silicon
   (C) breaking down organic molecules and releasing energy
   (D) warming air, which spins a turbine

10. A sample of radioactive material has a half-life of 20 years. It has an activity of 2 curies. How many years does it take for the material to have an activity level of 0.25 curies?

    (A) 20 years
    (B) 40 years
    (C) 60 years
    (D) 80 years

11. The term *vampire appliances* correctly refers to appliances that

   (A) generate more power than they consume
   (B) consume electricity even when they are not operating
   (C) are EnergyStar rated
   (D) are programmed to turn themselves off at midnight each night

12. All nonrenewable resource power plants use heat to

   (A) make hot air that generates power
   (B) create powerful magnetic fields that make electricity
   (C) create powerful water jets that spin turbines
   (D) produce steam to turn electric generators

13. The acidity of a lake would most likely increase because of

   (A) the construction of a hydroelectric dam in the region
   (B) increased power generation at a local wind farm
   (C) overfishing by commercial fisherman
   (D) the burning of coal by nearby factories

Free-Response Question

1. Nuclear power plants have been described as being part of the solution to the problem of the United States'
    dependency on foreign energy. Currently, some 20 percent of the electricity produced in the United States is
    generated by nuclear power.

    (a) **Describe** the key parts of a nuclear power plant. **Describe** the roles of the following: core, fuel rods, coolant,
        and heat exchanger.
    (b) **Describe** TWO practical methods of dealing with the long-term storage of the highly radioactive wastes
        produced by a power plant.
    (c) **Describe** ONE positive impact that a nuclear power plant might have on air pollution.
    (d) Opponents of nuclear power plants point out the problems caused by thermal pollution of nearby rivers.
        Describe how the thermal pollution occurs and ONE method to reduce this problem.

Summary

o   The Laws of Thermodynamics state that energy is neither created nor destroyed and tends toward entropy (disorder).

o   Nonrenewable energy is the primary source of energy production, using primarily fossil fuels (about 80% of the global demand), including:
*   oil
*   coal
*   natural gas

o   Nuclear energy is a non-fossil fuel, nonrenewable energy source.

o   Renewable energy sources account for only 14% of the global energy use. These include:
*   biofuels/biodiesel
*   solar
*   hydroelectric
*   wind
*   geothermal
*   hydrogen cells

o   *Renewable* does not always mean "clean" and *nonrenewable* does not always mean "dirty"; these terms refer to how long it takes to regenerate the fuel supply (generally more than 50 years or a human's lifespan).

o   Reduction in the use of both renewable and nonrenewable energy supplies can be done with greater energy conservation and improved technology efficiencies.

Chapter 9
Pollution

Topics in this chapter can be found in the "Atmospheric Pollution" unit and "Aquatic and Terrestrial Pollution" unit from the new Course and Exam Description.

We'll begin the chapter with a discussion of what it means for something to be toxic. We'll move on to discuss toxins in air pollution and then review the major aspects of thermal pollution, water pollution, and the problems that arise as a result of solid waste. As we go through each type of pollution, we'll also discuss the impact of pollution on the environment and human health, as well as some economic impacts. Let's begin!

TOXICITY AND HEALTH

A **toxin** is any substance that, when inhaled, ingested, or absorbed at sufficient dosages, damages a living organism, and the **toxicity** of a toxin is the degree to which it is biologically harmful. Almost any substance that is inhaled, ingested, or absorbed by a living organism can be harmful when it is present in large enough quantities—even water!

Substances are usually tested for toxicity using a **dose-response analysis.** In a dose-response analysis, organisms are exposed to a toxin at different concentrations, and the dosage that causes the death of the organism is recorded. The information from a set of organisms is graphed, and the resulting curve is referred to as a **dose-response curve**. The dosage of toxin it takes to kill 50 percent of the test animals is termed LD_{50}, and this value can be determined from the graph. A high LD_{50} indicates that a substance has a low toxicity; a low one indicates high toxicity. A **poison** is any substance that has an LD_{50} of 50 mg or less per kg of body weight.

The government regulates certain types of toxins in air, water, and food. **The Food and Drug Administration (FDA)** is the body that regulates food and related products; it was empowered to do so by the Federal Food, Drug, and Cosmetic Act of 1938. One important part of this act was the **Delaney Clause** (part of the Food Additives Amendment of 1958), which specifically bans any food additives found to cause cancer in humans or in animal testing. When the Delaney Clause was included, no one thought it would have a very broad application, but as scientists have identified more and more cancer-causing substances, its relevance has grown.

As they say, the dose makes the poison. When determining how harmful a substance is, all of the following must be considered:

- dosage amount over a period of time

- number of times of exposure

- size and/or age of the organism that is exposed

- ability of the organism to detoxify that substance

- organism's sensitivity to that substance (due, for example, to genetic predisposition or previous exposure)

- synergistic effect (when more than one substance combines to cause a toxic effect that's greater than any one component)

If just the negative health effects are plotted, instead of the level of the toxin at which death occurs, the resulting graph indicates the dosage that causes a change in the state of health. In this case, the ED_{50} is the point at which 50 percent of the test organisms show a negative effect from the toxin. The dosage at which a negative effect occurs is referred to as the **threshold dose**. Two more terms you should know for the test are acute effect and chronic effect. An **acute effect** is an effect caused by a short exposure to a high level of toxin; a snakebite, for example, causes an acute effect. A **chronic effect** is what results from long-term exposure to low levels of toxin; an example of this would be long-term exposure to lead paint in a house.

An **infection** is the result of a pathogen invading the body, and **disease** occurs when the infection causes a change in the state of health. For example, HIV, the virus that causes the disease AIDS, infects the body and typically has a long residence time. When it causes a change in a person's state of health, it has morphed into a disease called AIDS.

Pathogens are bacteria, viruses, or other microorganisms that can cause disease. There are five main categories of pathogens.

- Viruses (and other subcellular infectious particles, such as prions)
- Bacteria
- Fungi
- Protozoa
- Parasitic worms

Pathogens can attack directly or via a carrier organism (called a **vector**). One example of a pathogen that relies on a vector is the bacteria that causes Rocky Mountain spotted fever. It lives in the bodies of ticks, and when ticks bite humans, the ticks inject the bacteria, which causes the disease.

As you're probably well aware, other things besides pathogens can make people ill, including environmental factors such as tobacco smoke, UV radiation, or asbestos. Also, although you may be exposed to a toxin or an infectious agent and not experience a change in the state of your health, someone else who's exposed to the toxic agent or pathogen could become very ill.

The degree of likelihood that a person will become ill after exposure to a toxin or pathogen is called **risk.** Many environmental, medical, and public health decisions are based on potential risk. Calculating risk is referred to as **risk assessment,** and **risk management** means using strategies to reduce the amount of risk. The U.S. Department of Public Health and Public Services is an organization that makes use of risk assessment and management. For example, the department decides who can receive the flu shot each year. If the risk of getting the flu is high for a particular year, most of the population is encouraged to get the shot; however, if the risk seems small or the predicted flu strains are mild, only older people and the immunocompromised are advised to get the flu shot.

AIR POLLUTION

Substances that are considered contributors to air pollution have two sources: they can be natural releases from the environment or they can be created by humans. The effects of air pollution on humans can range in severity from lethal to simply aggravating. Some natural pollutants include pollen, dust particles, mold spores, forest fire smoke, and volcanic gases. One of the more recently described air pollutants from nature is produced by dinoflagellates, which, you might recall from Chapter 4, are the organisms that cause red tide. The toxins that are produced by these algae are caught in sea spray, in which they can be aerosolized and inhaled by humans, causing respiratory distress.

Although you may think that human-caused pollution is a relatively new phenomenon, people have added pollutants to the air throughout the history of humankind. Early humans' fire created pollutants, and the Romans' lead smelting resulted in air pollution that drifted thousands of miles from the source—and has even been discovered trapped in the ice of Greenland! It is true, however, that the large-scale production of pollutants began with the Industrial Revolution, and this is especially true of air pollution. The beginning of the Industrial Revolution marked the entrance of pollutants from fossil fuel into the atmosphere, for example, which has been environmentally disastrous.

Let's go over some terms used to describe pollution before we get into more specific details. **Primary pollutants** are those that are released directly into the lower atmosphere (remember the troposphere?) and are toxic; one example of a primary pollutant is carbon monoxide (CO). **Secondary pollutants** are those that are formed by the combination of primary pollutants in the atmosphere; an example of a secondary pollutant is acid rain. Acid rain is produced from the combination of **sulfur oxides** (such as SO_2 and SO_3) and water vapor.

> **Fast Fact**
>
> Fossil fuel emissions may also contain sulfur, which reacts during combustion:
>
> $$S + O_2 \rightarrow SO_2$$
>
> $$2SO_2 + O_2 \rightarrow 2SO_3$$
>
> $$SO_3 + H_2O \rightarrow H_2SO_4$$
> (sulfuric acid)

Pollutants can be released by **stationary sources,** such as factories or power plants, or they can be released by **moving sources,** like cars. **Point source pollution** describes a specific location from which pollution is released; an example of a point source location might be a factory or a site where wood is being burned. Pollution that does not have a specific point of release—for example, a combination of many sources, such as a number of cows releasing methane gas within a few square miles—is known as **non-point source pollution.**

The Major Culprits

The Environmental Protection Agency has determined that there are six pollutants (familiarly referred to as the dirty half dozen) that do the most harm to human health and welfare; the Environmental Protection Agency (EPA) refers to them as **criteria pollutants.**

The Six Criteria Pollutants

- carbon monoxide, CO

- lead, Pb

- ozone, O_3

- nitrogen dioxide, NO_2

- sulfur dioxide, SO_2

- particulates

In general, gases in the atmosphere are measured in units of parts per million, or ppm, when they are in relative abundance; when they are present in trace (very small) amounts, they are measured in parts per billion (ppb). For example, if in a certain geographic area, the carbon dioxide content of the air is 10 ppm, this would mean that there are ten molecules of CO_2 per one million molecules of air. These pollutants are monitored with the National Ambient Air Quality Standards (NAAQS). These standards were established by the Environmental Protection Agency (EPA) to protect human health.

Let's go back through the list above. Carbon monoxide (CO) is an odorless, colorless gas that's typically released as a by-product of incompletely burned organic material, such as fossil fuels. CO is hazardous to human health because it binds irreversibly to hemoglobin in the blood. Hemoglobin is the molecule that is responsible for transporting oxygen around the body from the lungs. Hemoglobin has a higher affinity for CO than it does for oxygen, which means that in the presence of both CO and O_2, CO will bind more readily than O_2. In our normal oxygen-rich environments, this competition is not a problem, but in areas where CO is present in large concentrations, it can be deadly. More than 60 percent of the CO released into the atmosphere comes from vehicles that burn fossil fuels.

Lead is an air pollutant that, as you now know, has been around since the time of the Roman smelters. It is generally released into the atmosphere as a particulate (a very small solid particle that can be suspended in the air), but then settles on land and water, where it is incorporated into the food chain and is subject to biomagnification. If it enters the human body, it can cause numerous nervous system disorders, including cognitive and developmental disabilities in children. At one time, lead entered the atmosphere primarily as a result of the burning of leaded gasoline. However, lead gas has been phased out, and now the primary source of lead is industrial smelting. Incidentally, the "lead" in your pencils is not the element lead; in fact, it's the mineral graphite. The graphite in pencils received the name "lead" because of its lead-like color when it's transferred to paper.

We began our discussion of ozone in Chapter 4 and have mentioned it several times since. Notice that the ozone the EPA calls one of the dirty half dozen is specifically—and only—the ozone that's formed as a result of human activity. All ozone is O_3 and is the same chemically. However, **stratospheric ozone** (which absorbs UV light from the sun and therefore protects life on our planet) is functionally very different from **tropospheric ozone** (a powerful respiratory irritant and precursor to secondary air pollutants). Up high, ozone helps us; down low, it hurts us. O_3 is a secondary pollutant; it is formed in the troposphere as a result of the interaction of **nitrogen oxides**, heat, sunlight, and volatile organic compounds (VOCs—more on these later). Tropospheric ozone is a major component of what we think of as smog (more on this later).

The next major culprit on the list, nitrogen dioxide (NO_2), is one in a family of nitrogen and oxygen gases. NO_2 and the other nitrogen oxides are formed when atmospheric nitrogen and oxygen react as a result of exposure to high temperatures; this type of reaction occurs in combustion engines, for example. In fact, more than half of the nitrogen oxides in the atmosphere are released as a result of combustion engines. Other sources of nitrogen oxides are utilities and industrial combustion. Nitrogen dioxide is also commonly found as a secondary pollutant and is a component of smog and acid precipitation.

Sulfur dioxide (SO_2) is a colorless gas with a penetrating and suffocating odor. It is a powerful respiratory irritant and is typically released into the air through the combustion of coal. As we mentioned in Chapter 8, the use of scrubbers in coal-burning plants has helped reduce the amount of SO_2

released into the atmosphere. However, there are other sources of sulfur dioxide, including metal smelting, paper pulping, and the burning of fossil fuels. Sulfur dioxide can also be a component of indoor pollution as a result of gas heaters, improperly vented gas ranges, and tobacco smoke. In the atmosphere, SO_2 reacts with water vapor to form acid precipitation. Here's one last note to help you with the test: both nitrogen and sulfur can combine with oxygen to make several different molecules. Rather than a list of all the possible molecules, you might see the terms NO_x and SO_x. These terms (O_x) mean that there are several sulfur- and nitrogen-containing compounds mixed together.

Particulate matter is the last on the EPA's list of the dirty half dozen. Like lead, it is not a gas, but exists in the form of small particles of solid or liquid material. These particles are light enough to be carried on air currents, and when humans breathe them in, the particles act as irritants. Examples include soot (black carbon) and sulfate aerosols.

There have been significant decreases in the atmospheric content of both lead and carbon monoxide since the 1970s, mostly because of the phasing out of lead gasoline and the introduction of car engines that burn more cleanly. However, there are other air pollutants that are a growing concern to environmentalists, including volatile organic compounds (VOCs), which are released as a result of various industrial processes including dry cleaning, the use of industrial solvents, and the use of propane. VOCs can react in the atmosphere with other gases to form O_3 and are a major contributor to smog in urban areas. Now, what exactly is smog?

Smog

As you might be aware, the setting for many of the Sherlock Holmes mysteries is the foggy, smoggy city of London. The smog that covered London throughout the 19th century, and well into the middle of the 20th century, was **industrial smog**—also known as **gray smog** or gray-air smog. As deadly as any of Holmes's adversaries in Sir Arthur Conan Doyle's stories, gray smog killed more than 2,000 people in a prolonged smog incident in 1911. However, the worst pollution-related incident in London occurred in 1952 and led to the death of about 10,000 city dwellers from pneumonia, tuberculosis, heart failure, and bronchitis. It was this disaster, resulting from the burning of large amounts of low-quality coal to heat homes and combat a cold fog, that prompted the **Clean Air Act** of 1956 in England.

> The **Clean Air Act** is part of the required environmental legislations to know for the exam, so make sure to memorize it!

Industrial smog is formed from pollutants that are typically associated with the burning of oil or coal. When **CO** and **CO_2** are released in the process of combustion, they combine with particulate matter in the atmosphere and produce smog. The production of smog can also be aided by weather conditions—air inversions, for example, which trap the pollutants; or fog, which holds the pollutants. As we mentioned above, sulfur dioxide may be another component in gray smog, combining with water vapor to form sulfuric acid that is suspended in the cloud of smog.

Photochemical smog—also known as **brown smog**—is a different type of smog, usually formed on hot, sunny days in urban areas. In photochemical smog, NO_x compounds, VOCs, and ozone all combine to form smog with a brownish hue. The intensity of sunlight on these days also promotes the formation of ozone from the combination of NO_x compounds. Los Angeles, California, and Athens, Greece, are two cities that are particularly susceptible to photochemical smog. Athens has enacted mandates that have already reduced the number of cars driven each day in the city and improved the quality of the air. For example, by law, cars with even numbered license plates can only be driven on even-numbered days—and cars with odd-numbered license plates can only be driven on odd-numbered days!

You should be familiar with the following chemical equations:

Photochemical Smog

$2NO + O_2 \rightarrow 2NO_2$ (causes brownish haze)

$NO_2 + UV$ light $\rightarrow NO + O$ followed by:

$O + O_2 \rightarrow O_3$ (O_3 is ozone and is very hazardous to plants, animals, and materials in the troposphere)

hydrocarbons $+ O_2 + NO_2 \rightarrow$ PANs (peroxyacyl nitrates—cause burning eyes and damage vegetation)

Ozone Depletion

While harmful in the troposphere, as you know, ozone in the stratosphere provides us with a much-needed defense against ultraviolet radiation. The ozone layer is responsible for blocking about 95 percent of the sun's ultraviolet radiation (UV), thus protecting surface-dwelling organisms from UV damage. Ozone is naturally created by the interaction of sunlight and atmospheric oxygen. The simplified reaction is

---

$$O_2 + UV \text{ (sunlight)} \rightarrow O + O$$
$$O + O_2 \rightarrow O_3$$

As early as the mid-1950s, a thinning of the ozone layer above the Antarctic was observed. In the 1970s, atmospheric scientists hypothesized, and later proved, that declining stratospheric ozone levels were due to a group of man-made chemicals known as **chlorofluorocarbons (CFCs)**. Invented in the 1930s, CFCs and many other related compounds (e.g., halons and hydrochlorofluorocarbons) were used in items such as propellants, fire extinguishers, and cans of hairspray.

Once released, CFCs migrate to the stratosphere through atmospheric mixing (they are very stable, which allows them to survive through the rise). In the upper stratosphere, intense UV radiation breaks the CFC molecules apart and releases chlorine atoms that form chlorine monoxide (ClO) while converting O_3 to O_2. Let's take a look at that reaction.

$$Cl + O_3 \rightarrow ClO + O_2$$

During the winter months, chlorine monoxide is concentrated on ice crystals that form in and around the Antarctic polar vortex. In early spring, the returning warmth of the sun frees the chlorine from the chlorine monoxide where it destroys more ozone. The reaction that frees the chlorine from chlorine monoxide is

$$ClO + O \rightarrow Cl + O_2$$

Ozone loss is greatest in the spring as the chlorine breaks down ozone into O_2. Remember that chlorine acts as a catalyst; it is not changed by its reaction with ozone and it can help break down another O_3 molecule immediately. As the air continues to warm, the natural production of ozone *increases* as more sunlight catalyzes the combination of oxygen back into ozone. This occurs in January and February (Antarctica's summer).

The Antarctic continent is the area exposed to the greatest amount of UV radiation, but prevailing winds can carry the ozone-depleted air to South America, Australia, and southern Africa. In 2006, the area of ozone thinness was over 26 million square kilometers. Reduced levels of ozone have been documented over the Arctic and even over some midlatitude regions.

The loss of ozone has serious implications for the Earth's ecosystems as well as for human health. The increased number of UV rays that reach Earth through the thin ozone layer can kill phytoplankton and other primary producers. The decrease in primary productivity of both marine and terrestrial ecosystems lowers the amount of available fish and crops. Human health issues from increased exposure to UV rays include eye cataracts, skin cancers, and the weakening of our immune systems.

The **Montreal Protocol** is part of the legislations required for the exam, so make sure to memorize it!

Now for the good news: there are several methods to manage the amounts of CFCs. In 1987, the **Montreal Protocol** was signed by more than 146 nations. The protocol calls for the worldwide end of CFC production. The United States stopped production in 1995. Since the institution of the Montreal Protocol, the release of ozone-depleting chemicals has been reduced by 95 percent. There are, however, many nations that still rely on CFCs, though work is being conducted to develop safe and effective substitutes.

Acid Rain

Acid precipitation—in the form of acid rain, acid hail, acid snow, etc.—occurs as a result of pollution in the atmosphere, primarily SO_2 and nitrogen oxides. These gases combine with water to form acids (typically nitric acid and sulfuric acid) that are deposited on the Earth through precipitation. Because this acid is highly diluted, acid precipitation isn't acidic enough to burn the skin upon contact, but it does have a significant, measurable effect on humans and the environment. How acidic is acid rain? Well, rain usually has a pH of about 5.6, but acid rain can have a pH as low as 2.3. The pH of rain is not 7.0 because of the carbonic acid in rainwater. Also, it's important for you to know that the pH scale is logarithmic, meaning that each whole pH value below 7 is ten times more acidic than the next higher value.

Acid precipitation can be a chronic and significant problem for large urban areas with many vehicles and areas that are downwind of coal burning plants. While **dry acid particle deposition** occurs two to three days after emission into the atmosphere, **wet deposition** is usually delayed for four to fourteen days after emission; therefore, pollution from wet deposition can travel in air currents to locations that are many miles downwind of the emission source.

Acid precipitation is responsible for the following effects:

- leaching of some minerals from soil (which alters soil chemistry)

- creating a buildup of sulfur and nitrogen ions in soil

- increasing the aluminum concentration in soil to levels that are toxic for plants

- leaching calcium ions from the needles of conifers

- elevating the aluminum concentration in lakes to levels that are toxic to fish

- lowering the pH of streams, rivers, ponds, and lakes, which may lead to fish kills

- causing human respiratory irritation

- damaging all types of rocks, including statues, monuments, and buildings

Some areas, like those with already acidic soils that were derived from granite, are particularly vulnerable to acid precipitation. Other areas that are particularly vulnerable to acid precipitation are those where the soil has been leached of its natural calcium content. This is because calcium acts as a natural buffer and tempers the effects of acid precipitation. Acid rain is a significant cause of sink holes in Florida due to acid dissolving the limestone rock that much of Florida is made of.

In some areas of the world, progress has been made toward controlling acid precipitation. The 1990 amendment to the Clean Air Act (CAA) has led to significant reductions in the amounts of SO_2 and NO_x that are emitted from industrial plants. Despite **National Ambient Air Quality Standards** (standards established by the United States Environmental Protection Agency), there is still considerable damage being done to soils and lakes in many areas, and these ecosystems will not be able to continue to tolerate significant lowering of their pH.

Motor Vehicles and Air Pollution

Today, all new vehicles sold in the United States must meet the EPA standards (in California, they must meet certain standards set by the state). Due to the Clean Air Act (the CAA) and its amendment (the CAAA), new cars (those produced after the year 1999) emit 75 percent fewer pollutants than cars made before 1970. The most significant device in controlling emissions in cars is the **catalytic converter**. This platinum-coated device oxidizes most of the VOCs and some of the CO that would otherwise be emitted in exhaust, converting them to CO_2. Newer models of catalytic converters also reduce nitrogen oxides, but not very successfully.

As mentioned in Chapter 8, in the Energy Policy and Conservation Act of 1975, the Department of Transportation was given the authority to set what's called **Corporate Average Fuel Economy (CAFE)** for motor vehicles. CAFE was intended to reduce both fuel consumption and emissions (not surprisingly, because burning less gas creates less air pollution).

Starting in 2011, the CAFE standards were expressed as mathematical functions depending on the vehicle's "footprint" (a measure of vehicle size determined by multiplying the vehicle's wheelbase by its average track width). That complicated 2011 mathematical formula was replaced starting in 2012 with a simpler formula with cut-off values.

In July 2011, President Obama announced an agreement with thirteen large automakers to increase fuel economy to 54.5 miles per gallon for cars and light-duty trucks by model year 2025.

All of these new standards will most likely result in higher purchase prices for vehicles, and they have certainly caused an outcry from auto manufacturers and oil refineries. However, the new standards are expected to reduce air pollutants by two million tons per year.

Vehicles of the Future

In 1990, the state of California passed a No-Pollution Vehicle Law mandating that, by 2003, 10 percent of the cars sold in the state would be pollution free. That law was later rescinded because of problems with the development of the zero pollution electric car, which looked promising at the time the bill was passed. In the past, electric cars were not widely adopted because they had a limited traveling range, they weighed more than their gasoline-burning counterparts, and they lacked amenities (like air conditioning). However, a renewed interest as well as new battery technology, which increases energy storage and reduces cost, has motivated car makers to produce a promising generation of fully electric cars. The Nissan Leaf, Chevrolet Bolt, and Tesla Model 3 are examples of some zero-pollution electric vehicles.

In addition to new electric cars, hybrid vehicles that run on a mixture of gas and electric power have been gaining popularity over the past few years, with the Toyota Prius leading the pack and many auto companies releasing hybrid versions of their standard vehicles. Government regulations, incentives, and public acceptance will probably determine how quickly the hybrid car moves into the mainstream vehicle market. One incentive that's been offered at a federal level is a full-dollar tax credit. The amount of the credit varies by car model. That incentive is scheduled to be reduced unless a new energy bill changes it. Some individual states also provide incentives to residents who purchase a hybrid vehicle.

A hydrogen fuel cell vehicle would produce even less pollution than a hybrid vehicle, but don't expect to see them on the market very soon. Mass producing the cells is still not cheap enough to make the cars economically viable, as we mentioned in Chapter 8.

It is highly unlikely that Congress will enact legislation that will provide real incentives for the purchase of hybrid vehicles or other alternatives that would reduce air pollution from vehicles. This is in part due to the fact that lobbying groups representing the oil companies and vehicle manufactures consistently lobby against these incentives. However, in the future, grassroots organizations that are backed by the voting public may influence legislation.

Indoor Air Pollution

The idea of air pollution that exists indoors and the concept of "sick building syndrome" are still relatively new, but it is now widely recognized that air pollutants are usually at a higher concentration indoors than outside. This makes sense if you consider that pollutants that exist outside can also move inside as doors and windows are opened. Once the pollutant is indoors, it remains trapped until air currents move it out the door or windows or through a ventilation system. Additionally, indoor spaces have certain pollutants that are unique to them. The World Health Organization (WHO) estimates that indoor air pollution is responsible for 4 million annual deaths worldwide. (That's one death every 8 seconds!) According to the Environmental Protection Agency (the EPA), indoor air pollution is one of the five major environmental risks to human health.

One reason that indoor air pollution has such a great impact on health is because humans spend a significant amount of time indoors. Especially in developed countries, people generally work and live in well-sealed buildings that have little air exchange. In developing countries, however, one of the worst indoor air pollutants is material that's used for fuel. Dung, wood, and crop waste are the primary fuels used by more than half the world's population in order to heat homes and cook food, and the particulate matter that results from burning these fuels can exceed acceptable levels by hundreds of times.

In developed countries, other pollutants play the biggest roles in the creation of indoor air pollution; the most abundant indoor pollutants are **volatile organic compounds (VOCs)**. VOCs are found in carpet, furniture, plastic, oils, paints, adhesives, pesticides, and cleaning fluids. Even dishwashers are responsible for the creation of VOCs, when chlorine detergent reacts with leftover foods. Another component of pollution in developed countries is CO; CO arises in indoor air as a result of gas leaks or poor gas combustion devices. CO detectors are available for homes, and can prevent CO poisoning.

Two of the most deadly and common indoor pollutants in developed countries are tobacco smoke and radon. Tobacco smoke affects not only the health of the smoker, but the health of those around the smoker, as well. Secondhand smoke causes many of the same symptoms in nonsmokers who simply breathe it in as smoking can cause to the smokers themselves. Secondhand smoke, which contains over 4,000 different chemicals, has been classified by the EPA as a Group A carcinogen (meaning that it causes cancer in humans). It's estimated that secondhand smoke causes 35,000–40,000 deaths per year from heart disease, and 3,000 deaths from lung cancer. In children younger than 18 months, it is responsible for 150,000–300,000 lower respiratory tract infections annually, and it increases the number and severity of asthma attacks in about one million asthmatic children.

Radon is the second leading cause of lung cancer (after smoking) in the United States. Radon is a gas that's emitted by uranium as it undergoes radioactive decay. It seeps up through rocks and soil and enters buildings. It is not found everywhere and must be tested for specifically. Homes that were built after 1990 have radon-resistant features.

The final indoor pollutants we'll review are actually living: certain living organisms, such as tiny insects, fungi, and bacteria are considered pollutants. Many people are allergic to mold spores, mites, and animal dander, but asthma attacks can also be triggered by these living pollutants. The water tanks for large air conditioning units are good places for certain types of bacteria to grow, and as air is distributed throughout the house, the bacteria are also distributed. Some bacteria can cause diseases; one example of this is *Legionella pneumophila*, which causes Legionnaires' disease.

Sick Building Syndrome

Sick building syndrome (SBS) is a term that's used when the majority of a building's occupants experience certain symptoms that vary with the amount of time spent in the building and for which no other cause can be identified. SBS is somewhat difficult to diagnose, and specific culprits are very difficult to identify. A condition is referred to as a **building-related illness** when the signs and symptoms can be attributed to a specific infectious organism that resides in the building. One example of a building-related illness is Legionnaires' disease. Some symptoms of SBS include the following:

- irritation of the eyes, nose, and throat

- neurological symptoms, such as headaches and dizziness; reduction in the ability to concentrate; or memory loss

- skin irritation

- nausea or vomiting

- a change in odor or taste sensitivity

There are many ways in which people can reduce the amounts of indoor pollutants that they're exposed to—for many people, simply quitting smoking or encouraging roommates to quit would make a huge difference. Other precautions that people can take are to limit the amount of exposure they have to certain chemicals, such as pesticides or cleaning fluids. Perhaps the most important step to take is making sure that buildings are as well ventilated as possible.

CLIMATE CHANGE

Scientists use very sophisticated computer models and make several thousand meteorological observations each day to monitor the daily temperature of the Earth's atmosphere. Over the last several years, their observations have shown that there has been a slow but steady rise in the Earth's average temperature. According to NASA, 2016 was the warmest year on record, and if El Niño and La Niña patterns were removed from the record, 2017 would have been the warmest year on record. Other qualified scientists have carefully documented a decrease in the size of glaciers and ice sheets, a slight rise in the average ocean level, and more severe rainstorms. In response to these phenomena, the Intergovernmental Panel on Climate Change (IPCC) gathered hundreds of scientists from around the world to study these problems. In a 2013 report, the IPCC stated that most of the observed increase in the global average temperature since the mid-20th century is very likely (greater than 95 percent) due to the observed increase in **anthropogenic greenhouse gas** concentrations. The three major gases are carbon dioxide (from pre-industrial levels of 280 ppm to 400 ppm in 2016), methane (from preindustrial levels of 715 ppb to 1,840 ppb in 2016), and nitrous oxide (from preindustrial levels of 270 ppb to 328 ppb in 2016). These gases absorb the infrared heat radiating from Earth and thus heat the lower atmosphere. This warming is in addition to the normal warming of the atmosphere by the greenhouse effect. Review the diagram on page 101.

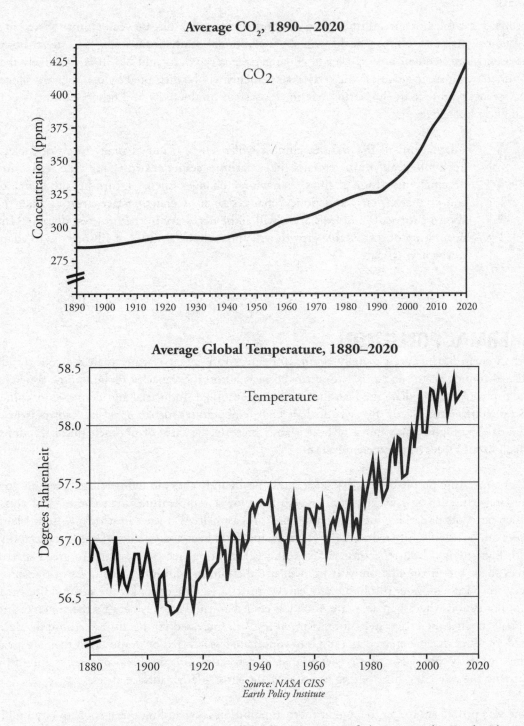

Average CO₂, 1890—2020

Average Global Temperature, 1880–2020

Source: NASA GISS
Earth Policy Institute

The increase in the Earth's temperature will lead to a variety of changes to the Earth. Physical changes on Earth include further lessening of glaciers and ice sheets, continued rising of average ocean levels (due mostly to the thermal expansion of water), changes in precipitation patterns (with wet areas getting more precipitation and dry areas getting less precipitation), an increase in the frequency and duration of storms, an increase in the number of hot days, and a decrease in the number of cold days.

Climate change will also affect biota. While there will be increased crop yields in cold environments, this is likely to be offset by loss of croplands as other areas suffer droughts and higher temperatures. Cold-tolerant species will need to migrate to cooler climates or they may become extinct. Heat-tolerant species (including mosquitoes and other disease vectors) may spread and invade new habitats. Human health will show additional deaths from water- and insect-borne diseases. More frequent heat spells will endanger the very young and old. It is very likely that commerce, transport facilities, and coastal settlements will be disrupted by ocean level changes and stronger, more frequent storms. Marine ecosystem productivity and fishery productivity is also likely to change.

> **Did you know?**
>
> **Carbon Sequestration** is the process of capturing and storing atmospheric carbon dioxide.

Adaptations to the warmer climate will need to occur at many levels of society. Technological improvements like **carbon sequestration** and the reduction of emissions from engines, behavioral changes such as turning off lights to conserve electricity, and policy changes such as enacting new treaties (like the **Kyoto Protocol**) and legislation will all be necessary in the next few decades. The promotion of sustainable growth will enhance the abilities of all societies to adapt to the new climate.

THERMAL POLLUTION

Urban environments are generally about 20 degrees Fahrenheit warmer than the countryside that surrounds them, and this is due to the heat absorbing capacity of buildings, concrete, and asphalt, which radiate the heat that they have absorbed. Industrial and domestic machines also directly warm the air. Because of their high temperatures, urban areas are known as **heat islands.** The high temperatures of heat islands increase the rates of photochemical reactions, which in turn leads to photochemical smog.

The temperature profile of an urban area shows peaks and valleys in temperature based on how the land is used. For example, green spaces have lower temperatures than commercial areas, which have lots of parking lots, cars, buildings, and asphalt. Two ways in which the heat island effect can be significantly reduced are: (1) replacing dark, heat-absorbing surfaces (such as roofs) with light-colored heat-reflecting surfaces, and (2) planting trees and adding to green spaces. Trees shade the urban environment from solar radiation; in addition, the process of transpiration (the release of water through plant leaves) creates a cooling effect for the surrounding area. Another reason why urban areas are often less cool than rural areas is because the concrete and asphalt in cities increase water runoff. Runoff leads to increased temperatures because the deep pools of water that are created as a result of runoff are less affected by evaporation than are areas where water is spread out thinly over a larger surface area. Green spaces can reduce runoff by trapping the water and distributing it more evenly across a larger surface area.

One way people are trying to combat thermal pollution is by adding green roofs to city buildings. A green roof, or living roof, is a roof that is fully or partially covered with plants, greenery, gardens, and other vegetation planted over some type of waterproofing material. Not only does this combat the heat island effect, but it also keeps the buildings cool in summer and warm in winter, reduces rainwater runoff, provides habitats for wildlife, and helps to clean the urban air.

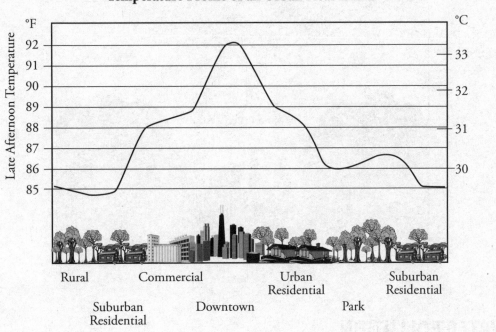

Temperature Profile of an Urban Heat Island

Another type of thermal pollution associated with many urban environments is **temperature inversion**. In this phenomenon, air pollutants become trapped over cities because they are not able to rise into the atmosphere. In normal atmospheric conditions, the warm, polluted air over a city rises into the cooler atmosphere. (Remember that warm air is less dense than the surrounding cool air, and less dense objects float!) In an inversion, the air above the city is warm and blocks the polluted air from rising. The polluted air remains hanging above the city and can cause respiratory problems. Inversions often occur in cities surrounded by mountains or cities bordered by mountains on one side and ocean on the other (for example, Los Angeles and Beirut). But thermal inversions can occur over any city where large masses of warm air can become stalled.

Thermal Inversion

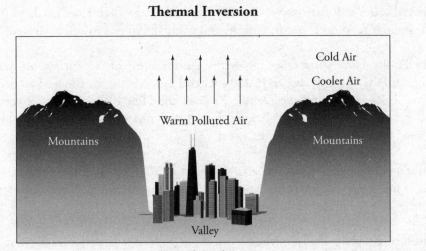

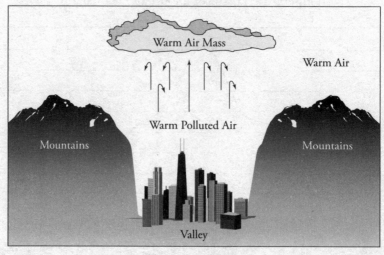

A Temperature Inversion

WATER POLLUTION

The **Clean Water Act** is part of the required environmental legislations to know for the exam, so make sure to memorize it!

When the Cuyahoga River near Cleveland, Ohio, caught fire in 1969, it became a symbol of polluted America. This fire, along with many other problems that began to arise with polluted bodies of water at that time, eventually resulted in the **Clean Water Act** (CWA) of 1972. The CWA had a dramatic effect on the quality of water in the United States. By 2016, 91 percent of community water systems met federal health standards—this number was up from the 79 percent that were considered clean by the government in 1993.

Experts say that Americans have some of the cleanest drinking (tap) water in the world. From the time of the passage of the CWA to 2017, 60 percent of the stream lengths that were tested were found to be sufficiently clean to allow fishing and swimming, while only 36 percent of the streams that were tested in 1972 were clean enough. Also as a result of the CWA, the annual loss of wetlands slowed significantly since 1972. The CWA has certainly had a positive effect on our water, but there are still plenty of water issues and bodies of water that need to be cleaned. Plus, the Clean Water Act needs to be constantly enforced, and the actions of specific citizens and companies need to be monitored.

For example, the Flint water crisis represents a breakdown of Americans' assurance in their clean water supply. The crisis began in 2014 when Flint, Michigan, changed its drinking water source from Lake Huron and the Detroit River to the Flint River. The water was not treated properly, and because of this, lead leached from water pipes into the water supply. Residents were exposed to dangerous levels of the toxin.

A federal state of emergency was declared for Flint in 2016, and Flint residents were told not to use the water supply for drinking, cooking, cleaning, or bathing. Water quality was deemed acceptable by 2017, and the lead pipes are being replaced.

One continual problem that contributes to water pollution is that runoff from land carries **excess nutrients** and pollutants to streams. This can result in large **dead zones**. For example, the dead zone in the Gulf of Mexico covers up to 5,000 square miles in the middle of what is the richest area for shellfish in the United States. This dead zone has caused the collapse of the shrimp and shellfish industries in that region.

The dead zone was created because the Mississippi River collects roughly 10,000 pounds of fertilizer and raw sewage pollution from 31 states and some of Canada as it travels south. Then it dumps all of this nutrient-rich water into the Gulf. The warm, nutrient-rich freshwater does not mix well with the colder saltwater, and this results in **eutrophication**, which allows phytoplankton to grow almost uncontrollably. In turn, the zooplankton that feed on them also experience a population explosion. When the phytoplankton and zooplankton die and sink to the bottom, bacteria metabolize the available dissolved oxygen as they decompose this detritus; the lack of oxygen creates a **hypoxic zone**, in which nothing that depends on oxygen can grow. This zone stays in place from May until September, when colder, wetter weather helps to break it up. To save this economically important fishery, a federal-state Hypoxia Task Force was formed in 1997 to reduce the size of the dead zone by two-thirds by 2015, but in 2015 it found that the zone was about the same size it was in 1994. The target date was extended to 2035.

You should review the diagram of the Mississippi dead zone given below. The black areas (near the coasts of Louisiana, Mississippi, and Texas) represent areas where the level of dissolved oxygen (DO) is very low.

Dead Zone

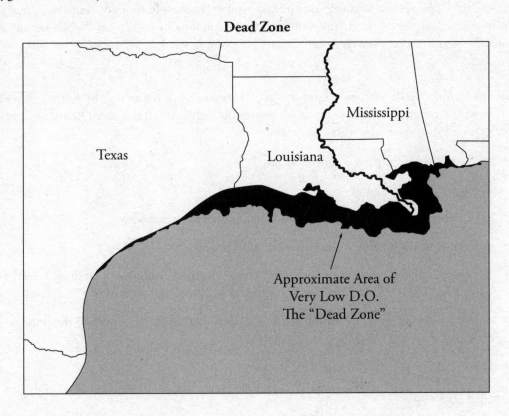

Sources

Like the terms that are used to describe sources of air pollution, particular sources that are responsible for water pollution, like paper mills, are called **point sources,** and pollution sources that do not have a definitive source (or result from contributions of many sources) are **nonpoint sources.**

Right now, the biggest source of water pollution is agricultural activities; the runners-up are industrial and mining activities. Unfortunately, standing bodies of water, such as ponds, reservoirs, and lakes, do not recover quickly from the addition of pollutants. The lack of water flow prevents the pollutants from being diluted, which means that they accumulate in the water and undergo biomagnification in the food chain. In a similar way, groundwater does not recover well from the addition of pollutants; this is again because there is very little movement of water and therefore very little flushing, mixing, or dilution. Furthermore, groundwater is generally very cold and low in dissolved oxygen, which makes recovery from degradable waste a slow process. The porous rock that surrounds the groundwater absorbs the pollutants, which makes them difficult to remove.

However, flowing streams and rivers can recover from moderate levels of pollutants if the pollutants are degradable. As illustrated by the implementation of long sewage pipes that once dumped raw sewage into the ocean off coastal areas, people thought that the ocean was able to dilute and recover from the addition of any amount of pollutants. While oceans can dilute, flush, and decompose large amounts of degradable waste, their capacity for recovery is unknown.

Water pollution is dealt with in two basic ways: (1) reducing or removing the sources of pollution, and (2) treating the water in order to remove pollutants or render them harmless in some way. Here's a list of the major water pollutants:

- excess nutrients (nitrogen, phosphate, etc.)

- organic waste

- toxic waste (pesticides, petroleum products, heavy metals, acids)

- sediments (soil washed with runoff water into streams)

- hot or cold water (hot water discharged from industrial facilities where it was used as a coolant; cold water from dam releases is discharged from the bottom of a reservoir)

- coliform bacteria (bacteria found in the intestines of animals that indicate the presence of fecal matter in water)

- invasive species (zebra mussels, for example)

- thermal pollution

Finally, if your AP Environmental Science class has performed water quality tests on water samples, you might have tested for the presence of various chemicals as well as insect larvae, which act as indicator species. Among the most important factors in judging water quality are:

- **pH,** which is a measure of acidity or alkalinity (normal for water is 6–8)

- **hardness,** which is a measure of the concentrations of calcium and magnesium

- **dissolved oxygen**—low levels of dissolved oxygen indicate an inability to sustain life (warm water holds less dissolved oxygen than cool water)

- **turbidity,** or the density of suspended particles in the water

- **BOD**, which is a measure of the rate at which bacteria absorb O from the water

Now, let's talk more specifically about a major water pollutant—wastewater.

Wastewater

Another group of water pollutants that are very dangerous to human health are infectious agents, such as those found in human and animal waste. Fecal waste not only contains the symbiotic bacteria that aid in the human digestive processes, but also contains disease-causing bacteria. Several human diseases, such as cholera and typhoid fever, are caused as a result of human waste entering the water source of a community. In fact, the major reason for the increase in the life span of humans was not modern developments in medicine; it was the introduction of cleaner drinking water and better ways of disposing of wastewater.

The term **wastewater** is used to refer to any water that has been used by humans. This includes human sewage; water drained from showers, tubs, sinks, dishwashers, and washing machines; water from industrial processes; and storm water runoff. Water that is channeled into storm drains, such as storm water, is generally dumped directly into rivers. (This is why storm drain covers in many locations have been stenciled with warnings about not dumping material down them.)

Today in the United States, wastewater that isn't storm water is moved through sewage pipes to a sewage treatment facility, but this was not always the case. Sewage water once was, and in developing countries still is, merely dumped into the nearest river or ocean. While some amounts of sewage can be diluted and broken down in these waters, too much waste poses serious risks to human health and the health of the aquatic ecosystems.

Now in the United States, sewage pipes deliver wastewater to a municipal sewage treatment plant, where it is first filtered through screens (in what's called a **physical treatment**) to remove debris such as stones, sticks, rags, toys, and other objects that were flushed down toilets. This debris is then usually separated and sent to a landfill. The remaining water is passed into a settling tank, where suspended solids settle out as sludge—chemically treated polymers may be added to help the suspended solids separate and settle out. This treatment is known as **primary treatment,** and it removes about 60 percent of the suspended solids and 30 percent of the organic waste that requires oxygen in order to decompose.

Secondary treatment refers to the biological treatment of the wastewater in order to continue to remove biodegradable waste. This treatment can be done using trickling filters, in which aerobic bacteria digest waste as it seeps over bacteria-covered rock beds. Alternately, the wastewater can be pumped into an activated **sludge processor,** which is basically a tank filled with aerobic bacteria. The solids in the water, including the bacteria, are once again left to settle out. The solids remaining are considered **sludge,** which is combined with the sludge from the primary treatment. Sludge used to be dumped into the ocean, but that practice has been banned. Instead, the sludge is further processed with anaerobic bacteria (to breakdown more organic material). This digestion also produces methane gas that can be used as an alternative fuel to run the treatment plant. After drying, this sludge cake can be processed and sold as fertilizer.

At the end of secondary treatment, 97 percent of the suspended solids; 95–97 percent of the organic waste; 70 percent of the toxic metals, organic chemicals, and phosphates; 50 percent of the nitrogen; and 5 percent of the dissolved salts have been removed from the wastewater. However, almost no persistent organic chemicals, such as pesticides, are removed, nor are radioactive isotopes. Generally, after secondary treatment, the wastewater is chlorinated to remove any remaining living cells and then discharged into a stream, the ocean, or water that's used to water lawns (called **gray water**). A negative effect of the final chlorination of the water is that trihalomethanes (potential carcinogens) can be formed when any organic matter left in the water reacts with the chlorine, and this is problematic. Two alternate processes to chlorination—ozonation and UV radiation—have been used to treat secondary-treatment water, but they have not proven to be as effective or long-lasting as chlorine and are also much more expensive.

Some municipal plants deposit wastewater directly into ground water; this is done in San Jose Creek in Los Angeles County. In these places, the water must be further treated by tertiary treatment. **Tertiary treatment** involves passing the secondary treated water through a series of sand and carbon filters and then further chlorination. At the San Jose Creek Plant, the tertiary treated water from the reclamation plants is discharged into percolation basins, where it replenishes groundwater, or it is used for irrigation and for watering lawns, golf courses, and plants in nurseries. Tertiary treatment is expensive, but in arid or semi-arid regions, every gallon that can be reclaimed is one that needs not come from rapidly depleting sources, such as diminished rivers or underground aquifers.

Private wastewater treatment in the form of septic tank systems is hallmarked by some as the most environmentally friendly type of waste disposal. Septic tanks act in a way that's similar to the primary and secondary treatments that take place in municipal treatment plants. The water is then discharged into leachate (drain) fields. In order to install these types of systems, the soil must be able to percolate the water—that is, the water must be made to move from the top of the soil though its various horizons. Some clay soils are not porous enough to allow percolation and thus are unsuitable for a septic field.

Water Quality Legislation

There are many pieces of federal law that cover water quality. The College Board now requires that you are familiar with the following:

Date	Name of Legislation	What It Did
1972	Clean Water Act	Used regulatory and non-regulatory tools to protect all surface waters in the United States. • sharply reduced direct pollutant discharges into waterways • financed municipal wastewater treatment facilities to manage polluted runoff • achieved the broader goal of restoring and maintaining the chemical, physical, and biological integrity of the nation's waters • supported "the protection and propagation of fish, shellfish, and wildlife, and recreation in and on the water"
1974, 1996, 2005, 2011, 2015	Safe Drinking Water Act	Established a federal program to monitor and increase the safety of the drinking water supply. It does not apply to wells that supply fewer than 25 people. Amendments in recent years have led to more stringent regulation of lead and algal toxins in drinking water.

SOLID WASTE (GARBAGE)

Solid waste can consist of hazardous waste, industrial solid waste, or municipal waste. Many types of solid waste provide a threat to human health and the environment.

The phrase "reduce, reuse, recycle" might seem simplistic, but it does outline the steps needed to reduce the amount of solid waste that must be dealt with. "Reduce," of course, refers to the minimizing of disposable waste. There are many types of packaging that are extremely wasteful—if you keep an eye out, you'll see them everywhere. "Reuse" applies to products that in some cases are disposable, but in other forms can be used over and over again, such as refillable bottles and tanks, reusable packing materials, secondhand goods, and cloth shopping bags. Reusing products prevents these high-quality goods from becoming waste. Finally, "recycling" is the reuse of materials. In **primary recycling,** materials such as plastic or aluminum are used to rebuild the same product—an example of this is the use of the aluminum from aluminum cans to produce more aluminum cans. Alternately, in **secondary recycling,** materials are reused to form new products that are usually lower quality goods—examples of this are old tires recycled to form carpet, and plastic bottles recycled to create decking material. Finally, another environmentally important process is composting. **Composting** allows the organic material in solid waste to be decomposed and reintroduced into the soil.

According to the EPA, one of the most effective steps in aiding the environment that occurred in the 20th century was the marked growth in the use of recycling and composting to deal with solid waste. Of the 268 million tons of municipal solid waste produced in the United States in 2017, over 90 million tons were recycled. According to the EPA, the following percent of each of these materials was recycled in the year 2017:

Material	Material Recycled in 2017 (Per 67.2 Million Tons)
Paper and paperboard	65.7%
Metals	12.4%
Glass	4.5%
Plastics	4.4%
Wood	4.5%
Rubber, leather & textiles	6.3%
Other	2.2%

In order to encourage people to reduce, reduce, and recycle, many communities have established Pay-As-You-Throw (PAYT) programs, which charge municipal customers for the amount of household garbage they throw away. As you can imagine, this has been a strong incentive for people to practice these good habits. An example of an incentive program that has worked beautifully is the bottle redemption bill. Ten states have enacted bottle bills and they have worked fantastically, especially in Michigan, which has a ten-cent redemption fee.

Landfills

In 1987, after it was discovered that landfills on Long Island were contaminating local groundwater, the barge *Mabro* left New York towing 3,186 tons of garbage in search of a dumping ground. However, it was barred from docking in several southern states, followed by the countries of Mexico, Cuba, and Belize. Three months and 6,000 miles later, it returned to New York, where it became a symbol for Americans who were concerned about the status of landfills in the United States. It was also at this time that the term NIMBY (which stands for Not In My Backyard) became popular. It was widely agreed upon that landfills were needed, but no one wanted a landfill close to their home.

Modern landfills are very different from the traditional caricature of a garbage dump filled with heaps of junked cars and rats foraging for food scraps. Federal regulations that protect human health and the environment have paved the way for **sanitary landfills**. For example, federal law prohibits landfills from being located near geological faults, wetlands, or flood plains. Additionally, landfill sites are periodically required to dig large holes in the ground and line them with geomembranes or plastic sheets that are reinforced with two feet of clay on the bottom and sides. Smoothing wet clay is much like making a clay pot; the layer that is created is virtually impermeable. Also, the waste in the landfill must be frequently covered with soil in order to control insects, bacteria, rodents, and odor; and the decomposed material that percolates to the bottom of the landfill (called **leachate**) is piped to the top of the site and collected in leachate ponds, which are closely monitored. Gases from the landfill, like methane, may even be piped up from the site and used to generate electricity. Sometimes the methane is burned in continuously flaming flares to avoid larger fires or explosions. To ensure that landfills do not contaminate the environment, they are required to be positioned at least six feet above the water table, and groundwater at the sites must be tested frequently for quality. When one site (hole) is full, it must be capped with an engineered cover, monitored, and provided with long-term care.

Waste may also be burned in municipal incinerators, which are generally capable of sorting out recyclables first. The energy released from the incineration can be used to generate electricity in what's called the **Waste-to-Energy (WTE) program.** This type of system is particularly effective in large municipal areas, where waste only needs to be transported short distances.

HAZARDOUS WASTE

Hazardous waste is any waste that poses a danger to human health; it must be dealt with in a different way than other types of waste. Hazardous waste includes such common items as batteries, cleaners, paints, solvents, and pesticides. Industry produces the largest amounts of hazardous waste, and most developed countries now regulate the disposal of these wastes. United States law mandates that hazardous materials be tracked "from cradle to grave" thanks to laws like **The Resource Conservation and Recovery Act (RCRA)**. The EPA breaks hazardous wastes down into four categories:

- **corrosive waste:** waste that corrodes metal

- **ignitable waste:** substances such as alcohol or gasoline that can easily catch fire

- **reactive waste:** substances that are chemically unstable or react readily with other compounds, resulting in explosions or causing other problems

- **toxic waste:** waste that creates health risks when inhaled or ingested, or when it comes into contact with skin

Hazardous wastes are disposed of in three main ways: in injection wells, in surface impoundments, and in landfills. Many communities have specific areas in their landfills that are designated for hazardous waste, and the standards for those areas of the landfills are higher than standards for non-hazardous waste areas. **Surface impoundment** is typically used for liquid waste; it involves the creation of shallow, lined pools from which the hazardous liquid evaporates. **Deep well injection** involves drilling a hole in the ground that's below the water table. These wells must reach below the impervious soil layer into porous rock, and waste is injected into the well. All three of these methods have their advantages, but none of them is satisfactory.

As you can probably imagine, radioactive waste must be contained in a different way than other hazardous wastes. For years, the United States has been trying to develop one major site for the disposal of all of our radioactive waste. Yucca Mountain, Nevada, was selected as the location of this site because of its remoteness and because nuclear testing had previously been done at the site. This decision is still controversial, in part because of NIMBY, but also because Nevada has no nuclear power plants. Some argue that because Nevada doesn't benefit from nuclear energy, the state should not be responsible for the spent fuel repository. It's a contentious subject. However, it is critical that some plan for the long-term storage of nuclear waste is made, because in the very near future the nation's nuclear power plants will be at maximum capacity for the storage of spent fuel. The Waste Isolation Pilot Plant in New Mexico is a new permanent site for **nuclear waste burial**. Waste that's left over from the construction of nuclear weapons is known as **transuranic waste**.

Some people define radioactive wastes that produce low levels of ionizing radiation as **low-level radioactive waste** and those that produce high levels of ionizing radiation as **high-level radioactive waste.** However, the EPA categorizes radioactive waste according to its place of origin. Therefore, in the EPA's classification system, some wastes that are considered high level may actually be less radioactive than certain low-level wastes. The EPA puts all radioactive wastes into six categories:

- nuclear reactor waste: high level

- waste from the reprocessing of spent nuclear fuel: high level

- waste from the manufacture of nuclear weapons: high level

- waste from the mining and processing of uranium ore: high level

- radioactive waste from industrial or research industries, including clothing, gloves, tubes, needles, animal carcasses, etc.: low level

- radioactive natural materials: not waste

In general, low-level waste is either stored on-site by licensed facilities until the radioactivity has degraded, or it is shipped to a low-level waste disposal facility. Mixed waste, containing both chemically hazardous waste and radioactive waste, is generally disposed of in the same manner.

In this book, we use the radioactive waste disposal terms used in the EPA classification system. However, on the AP Environmental Science Exam, a discussion of either system of classification would be considered correct—as long as you identify which classification system you're using.

Contaminated Waste Sites

After the 1970s, new regulations for the disposal of hazardous wastes solved many of the problems of how to add new wastes to landfills with minimal impact on the environment. At the same time, the issue lingered of what to do with sites that were already contaminated by hazardous waste or pollutants, known as **brownfield sites**. These sites had to be cleaned up and those who had acted irresponsibly had to be held accountable for the environmental problems they'd caused. For these reasons, the United States legislature created the **Superfund Program**, which was administered by the EPA.

Rocky Flats, Colorado, is a Superfund site where the party responsible for the damage happened to be the United States government. Starting in 1952 and continuing for almost 40 years, components of nuclear weapons, such as plutonium, uranium, beryllium, and stainless steel were all manufactured on this site. Now that the area has been significantly cleaned up, it is home to a variety of plants and animals, including bald eagles, and acts as a wind-power testing site.

Along with the Rocky Flats Plant, you should know the story of **Love Canal** near Niagara Falls, New York. The site was originally a canal built to bring power and employment to the surrounding community. After the canal's failure, the land was purchased by various companies that turned the canal into a landfill. After the town purchased the covered landfill area, 100 homes and a school were built on the site. In 1978, people saw rusting drums full of waste sticking up above ground. They also noticed dead and dying trees and gardens. Homeowners even reported having pools of smelly liquids in their basements; their children reported burning hands and faces after coming in from playing. Environmental Protection Agency employees soon came to the canal area, and by the end of August, 220 families had moved or said they would move out of the area. It was in response to the situation at Love Canal (and other sites in the United States) that laws like The Resource Conservation and Recovery Act (RCRA) and The Comprehensive Environmental Response, Compensation, and Liability Act (CERCLA) were passed.

Laws for Solid and Hazardous Wastes

There are many federal statutes that cover issues concerning solid and hazardous wastes. The College Board now requires that you are familar with the ones in the table below.

Date	Name of Legislation	What It Did
1976	The Resource Conservation and Recovery Act	• The solid waste program encouraged states to develop comprehensive plans to manage nonhazardous industrial solid waste and municipal solid waste, sets criteria for municipal solid waste landfills and other solid waste disposal facilities, and prohibits the open dumping of solid waste. • The hazardous waste program established a system for controlling hazardous waste from the time it is generated until its ultimate disposal—in effect, from "cradle to grave." • The underground storage tank (UST) program regulates underground storage tanks containing hazardous substances and petroleum products.
1980	The Comprehensive Environmental Response, Compensation, and Liability Act (CERCLA), commonly known as Superfund	• Created a tax on the chemical and petroleum industries and provided broad federal authority to respond directly to releases or threatened releases of hazardous substances that may endanger public health or the environment. • Established prohibitions and requirements concerning closed and abandoned hazardous waste sites. • Provided for liability of persons responsible for releases of hazardous waste at these sites. • Established a trust fund to provide for cleanup when no responsible party could be identified.

These are all required for the exam, so make sure to memorize them!

NOISE POLLUTION

Take your earphones off and think about this for a minute: the EPA considers noise to be a controllable pollutant. The **U.S. Noise Control Act** of 1972 gave the EPA power to set emission standards for major sources of noise, including transportation, machinery, and construction. Occupational Safety and Health Association (OSHA) has also set limits on the amount of noise that people can be exposed to in the workplace. Although the definition of noise pollution can be quite flexible, **noise pollution** in a broad sense is any noise that causes stress or has the potential to damage human health.

One concern about noise is that continued exposure to high levels can damage hearing. The louder the noise, the shorter the exposure it takes to damage inner ear cells and cause hearing impairment. Unfortunately, certain essential cells in the ear that are involved in hearing do not regenerate, so the loss of hearing is permanent. There are federal laws that regulate noise emissions from some equipment and modes of transportation, and OSHA is responsible for the regulation of noise in the workplace. In local communities, however, noise pollution is usually controlled by state or local laws. The **Quiet Communities Act** provides for the coordination of federal research and activities into noise control. Furthermore, this act authorized the use of Federal Aviation Administration funds for the development of noise abatement plans around airports.

CHAPTER 9 KEY TERMS
Don't *waste* any time; study these words now!

Toxicity and Health

toxin

dose-response analysis

dose-response curve

LD_{50}

ED_{50}

poison

threshold dose

acute effect

chronic effect

infection

disease

pathogen

vector

risk

risk assessment

risk management

Air Pollution

primary and secondary pollutants

sulfur oxides

nitrogen oxides

CO and CO_2

stationary sources

moving sources

point source pollution

non-point source pollution

criteria pollutants

tropospheric ozone vs. stratospheric ozone

industrial smog (gray smog)

photochemical smog (brown smog)

chlorofluorocarbons (CFCs)

ozone loss

acid precipitation

dry acid particle deposition

wet deposition

National Ambient Air Quality Standards

catalytic converter

Corporate Average Fuel Economy (CAFE)

Volatile Organic Compounds (VOCs)

sick building syndrome

building-related illness

Climate Change

methane

carbon sequestration

Kyoto Protocol

anthropogenic greenhouse gas

thermal pollution

heat island

temperature inversion

Water Pollution

excess nutrients

dead zone

eutrophication

hypoxic zone

point sources

nonpoint sources

pH

hardness

dissolved oxygen

turbidity

BOD

wastewater

physical treatment

primary, secondary, tertiary treatment

sludge

sludge processor

gray water

Solid Waste

solid waste

primary recycling

secondary recycling

composting

sanitary landfill

leachate

Waste-to-Energy programs

hazardous waste

Resource Conservation and Recovery Act

corrosive, ignitible, reactive, and toxic wastes

surface impoundment

deep well injection

nuclear waste burial

transuranic waste

low-level, high-level radioactive waste

brownfield site

Superfund program

Love Canal

Noise Pollution

U.S. Noise Control Act

noise pollution

Quiet Communities Act

Chapter 9 Drill

Directions: Each of the questions or incomplete statements below is followed by four suggested answers or completions. Select the one that is best in each case. For answers and explanations, see Chapter 13.

1. Which of the following is NOT a direct source of groundwater pollution?

   (A) Automobile exhaust
   (B) Wastewater lagoons
   (C) Underground storage tanks
   (D) Waste injected into deep wells

2. Which of the following cities has the greatest amount of gray-air smog?

   (A) New York, New York
   (B) Beijing, China
   (C) Los Angles, California
   (D) Chicago, Illinois

3. All of the following are true about sanitary landfills EXCEPT that they

   (A) have methods of monitoring leaks in the clay and plastic liners
   (B) pipe generated methane gas to storage tanks
   (C) pump leachate out of the landfill for treatment and disposal
   (D) are built so that trash sits on top of the land

4. Which of the following correctly explains what happens to the level of oxygen dissolved in water when organic waste is put in the water?

   (A) The level remains the same after the waste is added.
   (B) The level increases because of the availability of nutrients to animals that live in the water.
   (C) The level decreases because of the bacteria feeding off the waste and using the oxygen to live.
   (D) The level decreases because of the waste absorbing the oxygen.

5. An abundance of which of the following creatures (and a lack of other creatures) would indicate that water is polluted?

   (A) Trout and other game fish
   (B) Sludge worms, anaerobic bacteria, and fungi
   (C) Carp, gar, and leeches
   (D) Salamanders and turtles

6. Which of the following is the most common way of disposing of municipal solid waste?

   (A) Recycling
   (B) Composting
   (C) Placing in landfills
   (D) Burning

7. Which of the following gases involved in global climate change is increasing in the atmosphere at the fastest rate?

   (A) H_2O
   (B) Methane
   (C) Chlorofluorocarbons
   (D) CO_2

8. Which of the following choices gives the correct order of processing sanitary waste in a sewage treatment plant?

   (A) Disinfection—breakdown of organics by bacteria—solid separation
   (B) Solid separation—breakdown of organics by bacteria—disinfection
   (C) Solid separation—disinfection—breakdown of organics by bacteria
   (D) Breakdown of organics by bacteria—solid separation—disinfection

9. Which of the following is a secondary pollutant?

   (A) CO
   (B) Soot
   (C) VOCs
   (D) PANs

10. The United States hopes to build a nuclear waste disposal site in

   (A) Wheeling, West Virginia
   (B) Yucca Mountain, Nevada
   (C) Gallup, New Mexico
   (D) Hudson, New York

11. Oxides of nitrogen create the pollutant

   (A) nitric acid
   (B) nitrogen gas
   (C) sulfuric acid
   (D) carbonic acid

12. In comparison to the surrounding rural areas, cities are

 (A) cooler than the rural areas
 (B) the same temperature as the rural areas
 (C) hotter than the rural areas
 (D) incomparable to the surrounding areas as far as temperatures

13. Which of the following are two of the most deadly and common indoor air pollutants in developed countries?

 (A) Soot and VOCs
 (B) CO_2 and methane
 (C) Nitric acid and sulfuric acid
 (D) Radon and tobacco smoke

14. Which of the following is NOT a consequence of stratospheric ozone depletion?

 (A) UV damage that kills primary producers in the food chain
 (B) Fewer crops and fish
 (C) Human and animal respiratory irritation
 (D) Weaker human immune systems

15. The dosage at which a negative effect of a toxin occurs is called the

 (A) threshold dose
 (B) dose-response analysis
 (C) LD_{50}
 (D) ED_{50}

Free-Response Question

1. Photochemical smog is one of the most common forms of air pollution today.

   (a) **Identify** TWO primary pollutants that cause photochemical smog. **Describe** how they are produced.
   (b) **Identify** TWO secondary pollutants that make up photochemical smog. **Describe** how they are produced.
   (c) **Give** ONE reason why photochemical smog is more likely to be found in industrialized nations and gray-air smog is more likely to be found in non-industrialized nations.
   (d) **Give** ONE component of photochemical smog that affects human health. **Explain** its consequences.

Summary

o   Toxicity is a measurement of how harmful a substance is to living things. The toxicity of a dosage must take into account the following:
  - frequency and concentration of dosage
  - synergistic effect
  - age/size of organism
  - organism sensitivity

o   All pollution can either be traced back to a specific location, called point source pollution, or is derived from a combination of many sources, known as non-point source pollution.

Air Pollution

o   Sources of air pollution: point source and non-point source

o   The greatest production of air pollution since the Industrial Revolution has been from a drastic increase in the use of fossil fuels.

o   Major criteria air pollutants of concern in the troposphere are:
  - carbon monoxide (CO)
  - lead (Pb)
  - ozone (O_3)
  - nitrogen dioxides (NO_2)
  - sulfur dioxides (SO_2)
  - particulates

o   Ozone is an essential component of the stratosphere, blocking 95 percent of the sun's damaging ultraviolet radiation. Ozone depletion is largely due to a group of man-made chemicals known as chlorofluorocarbons (CFCs).

o   Acid rain is a secondary pollutant resulting from the combination of primarily SO_2 and nitrogen oxides combining with water vapor to form acids that alter the pH of precipitation.

Climate Change

o   Greenhouse gases (GHGs) are molecules that absorb heat and warm the lower atmosphere.

o   The major GHGs are:
  - carbon dioxide (CO_2)
  - nitrous oxide (N_2O)
  - methane (CH_4)

o   Heat islands occur in urban areas and other locations that are dominated by heat-absorbing materials, like concrete and asphalt. These increased temperatures can cause higher rates of photochemical reactions, which may lead to photochemical smog. Temperature inversions may also cause warm polluted air to not rise out of an urban area when a warm air mass stalls above the city.

Water Pollution

o   Sources of water pollution are either point (such as wastewater drainage pipes from factories) or non-point (for example, agricultural and urban runoff).

o   The greatest contributor to water pollution is excess nutrients from runoff, which can lead to eutrophication.

o   Major water pollutants include:
  - excess nitrogen and phosphate
  - organic waste
  - toxic waste
  - sediments and suspended particulates
  - thermal pollution
  - coliform bacteria
  - invasive species

o Important parameters for testing water quality are:
  - pH
  - hardness
  - dissolved oxygen (DO)
  - turbidity
  - biological oxygen demand (BOD)

o Managing and treating wastewater created by humans is essential to human health and increased life expectancy.

Waste

o Solid waste consists of many types that may each be a threat to human health and the environment.
  - Municipal waste (produced by households) and industrial waste (from commercial production) is either recycled, composted, buried in landfills, or incinerated for Waste-to-Energy programs.

o Hazardous waste poses dangers to human health and must be dealt with specifically via deep well injections, surface impoundments, and landfills, for example.

o United States legislation created the Superfund program to manage land previously used commercially and polluted with hazardous waste.

Required Environmental Legislation

o Clean Air Act of 1970 and the 1990 Amendment

o Clean Water Act 1972

o Delaney Clause

o Resource Conservation and Recovery Act (RCRA)

Chapter 10
Culture, Society, and Environmental Quality

> Conservation is a great moral issue, for it involves the patriotic duty of ensuring the safety and continuance of the nation.
>
> —Theodore Roosevelt

By now, you've probably almost completed your AP Environmental Science course. This means that you realize that what's good for the environment is usually good for people, because we live in an ecosystem on Earth just as all other plants and animals do. At this point, the idea of pursuing environmentally thoughtful courses of action, through government policy as well as our day-to-day actions, must seem like a great idea, right? Well, if environmental protection is good, why do we often get the message that it will cost us more, that people will lose jobs, or that it will infringe on our liberties in some way? The answer is that, to a certain extent, all of these things *will* happen. However, the health of humankind is dependent on the health of the Earth, and often certain sacrifices must be made for the greater good of the Earth.

In this chapter, we will first review the importance of **sustainability**. We will then move on to discuss public policy making, give you a brief history of environmental activism in the United States, and then go through the important acts and amendments that you'll need to know for the test. We'll wrap up the chapter with a discussion of green taxes—and then we'll be done. Let's begin!

THE IMPORTANCE OF BEING SUSTAINABLE

To environmentalists, sustaining environmental quality usually means working in the biotic and abiotic environments in a way that ensures they are capable of functioning sustainably. However, along with maintaining a sustainable environment, maintaining the health and happiness of the human species would also be a part of most environmentalists' goals for Earth. The human species cannot exist in an unsustainable environment; after all, humans are part of a larger ecosystem, just as all other species of living things are. However, our advantage—or rather, our responsibility—lies in the fact that we are the most technologically advanced and capable species on the planet. We are also the ones causing the most damage.

> What does the term "sustainable" mean to you? How much would you be willing to sacrifice in order to sustain environmental quality? These are questions that all citizens should ask themselves before entering the voting booth.

As environmentally literate, reasonable citizens, we know that we're sometimes obliged to make choices that may not make everyone happy, but we strive to make choices that will ultimately benefit the greatest possible number of people.

Highly developed countries comprise only 20% of the world population, but consume more than half of the world's energy resources. If every country consumed global resources to this extent, we would need more resources to live than Earth can supply. This is because most of the resources that we rely upon are limited—recall the fossil fuels we burn, the way that we use water, and the rate at which we produce and dispose of waste.

Public Policy

The exploitation of public resources has been the motivation behind environmental policy at the international, national, state, and local level for as long as public policies have been made. Strictly speaking, **policy** is defined as a plan or course of action—as of a government, political party, or business—intended to influence and determine decisions, actions, and other matters.

While policies that we make as a nation are usually fairly easy to enforce—because they often have our collective best interests as a nation in mind—international policy, as is established through the United Nations (UN), is only achievable and realistic if the affected countries all cooperate with decisions that are made collectively. For example, in the 1994 International Conference on Population and Development (which was sponsored by the UN), one of the goals agreed on by the participants at the conference was to enroll 90 percent of all boys and girls in primary school by 2010. However, this policy can be put into action only if the countries that signed the agreement are willing to carry out the necessary steps.

There are ways in which the UN can attempt to force countries to follow mandates that are agreed on by the majority. These include withholding borrowing power through the World Bank, trade rules, and withholding aid. However, there are often certain environmentally significant countries that don't belong to the UN, didn't sign whatever agreement is at issue, or that just don't have the infrastructure to enforce the objective—however worthy. Additionally, international agencies often don't have the power to control what happens inside a particular country.

Much more effective are international policies that are put into effect through treaties that the countries involved have all agreed to—their governments have all ratified the treaty. Obviously, policies that countries agree to are most often ones that benefit these countries in some way, so it isn't too surprising that they are more readily enforced. In other words, international laws that are not agreeable are not usually followed because they don't provide countries with incentives. Moreover, it is not possible to punish countries that don't follow these policies.

As we touched upon above, it's understandably much easier to enforce laws and policies in the United States than it is for us to police the other nations of the world. In the United States, state and local laws have an effect on the environment, but if there is a conflict between state or local law and federal law, most times federal law will take precedence. However, in some cases states have legislated controls that are even stricter than those the federal law requires. In these cases, the state laws are the ones enforced, rather than the more lenient federal law. Additionally, some laws are passed and enforced regionally because of particular geographic needs; one example of this is the difference in water laws to the east and west of the Mississippi River.

East of the Mississippi River, water laws are based on the principle that the upstream consumers control the water but, by law, cannot impede or reduce its flow or change its quality. A number of lawsuits based on this premise have been filed and are currently pending. One example is the diversion of water from the Apalachicola-Chattahoochee-Flint (ACF) and Alabama-Coosa-Tallapoosa (ACT) river basins. The state of Georgia would like to divert water from these basins to supply the growing needs of the urban area of Atlanta. The states of Alabama, Tennessee, and Florida, which are downstream from the diversion, are concerned about the flow and quality of water that will reach them if this diversion project is carried out. This controversy is still in the court system.

While we're on the topic, Atlanta is a good example of a city whose ecological footprint is far larger than the resources that are available on the land it occupies. (Remember that an ecological footprint is the amount of resources available to support a population and absorb its wastes.) City lawmakers are rightfully concerned about water shortages in the near future.

On the other hand, water laws west of the Mississippi are based on **water rights**. West of the Mississippi, it's held that the person who first files a claim on a water resource has rights to the use of the water. The amount of water an individual with a water right has claim to each year is determined by water flow that year (and also by how much water the individual wants!). But, regardless of the amount of water present or the place where the right was claimed, the oldest person who made the water claim gets to use his or her share of water before anyone else can partake. Obviously, water rights in the West do not require sharing, as they do in the East.

Environmental Policy in the United States—A Short History

Although the first laws of the United States, such as those contained in the Constitution, do not mention the environment specifically, the Bill of Rights includes the Fifth Amendment, which prohibits the taking of private property for public use without just compensation. This has been interpreted to include the "preventing of serious public harm" by those who wish to take private property and do something on it that will affect the environment or those around them in a negative way. Basically, this means that your neighbor cannot decide to build a small nuclear power plant on his or her residential property because this would violate zoning laws.

Early laws did not mention the environment because when these laws were written, there was so much land and so many resources in the United States that it was unimaginable that they could ever be in danger of being used up. In fact, many early laws, such as the Homestead Act of 1862, encouraged the settlement and exploitation of western lands. Others, such as the Mineral Lands Act of 1866, encouraged the use of resources, and, unfortunately, this exploitative act is still in effect for many mining regulations. A few years later, the General Mining Act of 1872, a federal law, was created to systematically oversee and control prospecting and mining for economic minerals, such as gold, platinum, and silver, on federal public lands.

Shortly after the Civil War, as people continued to migrate to the West, it was realized that the United States did not have an endless supply of land or resources. In fact, in order to preserve some of the lands in the West that were being very quickly settled, the first national park, Yellowstone National Park, was established in 1872. Further legislative action in 1891 created the forest reserves, which made these lands off-limits to logging in order to protect the land from being overharvested and to maintain the existing watersheds. This legislation marked the beginning of the federal government assuming an environmentally protective role.

Political and Cultural Activism

During this time period there were several men who stood out as early environmental activists, including Henry David Thoreau (1817–1862). Thoreau's book, *Walden*, describes his retreat from society and the quiet years that he spent living on Walden Pond studying nature.

Another important writer and scientist of this time period was George Perkins Marsh (1801–1882), whose book *Man and Nature* helped the American public understand that there are limits to natural resources. His plan for the conservation of resources is the basis for many of the resource conservation principles that we try to adhere to today. Another early environmental advocate was John Muir, a nature preservationist who founded the Sierra Club in 1892. He led a campaign for the protection of lands from human exploitation and advocated low-impact recreational activities such as hiking and camping. These ideas did not become popular until the 1960s.

As far as political leaders, arguably the most environmentally active president in the history of the United States was Theodore Roosevelt (1858–1919). Roosevelt was interested from an early age in the workings of the environment and even began his own natural history museum as a child. Interestingly enough, that collection became a part of the founding collection for New York's American Museum of Natural History.

Roosevelt's term as president has been called the **Golden Age of Conservation** because of the many environmentally friendly laws and policies he put into effect. During his presidency (1901–1909), he increased the area of national forest lands by 400 percent (up to 194,000,000 acres), establishing 150 new national forests and adding area to others. He established the first 51 bird reserves, signing the first one into existence by asking his advisors, "Is there any law which prohibits me from declaring this island a bird refuge?" When his advisors determined that there was not, he signed the bill with gusto, announcing, "Very well, then, I so declare it." Additionally, he established five national parks, including the Grand Canyon, four national games preserves, eighteen national monuments (established under the 1906 Antiquities Act), twenty-four reclamation projects, and seven conservation commissions.

Roosevelt also appointed the first chief of the United States Forest Service in the history of the United States, Gifford Pinchot (1865–1946). Pinchot applied the principles of sustainable harvest and multiple-use to wildlife protection, recreation, and resource extraction.

As you're probably well aware, the 1960s were a turbulent time in U.S. history, when the baby boomers born after World War II began to come of age and express their opinions to the world. The book *Silent Spring*, written by Rachel Carson in 1962, awoke in many Americans an awareness of the state of the environment. The air was dirty, the water was polluted, and hazardous wastes were collecting in landfills all over the country. Also at this time, the Apollo space missions allowed Americans to see planet Earth from afar for the first time, and this popularized the term "spaceship Earth." Paul Ehrlich's 1968 book, *The Population Bomb,* warned of the myriad problems that would arise along with the quickly increasing human population, and an entertainer named John Deutschendorf took the stage name of John Denver and began to popularize the environmental movement through song.

A multitude of environmental laws and policies were initiated during the presidency of Richard Nixon. For example, the first Earth Day was celebrated on April 22, 1970. Also in 1970, Nixon signed into law the National Environmental Policy Act (NEPA); this act created the Council on Environmental Quality and required the submission of an environmental impact statement before any major federal action could be taken. One of Nixon's major environmental contributions was

More Great Books
Walden is a great book to read if you are interested in philosophy or environmentalism. If you live in the Boston area, you can even go for a swim in Thoreau's beloved pond!

More Great Books
Head to your local bookstore or library and look over a copy of Rachel Carson's famous book—you'll be glad you did!

to consolidate two agencies that had environmental responsibilities into a bureau called the **Environmental Protection Agency (EPA).** Finally, two major legislative actions were enacted in this new era of environmental awareness in the United States. The first was the Clean Air Act of 1963, which we have mentioned many times in these pages, and the second was the Clean Water Act (introduced in 1972).

There is your brief history of environmental activity and activists in the history of the United States. Throughout this book, we've highlighted some other important environmental laws that you should be aware of for this exam. The ones in bold are those that have had a particularly significant impact. Make flashcards of all of these, as well as the Clean Air Act of 1963 and the Clean Water Act, so that you know them cold for test day!

Some Relevant Environmental Policy Acts

As you just learned, a few pieces of legislation helped form the environmental policy of the United States. The table below shows some of the legislation relevant to U.S. environmental policy, but are not required by the College Board. If you want to see required laws that deal with particular problems like endangered species, clear water, or clean air, go back to those chapters!

> **Relevant, But Not Required!**
> These acts are not required by the College Board to know for the exam, so no need to memorize them.

Date	Name of Legislation	What It Did
1970	National Environmental Policy Act	Created the Council on Environmental Quality that resulted in the creation of the Environmental Protection Agency (EPA) from the consolidation of various environmental agencies. It also mandates that federal agencies prepare environmental impact statements.
1990	Pollution Prevention Act	Designed to promote source reduction (stop pollution from being produced).

What Have We Done for Us Lately?

Although some environmental bills and amendments have been added since 1985, lately there has been a distinct anti-environmental movement influencing government actions. Large, established environmental groups, such as the Sierra Club, have declined in membership, which is an interesting litmus test of environmentalism in America. However, hope lies in the fact that new grassroots environmental organizations are currently growing throughout the United States. **Non-governmental organizations (NGOs)** like Greenpeace and the World Wildlife Fund also play a role in protecting the environment. Six environmental issues that are expected to take center stage in the 21st century include:

- climate change

- water shortages and water supplies

- population growth

- loss of biodiversity

- air and chemical pollution

- ocean acidification and pollution

Keep your eye out for discussions of these topics in the news, and listen critically to campaigning politicians to see where they stand on these issues. Such issues will prove to impact you personally in more and more ways as time goes by. The discussion of how to best reduce greenhouse gas emissions, for example, currently revolves around what's called **cap-and-trade policy**, an approach that provides economic incentives for limiting emissions of pollutants.

HOW ELSE DO WE MAKE ENVIRONMENTAL PROGRESS?

Most often, the United States government has approached environmental issues by passing "command and control" laws. These laws set limits on factors, such as the amounts of pollution that are allowable from various sources, and they establish penalties for those that go over the limits. Wildlife has always been protected by similar types of legislation. There is no doubt that these laws have led to cleaner air and water as well as the conservation of soil and other natural resources. Endangered and threatened species have also both been protected, and some extremely endangered species have even been able to recover somewhat.

However, there have always been problems with the "command and control" approach. For example, consider the Endangered Species Act. Red-cockaded woodpeckers are endangered because the open forests with big, old pine trees have been replaced by forests with younger, smaller pines. Also, periodic natural fires, which historically kept the pinewoods open, have been suppressed because humans have settled in these areas. Periodic fires are needed to control the brushy understory and keep the pinewoods open. Creating yet another problem for the endangered bird, timber owners have been known to kill them in order to avoid preserving their habitat. However, it's very hard to prove what happens to these birds—are they being exterminated by landowners, or are they simply migrating elsewhere or declining in number for other reasons?

Green Taxes

There are other approaches that are more successfully used to continue environmental improvement without forcing the enactment of other types of command and control. Over time, it has become clear that the act of punishing actions that hurt the environment are not nearly as effective as rewarding actions that help the

A green tax shift is a fiscal policy that lowers taxes on income, including wages and profit, and raises taxes on consumption, particularly the unsustainable consumption of nonrenewable resources. Some taxes that could be lowered by the implementation of a green tax shift are payroll and income taxes, and the following is a list of taxes that could be implemented or, if currently in existence, increased:

- carbon taxes on the use of fossil fuels

- taxes on the extraction of mineral, energy, and forestry products

- license fees for fishing and hunting

- taxes on technologies and products that are associated with substantial negative externalities

- garbage disposal taxes

- taxes on effluents, emissions, and other hazardous wastes

environment. Since the 1970s, the United States has substantially increased taxes on labor and modestly increased taxes on income, while allowing actions that create pollution and cause resource depletion to remain largely untaxed. The result is that the tax system of the United States encourages resource depletion and discourages investments in machinery and labor. A worldwide discussion is taking place about how to move away from taxing "goods," such as investments and employment (activities we should be encouraging), and toward taxing "bads," like pollution, which we would like to discourage. Pollution taxes have now been embraced by a growing number of mainstream economists and policy makers and are just one of a new group of taxes called **green taxes**.

Additionally, taxes on certain forms of consumption may occur through the "feebate" approach, in which additional fees are imposed on less sustainable products—such as sport-utility vehicles—and then pooled to fund rebates on more sustainable alternatives, such as hybrid electric vehicles.

In this scheme, taxes serve as policy tools as well as a way to protect the environment.

> The three main goals of green taxes are:
>
> * to generate revenue to correct past pollution damage and reduce future pollution
> * to change behavior
> * to use the funds received from pollution taxes for restoration

Market permits are also being used somewhat successfully to encourage reduction in pollutants. Market permits are cap-and-trade permits and they work in this way: companies are allowed to buy permits that allow them to discharge a certain amount of substances into certain environmental outlets. If they can reduce their discharge, they are allowed to sell the remaining portion of their permit to another company. Economically speaking, it is to a company's advantage to reduce its discharge and sell the remainder of its allowable discharge to another company. But perhaps a better idea is for the government to buy back the unused permits rather than have them sold to another industry; this would reduce the overall discharge.

Many people think that subsidies (which are giveaways or tax breaks on certain resources to encourage their use) are hurting the environment more than they're helping it; detractors think that subsidies only encourage the use of unsustainable products.

Note that all policies, treaties, and laws are important to our environment. However, for purposes of the AP Environmental Science Exam, you will probably only be asked questions concerning United States federal laws, which is why we suggested that you commit the different acts you saw throughout this book to memory. International treaties, summits, and policies such as those that are directed from the United Nations or one of its agencies will probably come up on future exams, but that isn't your problem.

Globalization

As you can imagine, our world is becoming more and more interconnected. Aircraft can fly around the world in about 24 hours; we have instant communication worldwide via phones, television, and the internet. This is called **globalization,** and it affects society, the economy, and the environment. Positive effects can be seen in new economic opportunities, our expanded access to information, and the interactions of many societies. For example, grapes can be grown in Chile, shipped north, and be sold in your supermarket in less than a week. There are also several negative impacts of globalization. In certain parts of China, large piles of unusable electronic components have been creating water pollution problems as rainwater leaches out heavy metals. The rapid spread of emerging diseases, increased levels of air pollution and hazardous waste, and the loss of marine fish stocks are just a few more examples of globalization's negative impacts. Not to mention those grapes from Chile required an enormous amount of energy (and fossil fuel) and probably added to the pollution during their trip north!

Remember when you read about the Commons, resources owned by no one but accessible by everyone? This concept is important when we consider global access to those resources. Fresh water, clean air, ample supplies of fish, and access to fertile croplands are all examples of the global Commons. It is important to use these resources sustainably because they are the foundations for economic and social development.

Poverty and greed can cause people to use resources in an unsustainable manner and damage the environment. Cutting down important rainforest habitats to raise crops and accepting companies that generate a lot of harmful pollutants are two examples of how people's hunger for money can lead to unsustainable practices. Unfortunately, the economically disadvantaged people who allow unsustainable practices to continue are also the ones most susceptible to environmental issues brought about by climate change and have the least amount of resources to combat the health and environmental problems that result.

International organizations such as the World Bank and the United Nations are two examples of institutions that are trying to ameliorate the poverty issue. The World Bank uses loans to reduce poverty and to help foster improvements in biodiversity, environmental policies, land management, pollution management, and water resources management. The United Nations, through its environmental program, seeks to promote international cooperation, develop regional programs to promote sustainability, and to assess global, regional, and national environmental trends.

There are several international agreements that cover pollution issues, but the following are the ones you need to know for the exam:

Date	Name of Agreement	What It Did
1987	Montreal Protocol	Cut the emissions of CFCs that damage the ozone layer. This was amended in Copenhagen (1992) to include other key ozone-depleting chemicals.
1997	Kyoto Protocol	Required the participating 38 developed countries to cut their greenhouse gas emissions back to 5% below 1990 levels. While the United States signed the agreement, it did not ratify the agreement. As a result, the United States is not bound to abide by the Kyoto Protocol.

Now, review the Chapter 10 Key Terms list and try the drill to make sure you've grasped all of this chapter's content.

CHAPTER 10 KEY TERMS

Study these terms and you will be sure to write excellent essays.

Policy
Golden Age of Conservation
sustainability
water rights
Environmental Protection Agency (EPA)
Non-governmental organization (NGO)
cap-and-trade policy
green taxes
globalization
market permits
Kyoto Protocol

People
Henry David Thoreau (*Walden*)
Rachel Carson (*Silent Spring*)
John Muir and the Sierra Club

Relevant Acts
National Environmental Policy Act
Pollution Prevention Act

Chapter 10 Drill

Directions: Each of the questions or incomplete statements below is followed by four suggested answers or completions. Select the one that is best in each case. For answers and explanations, see Chapter 13.

1. The release of CFCs was banned under an international treaty written in which of the following cities?

   (A) Montreal
   (B) New York
   (C) New Delhi
   (D) Kyoto

2. Which of the following policies prevents the harassment, capture, injury, or killing of all species of whales, dolphins, seals, and sea lions, as well as walruses, manatees, dugongs, sea otters, and polar bears?

   (A) National Fisheries Act
   (B) CITES
   (C) Clean Water Act
   (D) The Marine Mammal Protection Act

3. When a large federal project might have a significant impact on the environment, which of the following must be drafted?

   (A) A cost-benefit analysis
   (B) An interagency review
   (C) A report from the geographical information system
   (D) An environmental impact statement

4. Rachel Carson described which of the following problems to Americans?

   (A) Pesticide bioaccumulation and poisoning
   (B) The beauty of nature
   (C) Loss of sustainable forest practices
   (D) The problem of the "Tragedy of the Commons"

5. Which of the following established the "emissions trading policy"?

   (A) Clean Water Act
   (B) Resource Conservation and Recovery Act
   (C) Asbestos Hazard Emergency Response Act
   (D) Clean Air Act

6. All of the following deal with the study of environmental economics EXCEPT

   (A) the study of the impact of goods on the ecosystem
   (B) understanding the economic cost of pollution
   (C) developing policy alternatives to pollution-based industry
   (D) the location of new mineral reserves

7. The listing of threatened species and the purchase of land to protect their habitats is legislated in which of the following?

   (A) Federal Noxious Weed Act
   (B) Endangered Species Act
   (C) Convention on Biological Diversity
   (D) Fish and Wildlife Conservation Act

8. Which of the following philosophies would be held by someone with a holistic viewpoint on the management of the Earth's resources?

   (A) There are no problems; let's do whatever we want.
   (B) The free market works best, and the government should not interfere.
   (C) We can manage most problems with technology.
   (D) The biodiversity of the Earth is the most important issue, and we need it to sustain us.

9. Which of the following best describes the goal of environmentally sustainable economic growth?

   (A) Allowing rapid population growth so there will be more workers
   (B) Exploration to find more natural resources
   (C) Increasing the quality of goods without depleting the natural resources needed to make the goods
   (D) Cutting down forests and replacing them with rangeland

10. A cap-and-trade policy might be effective in controlling which type of the following pollutants?

   (A) Thermal pollution in rivers
   (B) Organic waste pollution in oceans
   (C) Underground water pollutants
   (D) Carbon dioxide in the atmosphere

11. Which of the following can be classified as an NGO?

   (A) World Wildlife Fund
   (B) Bureau of Land Management
   (C) Fish and Wildlife Commission
   (D) International Trade Commission

Free-Response Question

1. The town council of Hilltop Valley has just received a proposal from its advisory panel to build a new coal-fired power plant on the western edge of the county where Hilltop Valley is located. The plant will be located along the back of the county's major river so that water can easily be transported to the plant.

    (a) **Describe** TWO parts of the Clean Air Act that will impact the building of this plant. Give ONE example of a compound that would be impacted under the National Ambient Air Quality Standards (NAAQS) and discuss the regulation of hazardous air pollutant levels.

    (b) **Describe** TWO positive and TWO negative impacts that the plant might have on the region.

    (c) Opponents of the plant say that they will use the Endangered Species Act (ESA) to prevent the plant's construction. **Describe** how the ESA could be successfully used to stop construction.

    (d) **Describe** a goal of the Pollution Prevention Act of 1990 and give ONE example of how the plant can reduce a pollutant it might emit.

Summary

o   How individuals manage their resource use and sustain the environmental quality of their biotic and abiotic environments dictates the future existence of the human species on Earth.

Required Environmental Legislation

o   Major domestic and international environmental policies have been created to influence and determine individual and corporate actions.

o   International Policy (UN)
   • Kyoto Protocol

o   Carbon taxes and market permits have influenced industrial and commercial behavior through economic incentives or penalties.

o   Globalization has caused:
   • increased transportation pollution
   • unsustainable resource use
   • unequally distributed environmental issues
   • greater disparity between poverty and greed

Chapter 11
AP Environmental Science in the Lab

The College Board requires the AP Environmental Science course to have a laboratory and field investigation component that will complement students' learning about the environment. Although there is no formal lab manual for the course, teachers are expected to provide lab experiences that ultimately incorporate test concepts and principles learned throughout the course. These experiences must also help you design experiments, collect data, apply mathematical routines and methods, and refine testable explanations and predictions.

While the College Board allows a wide range of lab and field activities that can be conducted in the course, they all should encompass these elements:

- Connected to major concepts in science and to one or more areas of the course.

- Allow students to have direct experience with an organism or system in the environment.

- Involve observing systems, collecting data/information, and communicating observations/ or results.

They also should encourage the following:

- generate questions for investigation

- choose which variables to investigate

- design and conduct experiments

- design your own experimental procedures

- collect, analyze, interpret, and display data

- determine how to present your conclusions

The variety of different labs that can be performed gives students opportunities to explore various topics in environmental science; this also allows instructors the freedom to adapt their course laboratories to their geographic locations, which can make the course more interesting for students.

There are several abilities you will be expected to have gained through participating in laboratories in your course, and the College Board describes these abilities in the following way. You should be able to:

- critically observe environmental systems

- develop and conduct well-designed experiments

- utilize appropriate techniques and instrumentation

- analyze and interpret data, including appropriate statistical and graphical presentations

- think analytically and apply concepts to the solution of environmental problems

- make conclusions and evaluate their quality and validity

- propose further questions for study

- communicate accurately and meaningfully about observations and conclusions

On the AP Environmental Science Exam, you could be asked a number of different types of questions in which the lab component from your class will be helpful. However, you won't be able to get full credit simply by remembering the details of what you did in a particular lab. You will have to use your critical thinking skills to answer these questions.

A SAMPLE FREE-RESPONSE QUESTION INVOLVING A LAB

A lab-related free-response question on the exam might look a lot like the one below.

2. The county range management office has just received a federal grant to study the breeding success of hawks in your area and you—their student intern for the summer—have been chosen to design and perform the research. The range office wants to know how the hawk population at the state prairie reserve compares with the hawk population on the federal grazing land and the private ranch land. The state prairie reserve allows no grazing, cutting, or controlled burns. The antelope herd that grazed there until a year ago had to be destroyed because of an illness in the herd. Any lightning strikes are extinguished as quickly as possible. The federal grazing lands are rented out to local ranchers whose private lands border the federal land. The management of these lands allows for a specific number of grazing days per year from mid-spring until mid-summer. However, no cutting, even of damaged trees, is allowed, and accidental burns are extinguished promptly. Private lands, of course, can be managed as desired.

    (a) **Identify** and **describe** what factor or factors will enable you to judge the hawks' breeding success.

    (b) **Design** a plan of action for determining the hawks' breeding success in the three areas. What will be your control?

    (c) **Describe** how the land management plans of each of the three areas would affect the success of the hawk population.

    (d) **Discuss** why a predator such as the hawk would provide evidence of a successful land management plan.

Take a minute to reread the passage and then the questions. Look at part (a). How do you think you could determine the success of a bird population—or any population, for that matter? Well, one way to assess the health of a population is by seeing how many offspring they produce and how many of these survive to reproductive age. So, for hawks, the number of offspring produced could be determined by counting the number of eggs laid; the number of offspring that survive to reproductive age could be approximated by counting how many fledglings leave the nest. So, in answering this part of the question, you could cite these two methods, including any others you can think of.

In part (b), you're asked to design a plan of action. You know what your dependent variable is—the success of the hawk population—and you know how you are going to judge this success—by the number of eggs laid and the number of fledglings that leave the nest. The reasoning behind this proposed line of action is that the more eggs laid by individuals, the greater the health of the population. Rates of egg-laying also provide information about the health of the females of the population; specifically, about whether or not they are receiving adequate nutrition. However, keep in mind that most birds of the same species lay about the same number of eggs, but that the difference between clutch sizes may not be an indicator of health—it could be due to chance.

What is the independent variable in this experiment? It's the way in which the land is managed in each of the three areas. Remember that you cannot change the management practices; you have been given those.

What is the control in the experiment? This is a little less clear-cut. You're given three management plans, none of which operates on land that's left untouched. Therefore, you could set up a control at one of the reserves, but it might be difficult to attract hawks and start with an adequate base control number. Another option is to do a **library control.** In a library control, you could determine the maximum number of eggs and fledglings that could be expected based on data collected in previous years. Either one of these types of controls would be acceptable. Let's summarize the plan of action that you might propose.

You will grid off three random plots, each 10 acres in area, in each of the three management areas. You will then observe the hawks in these plots and determine the locations of their nests. You will examine each nest (using binoculars because the nests are usually high in trees) and determine the number of eggs in each clutch. You will continue to make observations throughout the summer to determine how many birds are hatched and how many live to the fledgling stage and leave the nest. You will statistically determine the percent success rate of each nest and the overall area to determine the success of the hawks. Then you will compare your data from the three areas.

So, the plan of action that you write in your exam booklet for answering the first parts of the question might include the following points:

Plan of Action

- Using library resources, determine the maximum success rate of offspring production in hawks (control).

- Plot out three random 10-acre plots on each of the three management areas.

- Survey plots for hawks and nests within each grid and record the locations of the nests.

- Count the number of eggs in each nest.

- Observe the nests and hatchlings and count the number of hatched birds. Band baby birds for tracking purposes.

- Count the number of fledglings that survive and leave the nest.

- Perform statistical analysis to determine the success of the hawks on each of the three management areas.

The plan of action is your longest answer in this section, but hopefully you can see that it wasn't too difficult. You took the information you learned this year and applied it to the situation, but you didn't need to know much about hawks.

Unfortunately, however, the next parts will require you to know a little more, rather than just asking you to use your imagination. For example, part (c) asks you to describe how the land management plan of each of the three areas would affect the success of the hawk population.

Let's go through them one by one. The state land would be heavy in undergrowth and might have damaged trees located on it. The diversity of the plant population in this area would be relatively low because the grass would predominate—it would be able to grow unchecked. New grass would grow, but probably not many other plant species. On the other hand, the federal land is grazed, so in this area there will be abundant growth of new grass. Also, due to this natural thinning of the grass, there will probably be some diversity in the vegetation, but without the fall grazing to clean up the dying grass and continue to thin it, the federal land might not be as diverse as the fall grazed land. The private land can be grazed at any time. Controlled fires can also be used on this land to remove dead grass, add nutrients back into the soil, and remove some of the plant species that compete with grass, such as cactus or sage brush. For these reasons, the private land may have more diversity than any of the other land management areas.

You could also theorize that the private land could be overgrazed and thus exhibit low diversity. Really, you could make either presumption and be counted as correct. However, since you are given data on the fire practices and grazing, the first choice might be better. Why is the vegetation diversity important? The more vegetation diversity there is, the more choices for the herbivores in food and habitats—and probably the greater the success of the herbivore. In any event, the more successful the plants are, the more successful the herbivores in the area will be. Because hawks are carnivores, the more successful the herbivores, the more successful the hawks will be. See how this line of reasoning naturally leads to your answer to part (c)?

You're almost done; let's think about part (d). The success of the management plan would be demonstrated by the success of the hawk because the hawk is the tertiary consumer in the food chain. It cannot be successful if the rest of the food chain isn't strong.

You're done, and you really did not need to know much about the prairie or hawks. What you did need to know was how to set up a field experiment, the concept of the food chain, and something about land management plans. Even the management plans could have been deduced with some critical thinking.

SOME COMMON AP ENVIRONMENTAL SCIENCE LABS

Below, we've listed some of the common labs performed during an AP Environmental Science course, a summary of the procedures you might follow, and the take-home message of each.

Remember that each AP Environmental Science course is different; you may have performed some of these labs, but you probably have not completed all of them. It's a good idea to review all of these labs and understand their basic workings as well as their intent.

Soil Analysis Lab

- **Soil Testing Laboratory:** In this lab, soil is tested for physical traits and chemical properties, which provide information about the soil's condition and suitability for crops, septic fields, or other purposes. All of the factors tested are listed below.

Chemical Properties

1. pH—Clay soil requires more lime (calcium oxide) or alum (aluminum sulfate) to alter its pH than do sandy or loam soils. Iron necessary for plant growth is unavailable when the soil becomes alkaline. Gymnosperms (pine, fir, etc.) grow better in mildly acidic soil.

2. Nitrogen—common plant nutrient component

3. Phosphorus—common plant nutrient component

4. Potash—common name for a compound that contains one of the potassium oxides

Physical Characteristics

1. Soil type—Sand, silt, clay. Use mesh screen, cheese cloth, and soil settling in water tubes to determine the percent of each type of particle in the sample.

2. Water-holding capacity—Because of the small pores between clay particles, water moves very slowly through clay. Therefore, clay has a greater holding capacity than silt or sand.

3. Permeability—the movement of gas or liquid through the soil

4. Friability—Good soil is rich, light, and easily worked with fingers, which is good for plant growth because roots can easily grow through it.

5. Percent humus—A measure of soluble organic constituents of soil; the higher the number, the better. Organic soil has qualities of both sand and clay. The small particles of organic soil come together to form larger clusters. Water can be retained inside a cluster, but can move between clusters to percolate. Organic material is also high in nutrients.

6. Buffering capacity—Resilience of different types of bedrock, such as marble, granite, and basalt, when exposed to acid. Marble has high calcium content and is a better buffer than other rocks.

Water Analysis Lab

- **Water Chemical and Physical Analysis:** Water can be tested from many different sources. Sample kits have tools for many different tests—for example, the LaMotte kit and the spectrophotometer type kit. Below are some commonly performed tests and some of the expected results. It is probably not necessary for you to memorize all the standards, but you should be familiar with them.

  1. pH—Normal pH of water is between 6.5–8.5.

  2. DO—Measure of dissolved oxygen in water. Warm-water fish require a minimum of 4 ppm and cold-water fish require 5 ppm.

  3. Turbidity—Measurement of water clarity. Higher turbidity means there will be low clarity, and little sunlight will be able to penetrate the water. A Secchi Disk may be used to measure turbidity, but a more accurate measure can be made with a turbidity unit.

  4. Phosphate—An important plant nutrient, typically found in fertilizer and runoff from agricultural lands. Too much leads to eutrophication of water, high BOD, and low DO levels. Levels should not exceed 0.025 mg/L in still water and 0.05 mg/L in flowing water.

  5. Alkalinity—Measure of compounds that shift the pH toward the alkaline. There are no EPA standards, but normal is between 100–250 ppm.

  6. BOD—Biological oxygen demand, which is required for the aerobic organisms in a body of water. Unpolluted natural waters have a concentration of 5 mg/L or less. High nutrient levels or lots of biological material ready for decomposition are associated with high BOD, and vice versa.

  7. Chlorine—EPA standards dictate that Cl cannot exceed 250 mg/L. NaCl is applied to roads and parking lots and can run off into streams. Other sources of excess Cl are animal waste, potash fertilizer (KCl), and septic tank effluent. Chlorine is associated with limestone deposits but is not common in other soils, rocks, or minerals.

  8. Hardness—A measure of salts composed of calcium, magnesium, or iron. Most water testing kits test for $CaCl_2$. Hard water is more than 121 ppm, and soft water is less than 20 ppm.

  9. Iron—Normal range is 0.1–0.5 ppm.

  10. Nitrates—As an important plant nutrient, nitrogen is typically found in fertilizer and is a component of runoff from agricultural lands. Too much leads to eutrophication of water, high BOD, and low DO levels. Over 0.10 mg/L is considered elevated, and the EPA limit is 10 mg/L.

  11. Total solids—Weight of the suspended solids and dissolved solids. All natural waters have some suspended solids, but problem solids are sewage, industrial waste, or excess amounts of algae.

12. Total dissolved solids—Naturally occurring in water, but may cause an objectionable taste in drinking water. They are also unsuitable for irrigation because they leave a salt residue on the soil. EPA standard is 500 mg/L, but dissolved solids may range from 20–2,000 mg/L.

13. Fecal coliform—Any bacteria that ferments lactose and produces gas when grown in lactose broth. New tests for this are performed by adding a water sample to a specialized media and observing color changes. Drinking water should show no colonies of growth from the water sample.

Air Quality Labs

- **Air Quality**: Air quality can be assessed using various methods.

- **Particulates**: Sticky paper can be used to collect air particulates from various sources, and then the paper can be examined under a microscope. It is not possible to see the smallest particulates, but they do color the white paper.

- **Ozone**: In this lab, an eco badge or a homemade potassium iodide gel sampler is hung or worn in order to collect data on tropospheric ozone. The badge or KI sample changes color in the presence of ozone and becomes more intensely colored as the amount of ozone increases.

- **Carbon Dioxide**: In this lab, a commercial sampling device is used to determine the amount of carbon dioxide in an air sample. Car exhaust, burning tobacco, or other pollutants can also be sampled.

Other Labs

- **Lichen**: A lichen survey can be used to judge air quality. Lichens are sensitive to air pollution, particularly sulfur dioxide. The most sensitive lichens are the fruticose types, followed by the foliose, and then the crustose.

- **Scrubber Model**: A model of a scrubber can be constructed to attempt to remove sulfur contaminates from burning coal. A calcium compound can be used to try to wash the contaminant from the air column.

- **Biodiversity of Invertebrates**: Insects can be counted in an area and then plotted to assess biodiversity. Traps such as fall traps or sticky traps can be set, and bait such as tuna or sugars can be used. The number of different insects captured is counted, and this number divided into the total number captured would give an idea of the biodiversity of the area. A taxonomic key can be used to determine the number of species and their taxonomic name. A similar setup can be used to determine the impact of an invasive species. In this process, native bugs can be trapped and set up in terrariums; the invasive species is then introduced and the effects documented.

- **Field Trips**: Field trips can be taken to various areas of interest, such as power plants, landfills, or municipal waste treatment plants. The possibilities are almost endless. Think about what you learned from these experiences and how they can help you on this exam.

- **Energy Audit**: In this lab, students are asked to use their homes as a laboratory and perform an energy audit, examining the amount of electricity used by their families over a set period of time and then using appliance standards to determine which is the largest energy consumer. A result of this lab is that students suggest how their family's electrical energy needs could be reduced.

- **Food Chain**: In this lab, students observe a natural ecosystem and examine the food chain. They identify the organisms that are producers, primary consumers, and secondary consumers and determine how many levels make up the food chain, what organisms act as decomposers, and the presence of symbiotic relationships.

- **Model Building**: In this lab, models are built to model land formations, coastlines, tectonic plates, mining operations, or any number of other physical formations. In modeling the tectonic plates, students can slide or bump together plate models that are lying on top of a viscous substance representing the magma. From this model, students see how subduction zones and volcanoes work. Other models also allow students to construct a representation of a physical structure in order to better understand that physical structure.

- **Mining**: Students can construct a model of a mine, representing the rock layers and the mineral deposits.

- **LC$_{50}$ (or LD$_{50}$)**: In this lab, students use a kit or a lab procedure to test the concentration (LC) or dose (LD) that would cause the death of 50 percent of a test organism. An example of this procedure is to test the effects of copper (common algaecide and fungicide ingredient) on Daphnia (a tiny species of freshwater crustacea). Daphnia are exposed to a range of Cu levels and then fed fluorescently tagged sugar. Healthy Daphnia show up blue under UV light.

- **Population Density and Biomass**: This experiment can be performed in a number of different ways. One method is to mark off two plots and then remove all the vegetation from Plot 1. Plot 2 is left to grow for an additional period of time. This investigation may also be performed using grass squares in the lab or bottles of algae. The vegetation from Plot 1 is dried and weighed. This provides the baseline data for the amount of biomass present at the beginning of the experiment. After the time period of growth for the second plot, the vegetation is removed from the second plot and treated as the first was. This provides information about the increase in biomass for the experimental time period. Another way this can be expressed is as the productivity of vegetation. By subtracting the initial growth (Plot 1) from the experimental growth (Plot 2), productivity is calculated. This would constitute the net productivity for the plot. The grass in the second plot produces X amount of glucose (photosynthate). However, some of this photosynthate must be used to support the needs of the plant, so the number you calculate is the net, not the gross, productivity.

- **Gross and Net Productivity**: If an experiment is performed using bottles of growing algae or duckweed, the gross productivity can also be determined. One of the bottles in this experiment contains the starting plant material, but instead of being exposed to light, it is covered with foil to prevent light from entering the bottle. In this bottle, the plants are only able to perform respiration. Therefore, at the end of the experimental time, this bottle should have less biomass than the initial plant material because the plants have to use stored sugar for metabolic process. This provides you with the metabolic rate information, which, when added to the net productivity, gives you the gross productivity.

- **Other Uses for Vegetation Plots**: Sample plots may be used to examine the biodiversity of vegetation or the patterns of plant growth or dispersal. For example, one species of plant can be marked in a plot, and unless a garden was used as the sample plot, the target plant will probably be dispersed in clumps. Remember that random dispersal is not typical in nature, and even distribution typically only occurs in plots planted by humans.

- **Population Growth**: Population growth experiments can involve fast-growing populations such as bacteria, duckweed (*Lemma minor*), roly-polies (sowbugs), or fruit flies (*Drosophila*) and can involve graphing, as does the analysis of human population data. Population growth can be graphed, with time along the *x*-axis and population growth on the *y*-axis. Initially, the curve is a J shape, but it becomes S-shaped. Bacterial growth curves have an extended hook on the J due to the lag time in growth as the bacteria acclimate to the new media.

- **Variables**: In population growth experiments, often other variables are added to samples of bacteria to compare the growth of a normal population to one that has been altered in some way. This can be used to assess the effect of extra nutrients, such as nitrogen or phosphorus, or the negative effects of certain substances on the organism.

- **Turbidity and Bacteria**: When studying bacteria, turbidity is commonly observed and recorded. The more turbid (cloudy) the tube, the more growth has taken place. Turbidity can be observed with a spectrophotometer. When the sample tube is inserted into the spectrophotometer, the instrument passes light through the sample tube and measures the amount of light that passed through or was absorbed.

- **Population Size**: In this type of lab, the size of a population of species such as the gypsy moth, caterpillar, or other insect is studied. Remember that a population is a group of individuals of the same species located in a given area. The experiment can involve a collection box and colored stickers that attract caterpillars. The first day, caterpillars are captured and marked. On the second day, the caterpillars are captured and the new and recaptured individuals are marked. By dividing the number of recaptured insects by the size of the population on day one, an estimate of the population that was originally captured can be obtained. On day three, the total number of caterpillars captured is counted. The total number of caterpillars captured on day three is divided by the calculated number from day two—this gives an estimate of the total population.

- **Salinization**: In this lab, students determine if and how salt levels retard seed germination. Students use various dilutions of salts (a variety of salt types may be used) to saturate a growing surface. Seeds are sown to determine levels of salt that inhibit germination. This lab mimics what happens in irrigated farm land with salinization.

- **Note:** If you are asked to graph data on this exam, make sure you use reasonable units! Count the graph blocks you're given; this may help you determine an appropriate scale. Also, remember that time is usually plotted on the *x*-axis.

Well, this marks the end of the content review for this exam—congratulations, you've finished! Study the words in the Glossary, which follows, and then take Practice Tests 2 and 3.

Good luck!

Chapter 12
Glossary

GLOSSARY

These are all AP Environmental Science terms you should know for the exam, so make flashcards or write down these words in a notebook, and study them any chance you get. Do whatever you have to do to commit them to memory before exam day!

Chapter 4: Earth's Interdependent Systems

abiotic—related to factors or things that are separate and independent from living things; nonliving

acid—any compound that releases hydrogen ions when dissolved in water; also a water solution that contains a surplus of hydrogen ions

air mass—enormous bodies of air that move as a unit

A horizon—a soil horizon; the layer below the O horizon is called the A horizon, which is formed of weathered rock with some organic material; often referred to as topsoil

alkaline—a basic substance; chemically, a substance that absorbs hydrogen ions or releases hydroxide ions; in reference to natural water, a measure of the base content of the water

aquifer—an underground layer of porous rock, sand, or other material that allows the movement of water between layers of nonporous rock or clay; frequently tapped for wells

arable—land that is fit to be cultivated

asthenosphere—the part of the mantle that lies just below the lithosphere

atmosphere—the gaseous mass or envelope surrounding a celestial body—especially the one surrounding Earth—that is retained by the celestial body's gravitational field

barrier island—a long, relatively narrow island running parallel to the mainland, built up by the action of waves and currents and serving to protect the coast from erosion by surf and tidal surges

biological weathering—any weathering that's caused by the activities of living organisms

biotic—living or derived from living things

B horizon—a soil horizon that receives the minerals and organic materials that are leached out of the A horizon

chemical weathering—the result of chemical interaction with the bedrock that is typical of the action of both water and atmospheric gases

C horizon—a soil horizon made up of larger pieces of rock that have not undergone much weathering

clay—the finest soil, made up of particles that are less than 0.002 mm in diameter

climate—weather conditions, especially temperature and precipitation, that remain constant over 30 years or more

conduction—the transmission or conveying of something through a medium or passage, especially the transmission of electric charge or heat through a conducting medium without perceptible motion of the medium itself

convection—the vertical movement of a mass of matter because of heating and cooling; can happen in both the atmosphere and Earth's mantle

convection currents—air currents caused by the vertical movement of air due to atmospheric heating and cooling

convergent boundary—a plate boundary where two plates are moving toward each other

coral reef—an erosion-resistant marine ridge or mound consisting chiefly of compacted coral together with algal material and biochemically deposited magnesium and calcium carbonates

Coriolis effect—the observed effect of the Coriolis force, especially the deflection of an object moving above the Earth, rightward in the Northern Hemisphere, and leftward in the Southern Hemisphere, as away from the equator

crop rotation—the practice of alternating the crops grown on a piece of land to replenish soil nutrients; e.g., corn one year, legumes for two years, and then back to corn

delta—a usually triangular alluvial deposit at the mouth of a river

divergent boundary—a plate boundary at which plates are moving away from each other; causes an upwelling of magma from the mantle to cool and form new crust

doldrums—a region of the ocean near the equator, characterized by calms, light winds, or squalls

drip irrigation—a method of supplying irrigation water through tubes that drip water onto the soil at the base of each plant

earthquake—the result of vibrations that release energy from within the Earth; they often occur as two plates slide past each other at a transform boundary

El Niño—a climate variation that takes place in the tropical Pacific about every three to seven years, for a duration of about one year

erosion—the process of soil particles being carried away by wind or water; moves the smaller particles first and hence degrades the soil to a coarser, sandier, stonier texture

estuary—the part of the wide lower course of a river where its current is met by the tides

fault—the place where two tectonic plates abut each other

genetically modified organism (GMO)—any organism whose genome has been engineered at the molecular level to favor the expression of desired traits, usually by the inclusion of genes from unrelated species of organisms

Green Revolution—the time after the Industrial Revolution when farming became mechanized and crop yields in industrialized nations boomed as farmers began using large amounts of chemical fertilizers and pesticides

greenhouse effect—the phenomenon whereby Earth's atmosphere traps solar radiation, caused by the presence in the atmosphere of gases such as carbon dioxide, water vapor, and methane that allow incoming sunlight to pass through but absorb heat radiated back from Earth's surface

Hadley cell—a system of vertical and horizontal air circulation that creates major weather patterns, predominately in tropical and subtropical regions

headwaters—the water from which a river rises; a source

horizon—a layer of soil

humus—the dark, crumbly, nutrient-rich material that results from the decomposition of organic material, which is also a product of composting organic waste

hurricane (typhoon, cyclone)—a severe tropical storm originating in the equatorial regions of the Atlantic Ocean or Caribbean Sea or eastern regions of the Pacific Ocean, that travels north, northwest, or northeast from its point of origin and usually involves high-speed winds and heavy rains

inner core—the molten core of the Earth

insolation—the delivery rate of solar radiation per unit of horizontal surface

jet stream—a high-speed, meandering wind current, generally moving from a westerly direction at speeds often exceeding 400 km (250 miles) per hour at altitudes of 15 to 25 km (10 to 15 miles)

land degradation—deterioration of land quality (topsoil, organisms, vegetation, water quality), usually caused by its exploitation

La Niña—a cooling of the ocean surface off the western coast of South America, occurring periodically every 4 to 12 years and affecting Pacific and other weather patterns

lithosphere—the outer part of the Earth, consisting of the crust and upper mantle, approximately 100 km (62 miles) thick

loam—soil composed of a mixture of sand, clay, silt, and organic matter

mantle—the layer of the Earth between the crust and the core

monoculture—the cultivation of a single crop on a farm or in a region or country; a single, homogeneous culture without diversity or variety

monsoon—a wind system that influences large climatic regions and reverses direction seasonally

O horizon—the uppermost horizon of soil; primarily made up of organic material, including waste from organisms, the bodies of decomposing organisms, and live organisms

physical (mechanical) weathering—any process that breaks rock down into smaller pieces without changing the chemistry of the rock; typically wind and water

plate boundaries—the edges of tectonic plates

polyculture—the practice of planting several crops on the same plot of land simultaneously to increase biodiversity and sustainability

prior appropriation—when water rights are given to those who have historically used the water in a certain area

rain shadow effect—the low-rainfall region that exists on the leeward (downwind) side of a mountain range; the result of the mountain range's causing precipitation on the windward side

red tide—a bloom of dinoflagellates that causes reddish discoloration of coastal ocean waters; certain dinoflagellates of the genus *Gonyamlax* produce toxins that kill fish and contaminate shellfish

R horizon—the bedrock that lies below all of the other layers of soil

riparian right—the right, as to fishing or to the use of a riverbed, of one who owns riparian land (the land adjacent to a river or stream)

salinization—occurs when soil becomes waterlogged from excess irrigation and then dries out; as the water evaporates, the salt crystallizes and forms a layer on the soil surface, which prevents the growth of plants

sand—the coarsest soil, with particles 0.05–2.0 mm in diameter

silt—soil with particles 0.002–0.05 mm in diameter

Southern Oscillation—the atmospheric pressure conditions corresponding to the periodic warming of El Niño and cooling of La Niña

subduction zone—in tectonic plates, the site at which an oceanic plate is sliding under a continental plate

thermocline—a layer in a large body of water, such as a lake, that sharply separates regions differing in temperature, so that the temperature gradient across the layer is abrupt

thermosphere—the outermost shell of the atmosphere between the mesosphere and outer space, where temperatures increase steadily with altitude

topsoil—the A horizon of soil is often referred to as topsoil and is most important for plant growth

trade winds—the more or less constant winds blowing in horizontal directions over Earth's surface, as part of Hadley cells

transform boundary—also known as transform faults, boundaries at which plates are moving past each other, sideways

tropical storm—a cyclonic storm having winds ranging from approximately 48 to 121 km (30 to 75 miles) per hour

upwelling—a process in which cold, often nutrient-rich, waters from the ocean depths rise to the surface

volcanoes—an opening in the Earth's crust through which molten lava, ash, and gases are ejected

watershed—the region draining into a river system or other body of water

water-scarce—countries that have a renewable annual water supply of less than 1,000 m^3 per person

water-stressed—countries that have a renewable annual water supply of about 1,000–2,000 m^3 per person

weather—the day-to-day variations in temperature, air pressure, wind, humidity, and precipitation mediated by the atmosphere in a given region

Memorization Tip!
Take these Key Terms and make your own flashcards by hand. The act of transcribing the words and definitions will help you memorize the content. Then keep your handy flashcard deck on you at all times for quick review on the go.

weathering—the gradual breakdown of rock into smaller and smaller particles, caused by natural chemical, physical, and biological factors

wetlands—a lowland area, such as a marsh or swamp, that is saturated with moisture, especially when regarded as the natural habitat of wildlife

Chapter 5: The Inhabitants of Planet Earth and Their Relationships

ammonification—the production of ammonia or ammonium compounds in the decomposition of organic matter, especially through the action of bacteria

assimilation—the process in which plants absorb ammonium (NH_3), ammonia ions (NH_4^+), and nitrate ions (NO_3) through their roots

autotroph—producers; organisms that can produce their own organic compounds from inorganic compounds; they use energy from the sun or from the oxidation of inorganic substances

bioaccumulation—the accumulation of a substance, such as a toxic chemical, in various tissues of a living organism

biological extinction—true extermination of a species; no individuals of this species left on the planet

biomagnification—the process by which the concentration of toxic substances increases in each successive link in the food chain

biosphere—the part of the Earth and its atmosphere where living organisms exist or that is capable of supporting life

carnivore—an animal that consumes only other animals

chemotroph (chemoautotroph)—an organism, such as a bacterium or protozoan, that obtains its nourishment through the oxidation of inorganic chemical compounds, as opposed to photosynthesis

climax community—a stable, mature community in a successive series that has reached equilibrium after having evolved through stages and adapted to its environment

combustion—the process of burning

commercial or economic extinction—a few individuals exist but the effort needed to locate and harvest them is not worth the expense

community—formed from populations of different species occupying the same geographic area

competitive exclusion—the process that occurs when two different species in a region compete and the better adapted species wins

consumer—an organism that must obtain food energy from secondary sources by, for example, eating plant or animal matter

decomposer—bacteria or fungi that absorb nutrients from nonliving organic matter, like plant material, the wastes of living organisms, and corpses, and convert these materials into inorganic forms

denitrification—the process by which specialized bacteria (mostly anaerobic bacteria) convert ammonia to NO_3, NO_2, and N_2, which are released back into the atmosphere

detritivore—organisms that derive energy from consuming nonliving organic matter, such as dead animals or fallen leaves; Earthworms and many species of fungi are detritivores

ecological extinction—the condition in which there are so few individuals of a species that the species can no longer perform its ecological function

ecological succession—the transition in species composition of a biological community, often following ecological disturbance of the community; the establishment of a biological community in any area virtually barren of life

edge effect—the condition in which there is greater species diversity and biological density at ecosystem boundaries than there is in the heart of ecological communities

energy pyramid—the structure obtained if we organize the amount of energy contained in producers and consumers in an ecosystem by kilocalories per square meter, from largest to smallest

evaporation—to convert or change into a vapor

evolution—change in the genetic composition of a population during successive generations as a result of natural selection acting on the genetic variation among individuals and resulting in the development of new species

extinction—the death of an entire species; permanent inactivity

food chain—a succession of organisms in an ecological community that constitutes a continuation of food energy from one organism to another as each consumes a lower member and, in turn, is preyed upon by a higher member

food web—a complex of interrelated food chains in an ecological community

Gause's principle—states that no two species can occupy the same niche at the same time, and that the species that is less fit to live in the environment will either relocate, die out, or occupy a smaller niche

Gross Primary Productivity—the amount of sugar that the plants produce in photosynthesis minus the amount of energy the plants need for growth, maintenance, repair, and reproduction

habitat—the area or environment where an organism or ecological community normally lives or occurs

habitat fragmentation—when the size of an organism's natural habitat is reduced, or when development occurs that isolates a habitat

heterotroph—an organism that cannot synthesize its own food and is dependent on complex organic substances for nutrition

indigenous species—species that originate and live, or occur naturally, in an area or environment

invasive species—an introduced, nonnative species

keystone species—a species whose very presence contributes to an ecosystem's diversity and whose extinction would consequently lead to the extinction of other forms of life

Law of Conservation of Matter—states that matter can neither be created nor destroyed

mutualism—a symbiotic relationship in which both species benefit

natural selection—the process by which, according to Darwin's Theory of Evolution, only the organisms best adapted to their environment tend to survive and transmit their genetic characteristics in increasing numbers to succeeding generations, while those less adapted tend to be eliminated

Net Primary Productivity (NPP)—the amount of energy that plants pass on to the community of herbivores in an ecosystem

niche—the total sum of a species' use of the biotic and abiotic resources in its environment

nitrification—the process in which soil bacteria convert ammonium (NH_4^+) to a form that can be used by plants; nitrate, or NO_3

nitrogen fixation—the conversion of atmospheric nitrogen into compounds, such as ammonia, by natural agencies or various industrial processes

omnivores—organisms that consume both producers and primary consumers

parasitism—a symbiotic relationship in which one member is helped by the association and the other is harmed

photosynthesis—the process in green plants and certain other organisms by which carbohydrates are synthesized from carbon dioxide and water using light as an energy source: most forms of photosynthesis release oxygen as a byproduct

pioneer species—organisms in the first stages of succession

population—a group of organisms of the same species that live in the same area

predation—when one species feeds on another

predator—a species that feeds on another species

prey—a species that's subject to predation by another species

primary consumers—organisms that consume producers (plants and algae)

primary succession—when ecological succession begins in a virtually lifeless area, such as the area behind a moving glacier

producer—an organism that is capable of converting radiant energy or chemical energy into carbohydrates

realized niche—when a species occupies a smaller niche than it would in the absence of competition

reservoir—a place where a large quantity of a resource sits for a long period of time

respiration—the process in which animals (and plants!) breathe and give off carbon dioxide from cellular metabolism

residency time—the amount of time a resource spends in a reservoir or an exchange pool

secondary consumers—organisms that consume primary consumers

species—organisms that are capable of breeding with one another and incapable of breeding with other species

species richness—the number of different species found in an ecosystem

symbiotic relationships—close, prolonged associations between two or more different organisms of different species that may, but do not necessarily, benefit the members

tertiary consumers—organisms that consume secondary consumers or other tertiary consumers

transpiration—the act or process of transpiring, or releasing water vapor, especially through the stomata of plant tissue or the pores of the skin

trophic level—each of the feeding levels in a food chain

Chapter 6: Population Ecology

age-structure pyramids—graphical representations of populations' ages

albedo—the fraction of solar energy that is reflected back into space

biotic potential—the amount that the population would grow if there were unlimited resources in its environment

birth rate (crude birth rate)—the number of live births per 1,000 members of the population in a year

brownfield site—an industrial or commercial site that is idle or underused because it is contaminated by hazardous waste or pollutants

carrying capacity—the maximum population size that can be supported by the available resources in a region

collapse—severe to catastrophic dieback of a population due to overshoot

death rate (crude death rate)—the number of deaths per 1,000 members of the population in a year

demographic transition model—a model that is used to predict population trends based on the birth and death rates as well as economic status of a population

ecological footprint—the amount of Earth's surface required to supply the needs of and dispose of the waste from a particular population

food desert—an area, usually low-income, in which many residents cannot access affordable, healthy food

emigration—the movement of individuals out of a population

genetic drift—the random fluctuations in the frequency of the appearance of a gene in a small, isolated population, presumably owing to chance, rather than natural selection

immigration—the movement of individuals into a population

K-selected—organisms that reproduce later in life, produce fewer offspring, and devote significant time and energy to the nurturing of their offspring

logistic population growth—the condition in which a population is well below the size dictated by the carrying capacity of its region, so it will grow exponentially; but as it approaches the carrying capacity, its growth rate will decrease and the size of the population will eventually become stable

overshoot—when a population exceeds its carrying capacity

population density—the number of individuals of a population that inhabit a certain unit of land or water area

population momentum—the condition in which a large proportion of women are in child-bearing years, so that a country's population will continue to grow even if the fertility rate falls to replacement levels

replacement birth rate—the number of children a couple must have in order to replace themselves in a population

r-selected—organisms that reproduce early in life and often and have a high capacity for reproductive growth

total fertility rate—the number of children an average woman will bear during her lifetime; this information is based on an analysis of data from preceding years in the population in question

Chapter 7: Resource Utilization

agroforestry—when trees and crops are planted together, creating a mutualistic symbiotic relationship between them

aquaculture—the raising of fish and other aquatic species in captivity for harvest

bottom trawling—a fishing technique in which the ocean floor is scraped by heavy nets that smash everything in their path

by-catch—any other species of fish, mammals, or birds that are caught that are not the target organism

capture fisheries—fish production in which fish are caught in the wild and not raised in captivity for consumption

clear-cutting—the removal of all of the trees in an area

conservation—the management or regulation of a resource so that its use does not exceed the capacity of the resource to regenerate itself

consumption—the day-to-day use of environmental resources such as food, clothing, and housing

contour plowing—a process in which rows of crops are plowed across the hillside; this prevents the erosion that can occur when rows are cut up and down on a slope

deforestation—the removal of trees for agricultural purposes or purposes of exportation

drift nets—nets that drift free in the water and indiscriminately catch everything in their path

ecosystem capital (natural capital)—the value of natural resources

fishery—the industry or occupation devoted to the catching, processing, or selling of fish, shellfish, or other aquatic animals

greenbelt—open or forested areas built at the outer edge of a city

ground fires—smoldering fires that take place in bogs or swamps and can burn underground for days or weeks; originating from surface fires, ground fires are difficult to detect and extinguish

intercropping (strip cropping)—the practice of planting bands of different crops across a hillside

long lining—in fishing, the use of long lines with baited hooks, which will be taken by numerous aquatic organisms

malnutrition—poor nutrition that results from an insufficient or poorly balanced diet

mineral deposit—an area where a particular mineral is concentrated

mining—the excavation of the Earth for the purpose of extracting ore or minerals

monoculture—when just one type of plant is planted in a large area

natural resources—biotic and abiotic natural ecosystems

nonrenewable resources—resources that are often formed by very slow geologic processes, and therefore considered incapable of being regenerated within the realm of human existence

no-till methods—refers to when farmers plant seeds without using a plow to turn the soil

old growth forest—a forest that has never been cut; they have not been seriously disturbed for several hundred years

overgrazed—when grass is consumed by animals at a faster rate than it can regrow

preservation—the maintenance of a species or ecosystem in order to ensure its perpetuation, with no concern as to their potential monetary value

production—the use of environmental resources for profit

renewable resources—refers to resources, such as plants and animals, which can be regenerated if harvested at sustainable yields

second growth forests—areas where cutting has occurred and a new, younger forest has arisen

selective cutting—the removal of select trees in an area, leaving the majority of the habitat in place and therefore having less of an impact on the ecosystem

shelter-wood cutting—when mature trees are cut over a period of time (usually 10–20 years), leaving mature trees, which can reseed the forest, in place

silviculture—the management of forest plantations for the purpose of harvesting timber

slash and burn—when an area of vegetation is cut down and burned before being planted with crops

surface fires—fires that typically burn only the forest's underbrush and do little damage to mature trees; these fires actually serve to protect the forest from more harmful fires by removing underbrush and dead materials that would burn quickly and at high temperatures

tailings—piles of gangue, which is the waste material that results from mining

traditional subsistence agriculture—when each family in a community grows crops for themselves and relies on animal and human labor to plant and harvest crops

terracing—creating flat platforms in the hillside that provide a level planting surface, which reduces soil runoff from the slope

tree farms—planted and managed tracts of trees of the same age that are harvested for commercial use; also known as plantations

uneven-aged management—the broad category under which selective cutting and shelter-wood cutting fall; selective deforestation

Chapter 8: Energy

active collection—the use of devices, such as solar panels, to collect, focus, transport, or store solar energy

anthracite—the cleanest-burning coal; almost pure carbon

barrels—the unit used to describe the volume of fossil fuels

bituminous—the second purest form of coal

breeder reactor—a nuclear reactor that generates more fissionable material than it consumes

crude oil—the form petroleum takes when in the ground

energy—the capacity to do work

fission—a nuclear reaction in which an atomic nucleus, especially a heavy nucleus such as an isotope of uranium, splits into fragments, usually two fragments of comparable mass, releasing from 100 million to several hundred million electron volts of energy

fossil fuel—a hydrocarbon deposit, such as petroleum, coal, or natural gas, derived from living matter of a previous geologic time and used for fuel

First Law of Thermodynamics—the law that states that energy can neither be created nor destroyed; it can only be transferred and transformed

fly ash—a waste product produced by the burning of coal

half-life—the amount of time it takes for half of a radioactive sample to degrade

Hubbert peak (peak oil)—an influential theory that concerns the long-term rate of conventional oil (and other fossil fuel) extraction and depletion; it predicts that future world oil production will soon reach a peak and then rapidly decline

hydroelectric power—power generated using water

kinetic energy—the energy of motion

lignite—the least pure coal

nuclear fusion—a nuclear reaction in which two nuclei are fused to form one or more different atomic nuclei (and subatomic particles)

overburden—the rocks and Earth that are removed when strip mining for a commercially valuable mineral resource

passive solar energy collection—the use of building materials, building placement, and design to passively collect solar energy that can be used to keep a building warm or cool

peak oil (see **Hubbert peak**)

petroleum—oil; a hydrocarbon that forms as sediments are buried and pressurized

photovoltaic cell (PV cell)—a semiconductor device that converts the energy of sunlight into electric energy

potential energy—energy at rest, or stored energy

proven reserve—an estimate of the amount of fossil fuel that can be obtained from reserve

radiant energy—sunlight

radioactive—a material that undergoes radioactive decay, a process by which unstable nuclei emit energy over time as particles or photons

scrubbers—devices containing alkaline substances that precipitate out much of the sulfur dioxide from industrial plants' air effluent

Second Law of Thermodynamics—the law that states that the entropy (disorder) of the universe is increasing; one corollary of the Second Law of Thermodynamics is the concept that, in most energy transformations, a significant fraction of energy is lost to the universe as heat

strip mining—involves the removal of Earth's surface all the way down to the level of the mineral seam

subbituminous—the third purest form of coal

underground mining—involves the sinking of shafts to reach underground deposits; in this type of mining, networks of tunnels are dug or blasted, and humans enter these tunnels in order to manually retrieve the coal

wind farm—a group of modern turbines

Chapter 9: Pollution

acid precipitation—acid rain, acid hail, acid snow; all of which occur as a result of pollution in the atmosphere

acute effect—the effect caused by a short exposure to a high level of toxin

catalytic converter—a platinum-coated device that oxidizes most of the VOCs and some of the CO that would otherwise be emitted in exhaust, converting them to CO_2

closed-loop recycling—when materials, such as plastic or aluminum, are used to rebuild the same product; e.g., the use of the aluminum from aluminum cans to produce more aluminum cans

composting—a process that allows the organic material in solid waste to be decomposed and reintroduced into the soil, often as fertilizer

building-related illness—when the signs and symptoms of an illness can be attributed to a specific infectious organism that resides in the building

chronic effect—an effect that results from long-term exposure to low levels of toxin

Delaney Clause—a part of the Food Additives Amendment of 1958 that specifically bans any food additives found to cause cancer in humans or in animal testing

deep well injection—drilling a hole in the ground below the water table to hold waste

disease—occurs when infection causes a change in the state of health

dose-response analysis—a process in which an organism is exposed to a toxin at different concentrations, and the dosage that causes the death of the organism is recorded

dose-response curve—the result of graphing a dose-response analysis

ED_{50}—the point at which 50 percent of the test organisms show a negative effect from a toxin

Food and Drug Administration (FDA)—the governmental body in the US that regulates food and related products

global warming—an intensification of the greenhouse effect due to the increased presence of heat-trapping gases in the atmosphere

gray smog (industrial smog)—smog resulting from emissions from industry and other sources of gases produced by the burning of fossil fuels, especially coal

hazardous waste—any waste that poses a danger to human health and therefore must be dealt with in a different way from other types of waste

heat islands—urban areas that heat up more quickly and retain heat better than nonurban areas

high-level radioactive waste—radioactive wastes that produce high levels of ionizing radiation

industrial smog—(see **gray smog**)

infection—the result of a pathogen invading a body

LD_{50}—the point at which 50 percent of the test organisms die from a toxin

leachate—the liquid that percolates to the bottom of a landfill

low-level radioactive waste—radioactive wastes that produce low levels of ionizing radiation

noise pollution—any noise that causes stress or has the potential to damage human health

non-point source pollution—pollution that does not have a specific point of release

open-loop recycling—when materials are reused to form new products

ozone holes—the thinning of the ozone layer over Antarctica (and to some extent, over the Arctic)

pathogens—bacteria, virus, or other microorganisms that can cause disease

photochemical smog—usually formed on hot, sunny days when NO_x compounds, VOCs, and ozone combine to form smog with a brownish hue

point source pollution—a specific location from which pollution is released; e.g., a factory where wood is being burned

poison—any substance that has an LD_{50} of 50 mg or less per kg of body weight

physical treatment—in a sewage treatment plant, the initial filtration that is done to remove debris such as stones, sticks, rags, toys, and other objects that were flushed down the toilet

primary pollutants—pollutants that are released directly into the lower atmosphere

primary treatment—when physically treated sewage water is passed into a settling tank, where suspended solids settle out as sludge; chemically treated polymers may be added to help the suspended solids separate and settle out

risk assessment—calculating risk, or the degree of likelihood, that a person will become ill upon exposure to a toxin or pathogen

risk management—using strategies to reduce the amount of risk; e.g., the likelihood that a person will become ill upon exposure to a toxin or pathogen

secondary pollutants—pollutants that are formed by the combination of primary pollutants in the atmosphere

secondary treatment—the biological treatment of wastewater in order to continue to remove biodegradable waste

sick building syndrome—a condition in which the majority of a building's occupants experience certain symptoms that vary with the amount of time spent in the building, without being able to identify a specified cause or illness

sludge—the solids that remain after the secondary treatment of sewage

sludge processor—a tank filled with aerobic bacteria that's used to treat sewage

solid waste—can consist of hazardous waste, industrial solid waste, or municipal waste; many types of solid waste provide a threat to human health and the environment

stationary sources—non-moving sources of pollution, such as factories

Superfund program—a program funded by the federal government and a trust funded by taxes on chemicals; identifies pollutants and cleans up hazardous waste sites

threshold dose—the dosage level of a toxin at which a negative effect occurs

toxicity—the degree to which a substance is biologically harmful

toxin—any substance that is inhaled, ingested, or absorbed at dosages sufficient to damage a living organism

tropospheric ozone—ozone that exists in the troposphere

U.S. Noise Control Act—gave the EPA power to set emission standards for major sources of noise, including transportation, machinery, and construction

vector—the carrier organism through which pathogens can attack, such as a tick

wastewater—any water that has been used by humans; this includes human sewage, water drained from showers, tubs, sinks, dishwashers, washing machines, water from industrial processes, and storm water runoff

Waste-to-Energy (WTE) program—when the energy released from waste incineration is used to generate electricity

Chapter 10: Culture, Society, and Environmental Quality

green tax—a fiscal policy that lowers taxes on income, including wages and profit, and raises taxes on consumption, particularly the unsustainable consumption of nonrenewable resources

market permits—when companies are allowed to buy permits that allow them a certain amount of discharge of substances into certain environmental outlets; if they can reduce their amount of discharge, they are allowed to sell the remaining portion of their permit to another company

Chapter 13
Chapter Drills:
Answers and
Explanations

CHAPTER 1 DRILL

1. **C** This is a calculations question. In the 28-year period, the population doubled twice (400 → 800 → 1,600). Therefore, the doubling period is 14 years. Using the Rule of 70 formula, the correct calculation is 70 ÷ 14 = 5 percent.

2. **D** All three options are related to the destruction of the barrier island. It doesn't matter that some are natural events and some are man-made causes because the question is really concerned with the inertia of the barrier island habitat. Since you are unable to eliminate any of them, the correct answer is all of them.

3. **A** This is a question where using Process of Elimination may be the most useful. You may recognize that the relationship described in this scenario is a predator-prey relationship, so you should be able to eliminate non-predatory relationships right away. Keystone species also need to impact more than just one other species, so (C) is not viable. Get rid of (D) because you have never heard of such a thing before. This leaves (A) and (B), which are related to species relationships, so you just need to determine who is controlling who. In this case, the coyote population is controlling the size of the house cat population, so the answer is (A).

4. **D** This is a question where you must identify if just one or all of the answers are correct. Choices (A) and (C) are both related to the excess nutrients that are important for the algae blooms to occur. Choice (B) is the result of the decomposing algae. Therefore, since none prove to be incorrect, (D) is the best answer.

5. **D** This question prompts you to refer to the mnemonic device HIPPCO. Choice (B) is not part of the HIPPCO mnemonic, so it can be eliminated. Choices (A), (C), and (D) are all related to HIPPCO, so you then must resort to your knowledge about fisheries. Since harvesting fish at such great rates is more influential on the population sizes than invasive species, habitat loss, or pollution, (D) is the correct answer.

6. **B** This is an EXCEPT question, so you must identify the only incorrect answer. Shipping trucks in (A) are a non-point source of air pollution because they introduce contaminants over a widespread area. Choices (C) and (D) are both examples of non-point water pollution because they are diffused sources. Choice (B) is the only point source because the smokestack is an identifiable source of the pollution, and therefore does not qualify it as a non-point source pollution.

7. **D** This is a calculations question. Choice (D) uses the correct growth rate formula (birth rate − death rate) to get 5%. This rate is then used to calculate the number of years for doubling using the Rule of 70 where N= 14 years. The population has to double twice to get from 10 million to 40 million (10 mil → 20 mil → 40 mil). Therefore, it will take two doubling periods, or 28 years. This brings us to the year 2028.

8. **C** There are several answer options that use absolute terms (examples of which include "always" and "never"). Choices (A) and (B) use words that make the statement too extreme because the Dust Bowl was not *solely* created by a drought, and it didn't *only* affect farmers, as other people were affected by the Dust Bowl. Choice (D) is incorrect because it was actually the poor price of wheat that drove the event, not prevented it. Narrowed down to two answers, you may have to just guess at this point, but the correct answer is (C).

9. **A** This is a word association question. You will hopefully remember that (B), (C), and (D) are all part of the natural process of the nitrogen cycle. If you find the nitrogen cycle complicated, though, you should notice that respiration is only associated with the carbon cycle.

CHAPTER 2 DRILL

1.

(a) This is a Rule of 70 calculation question. The population has doubled twice from 1972 to 2012. Therefore, the doubling period is 20 years. Using the Rule of 70 formula, the correct calculation is 70 ÷ 20 = 3.5 percent.

(b) This is an endangered species question, so consult your HIPPCO acronym. Prairie dogs declined due to:
- **H**abitat loss to livestock and agriculture competing for land
- **P**ollution to their habitat through a deliberate poisoning eradication program, because farmers thought the tunnels posed threats to horse and cow legs
- **C**limate—massive drought (Dust Bowl fell between 1880 and 1972)

(c) This is a keystone species question. The burrow colonies developed by the prairie dogs contribute to the success of other species by providing shelter for animals such as burrowing owls, ferrets, and snakes; providing a food source for predators; and their burrowing churns the soil to enable the Earth to better sustain plant life and infiltrate water.

(d) This is an economic/ecological pros question.

Economic

For	Against
• Provides food sourse for economically important species • Eradication plans are expensive • Eco-tourism to see organisms that rely on the prairie dog	• Produces holes that can injure livestock • Dangerous or destructive to property, infrastructure, and development

Ecological

For	Against
• Provides food sourse for other species • Provides burrows for other species • Churned soil is better for plants	• Overpopulation of prairie dogs could alter ecological dynamics

(e) This is a policy question. You should consider any policy that is used to protect an endangered species or its habitat.

- Endangered Species Act (EPA)

- Convention on International Trade in Endangered Species (CITES)—International legislation banning

- Marine Mammals Protection Act (MMA)

- Migratory Bird Conservation Act

2.

(a) This question requires you to read the DO and BOD graphs of the oxygen sag curve, which show that the biological oxygen demand is greater than the oxygen available. This is because the bacteria and organisms that gravitate to the waste and begin to consume it require great amounts of oxygen and deplete the supply. Therefore, the limiting factor in the second, third, and fourth zones is oxygen. Organisms that require greater levels of oxygen will not be able to survive in these areas, but organisms with wide ranges of tolerance for oxygen availability will survive here.

(b) Again, this question asks you to read the diagram for information. The health of the stream diminishes right after the point source pollution because the pollution eliminates what species are normally found there. You may also reference specific indicator species as those require specific levels of water quality being absent or present in a zone.

(c) This question asks you to explore your knowledge of the wastewater treatment process. There are two different ways to think about how both wastewater treatment plants and wetlands improve water quality: (1) removing solids in primary treatment, and (2) biological treatment during secondary treatment. The physical removal of solids can be done through processes such as filtering, screening, skimming, and settling out suspended solids. The biological treatment is using bacteria and microbes to digest and decompose any waste in the water.

(d) This question also asks you to explore your knowledge of the final stage of the wastewater treatment process. Possible contaminates are: microbes/bacteria used during secondary treatment, pathogens, *E.coli*, coliform bacteria, *Giardia*, cholera. Methods for disinfection are chlorine, UV light, lime, or microfiltration.

3. This question is designed to have you consider the many issues associated with waste management. This is also a computational question.

(a) Be aware of the fact that this question is only looking for options that have an environmental impact. If you choose *sanitary landfill,* acceptable environmental reasons include the following:
- Sanitary landfills produce lower air pollution and little odor.
- After retirement, methane can be recovered from sanitary landfills to be used as a fuel source.
- Sanitary landfills still encourage waste reduction and encourage recycling.

If you choose *incineration,* options for acceptable environmental reasons include the following:
- Incineration requires less land space.
- Trash volume is reduced.
- Incineration reduces the chance of water pollution.

(b) Be aware of the fact that this question is only looking for one economic impact. If you choose *sanitary landfill,* acceptable economic reasons include the following:
- Sanitary landfills can be built quickly and cheaply, especially if the city has a lot of space.
- After retirement, methane can be recovered from sanitary landfills to be used as a fuel source and/or sold.

If you choose *incineration,* acceptable economic reasons include the following:
- The process of incineration produces electricity, which may produce a profit for the city.
- Incineration frees up land that can then be used for other businesses.

(c) This is a calculations question.

(i) A \$350/week profit means you are earning \$50/day. You have two major costs: gas and the driver's salary. Gas = 10 gals/day x \$2/gal = \$20/day. Driver = \$100/day. Total cost per day = \$120/day. You'll need to make \$170/day for a \$50/day profit, and therefore you'll need to make \$1,190/week to earn a \$350 profit.

(ii) You have already determined from (i) that you'll need to earn \$170/day. Since each can is worth 2 cents, or \$0.02, you can solve for x in \$0.02x = \$170, so x = 8,500 cans per day.

(d) There are several methods that could be described here.

- Educate people on the concept that trash doesn't simply just "go away" as well as on consumption patterns.

- Promote Pay-as-You-Throw programs.

- Develop a compostable waste collection program.

- Limit the amount people are allowed to throw away, or provide disincentives with fees.

- Encourage manufacturers to produce less packaging.

- Outlaw plastic shopping bags or charge a fee for usage.

- Encourage low-waste societies.

CHAPTER 4 DRILL

Multiple Choice

1. **B** A seismograph measures the magnitude of earthquakes by recording Earth's movement, or seismic waves. The Richter Scale is used to measure the amplitude of the highest S-wave of an earthquake.

2. **B** The aurora borealis occurs in the thermosphere, or ionosphere. This is the highest level of the atmosphere, extending 60 miles and higher. The aurora borealis is caused by electrons from the sun striking oxygen atoms in the thermosphere.

3. **D** Climate is the 30-year average of temperature and precipitation in a certain area, and the temperature and precipitation of a region are the most important factors in determining the rate of plant growth. Finally, because the amount of plant matter determines the amount of animal life, the communities found in a particular habitat are directly related to its temperature and precipitation.

4. **B** As the sun's light passes through the atmosphere, it strikes the solid Earth. The Earth, with its soil, water, buildings, asphalt, and concrete, absorbs this radiant energy. This energy is then radiated back into the atmosphere as infrared radiation. This radiation can be reflected back into the atmosphere (the greenhouse effect), or it can pass back into space.

5. **C** All of the answer choices listed are parts of the hydrosphere except for (C), parent rock.

6. **A** Lakes are divided into various zones depending on light and temperature. The littoral zone is where rooted plants live, while the limnetic zone is open, sunlit water that's generally warm. The profundal zone is the deep open water where no photosynthesis occurs, and the benthic zone is the cold, dark zone at the bottom. Only organisms that can tolerate cold and low oxygen levels can live in the benthic zone.

7. **D** Approximately 75 percent of Earth's surface is covered by water; this includes both saltwater and freshwater.

8. **D** An area where saltwater and freshwater mix that has a very high level of productivity is correctly called an estuary. Estuaries are sites where the "arm" of the sea extends inland to meet the mouth of a river. Estuaries are often rich with many different types of plant and animal species, because the fresh water in these areas usually has a high concentration of nutrients and sediments.

9. **A** Both freshwater and saltwater bodies experience a seasonal movement of water from the cold and nutrient-rich bottom to the surface; this is called an upwelling. Upwellings are composed of cold, nutrient-rich waters.

10. **B** Unconfined aquifers are ones where water is free to move vertically or horizontally through an area.

11. **C** Monsoons, which develop mainly in coastal areas, occur when land heats up and cools down more quickly than water does. In a monsoon, hot air rises from the heated land, and a low-pressure system is created.

Free Response

1. You might need some information from Chapter 4 to answer this question. This will give you an opportunity to see how prepared you are for this exam before you review all of the major topics you'll need to know. How does your answer compare with ours?

(a) The runoff volume would be much greater in the clear-cut plot than the forested plot. In the forested area, the trees help hold water in the soil. Also, the leaves slow down the speed of the raindrops, lessening their impact on the soil. The roots also help bind the soil and hold it in place, so there is no erosion. In the cut area, the water stays closer to the surface, so there is more water for runoff. The rain can fall onto the soil with full force and erosion can take place. The soil is not held together, and the particles have a greater likelihood of moving.
(2 points maximum—1 point for the correct answer and 1 point for a correct explanation)

(b) The phosphate levels would be much higher in the runoff from the cut plot. The phosphate would be leached out of the soil in the clear-cut plot because of all the water running off the soil. Another explanation would be that the trees absorb most of the phosphorus out of the soil. There would be less phosphate in the soil of the forested land, so the runoff would contain less.
(2 points maximum—1 point for the correct answer and 1 point for a correct explanation)

(c) There are several possible negative effects. First, the added nutrients could cause an algal bloom in the stream. The bloom might make the stream less habitable for fish or insect larvae. As the algae decompose, the amount of dissolved oxygen would go down. The sediment might increase the water's turbidity, making it cloudier, and thus lowering the ability of producers to live in the stream. Also, the increased water volume might cause more erosion or possibly flooding farther downstream. Finally, the lack of shade would increase the water's temperature. This increase would lower the dissolved oxygen (DO) levels.

(2 points maximum—1 point for the correct answer and 1 point for a correct explanation)

(d) Because of the very rapid rate of decay and high metabolism of the living plants, there is little organic material in rain forest soil. Anything that falls to the forest floor is quickly decomposed, and the remains are rapidly absorbed by plants. When the forest is cut down, this soil is directly exposed to the large amounts of rain that falls in these forests. The rain quickly washes away the remaining organic matter, leaving even fewer nutrients in the soil.

(4 points maximum—2 points for the correct explanation of why there are few nutrients and 2 points for the correct explanation of the rapid loss of the remaining nutrients)

CHAPTER 5 DRILL

Multiple Choice

1. **C** The relationship between the bird and the tick is best described as parasitism, which is one type of symbiotic relationship. In this case, the parasite is the tick; it benefits from the relationship, while the bird is harmed.

2. **B** This concept is called competitive exclusion. The idea behind competitive exclusion is that two species that share the same niche cannot infinitely exist in the same ecosystem; eventually one will prove to be more fit and outcompete the other.

3. **D** In resource partitioning, different species can use slightly different parts of the habitat to avoid direct competition.

4. **D** One aspect of the roles of both predators and parasites in an ecosystem is that they generally eliminate the weakest members of a population. The weakest individuals are those who are young, sick, or old; these individuals are eliminated, leaving the best-adapted organisms to survive and reproduce.

5. **A** Choice (A) is false. Small adult plants are found in early succession stages. The other characteristics are all true of climax communities.

6. **C** Food chains show feeding relationships. The energy in a food chain flows from the sun, through producers (plants), to primary consumers, to secondary consumers, and then to tertiary consumers. In other words, energy flows from prey to predator: (C).

7. **A** Sulfur is stored as sulfur salts in rocks and as sulfur dioxide in the atmosphere.

8. **D** Microevolution is the process where a population shows small-scale genetic changes over a short period of time. Antibiotic resistance in bacteria occurs when overuse of antibiotics kills off the susceptible cells and the resistant ones are able to thrive.

9. **D** A grassland can support large numbers of animals, and the herd helps with protection from predators.

10. **D** Deer are herbivores feeding on plants, not members of the decay food web found in the soil.

Free Response

1. **(a)** Compound "X" can enter the pond from surface water runoff that carries the compound. It could also enter the pond by being carried in by rain, snow, or dust that falls into the water. The substance might get carried in by ground water from a polluted aquifer.

    (2 points maximum—1 point for the correct name and 1 point for a correct description of the process)

    (b) Water (0.1 ppb) → Zooplankton (0.2 ppb) → Small fish (0.1 ppm) → Predatory fish (1.0 ppm) → Hawk (3.0 ppm)

    (2 points maximum—1 point for the correct order and 1 point for the concentrations)

    (c) Bioaccumulation and biomagnification are two of the most important processes to know for the test. In bioaccumulation, fat-soluble molecules accumulate and stay in the fatty tissues of animals since they cannot dissolve in water. Biomagnification occurs when compounds are passed from prey to predator. Since a predator needs to eat a lot of prey, each of the prey organisms gives some of the compound to the predator. The compounds accumulate, and the concentration becomes much higher than you would expect to be in the environment.

    (4 points maximum—2 points for a correct description of each process)

    (d) Mercury and PCBs (polychlorinated biphenyl) are two very common molecules that bioaccumulate and biomagnify. Negative effects of PCBs include skin conditions such as chloracne and rashes, changes in blood and urine that may indicate liver damage, dermal and ocular lesions, irregular menstrual cycles, lowered immune response, fatigue, headache, cough, and poor cognitive development in children. Negative effects of mercury include itching, burning or pain, skin discoloration (pink cheeks, fingertips, and toes), swelling, desquamation (shedding of skin), sweating, tachycardia, increased salivation, and hypertension. Affected children may show red cheeks, nose, and lips; loss of hair, teeth, and nails; transient rashes; muscle weakness; and increased sensitivity to light. Other symptoms may include kidney disfunction, emotional lability, memory impairment, and insomnia.

    (2 points maximum—1 point for the compound and 1 point for a correct symptom associated with the compound)

CHAPTER 6 DRILL

Multiple Choice

1. **C** Remember that with EXCEPT questions, you're looking for the answer that does not fit the statement. If you read through them, you'll see that (C) is the only factor that does not measure some characteristic of populations. Habitat is important in determining a population's size, but it is not a way to measure a population.

2. **D** Emigration refers to the movement of individuals that are leaving, or emigrating from, a population. Birth rate is the number of births per thousand; carrying capacity is the maximum number that can live sustainably in a habitat; immigration is the movement of individuals into a population; and environmental resistance is all of the factors in a habitat that limit a population's growth.

3. **C** To do this problem, remember the Rule of 70, which approximates the time it takes for a population to double (called its doubling time). Take the rate of change (2 percent) and divide that number into 70. So, $\frac{70}{2} = 35$, which is the number of years that it will take this population to double in size.

4. **D** Age-structure pyramids are often constructed using data about the number and gender of various age groups in a population. Generally, the female population is shown on the right side of the pyramid, and the male population is situated on the left side. The length of each age group indicates how the total number of individuals in that age group compares to the other age groups and the population as a whole.

5. **D** *K*-selected organisms reproduce later in life, produce fewer offspring, and devote significant time and energy to the nurturing of their offspring. For these species, it is important to preserve as many members of the offspring as possible because they produce so few; parents have a tremendous investment in each individual offspring. Some examples of *K*-selected species are humans, lions, and cows.

6. **A** Organisms like the hare and the lynx (predator and prey) exhibit regular changes in their population (every 10 years) in a pattern known as boom-and-bust. For the other choices, an irruptive population is very large and then very small; an irregular population behaves in a chaotic manner; a logistic population doubles in a short time; and a stable population varies only slightly above and below its carrying capacity over time.

7. **B** The demographic transition model is used to study countries' transitions from one type of economy to another; specifically, how the transition affects the population. The four stages of the transition are preindustrial (high birth and death rates), transitional (high birth rates and low death rates), industrial (declining birth and death rates), and finally postindustrial (very low birth and death rates).

8. **C** HIV/AIDS is correct. The other diseases are seen more commonly in Western Europe and North America. HIV/AIDS spreads very rapidly because it is caused by an easily transmittable virus; the other diseases listed are not communicable—they cannot be passed from person to person.

9. **A** 1.3 billion is the figure that is closest to the current population in India.

10. **D** Choice (D) is the only answer choice that lists a factor that is density independent. Destruction by humans (or a natural event) would occur whether the population density was low or high. Density dependent factors only influence a population when the density is high. You might have thought that this answer choice was incorrect because you reasoned that, as a population increases in size, it could get so large that it would degrade its environment. However, remember carrying capacity! Habitats only tolerate the existence of a certain number of individuals in a population.

11. **C** Environmental resistance factors slow a population's growth. Factors might include competition, parasites, or a lack of resources.

12. **A** Births and immigration add individuals to a population, whereas deaths and emigration remove individuals. The difference between gain and loss is the growth.

13. **D** Habitat conservation is the only one of these factors that promotes species growth. All the other factors cause a population to decline.

14. **C** In developing countries experiencing poverty, women tend to have more children than women in developed nations. While this is in part the result of a lack of available birth control, this cultural phenomenon is also the result of the need for these children to go to work and provide an economic "safety net" for families. Also, infant mortality rates are higher at poverty level, which means that women must have more children to even attain replacement rate.

Free Response

1. **(a)** The carrying capacity is the maximum number of individuals that a habitat can sustain for a long period of time. If a population exceeds the carrying capacity, there will be a die-off of individuals until the population dips below the carrying capacity. When the population is lower than the carrying capacity, the population can begin to increase. Factors that can limit population in a habitat are physical factors (temperature, nutrient availability, amount of light, amount of dissolved oxygen, or pH) and biotic factors (parasites, predators, competitors).
    (4 points maximum—2 for the definition and 1 for each correct example)

    (b) One example of how nature limits consumption is competition that occurs between two populations for the same habitat. For example, if two different species of animals prey on the same species in the same habitat, they are said to be in direct competition. Sooner or later, the population of prey will be small enough that one predator in the population would not have enough resources. This might cause them to become extinct, to leave that area, or to switch to another food source, thus ending the competition. Some examples of competition are two raptor birds that compete for mice or fish; hunting cats like cheetahs and lions, which compete for grazing animals; or two species of birds that compete for insects.
    (4 points maximum—2 points for correctly explaining how consumption can be controlled and 1 point each for 2 correct examples)

    (c) Human activities that violate the limits of population growth can include examples of how we harvest natural resources to help grow food; this affects the habitat of certain plant and animal species. Examples of harvesting more natural resources might include irrigation to increase water availability; fertilizers to overcome a lack of certain minerals in the soil; or turning the natural biome into farmland to raise more food crops. We eliminate competitors for our food supply and we use medicines to kill parasites. There are numerous possible correct answers to this question.
    (2 points maximum—1 point for each correct explanation of how humans remove competition and exploit resources)

CHAPTER 7 DRILL

Multiple Choice

1. **A** Smelting is a process that separates a desired ore from other materials in mined ore. It is usually accomplished by heating the ore and ladling off the desired molten element.

2. **C** Choice (A), cost-benefit analysis, is the comparison of the benefit of an action relative to the costs of that action. Choice (B), external costs, refers to the costs that occur after someone purchases something. For example, after you buy a car, the cost of gasoline is an external cost. Marginal benefit, (D), is the tradeoff between how much is gained by buying forest land (for example) or using the money to do some other beneficial activity. So, if you drive a car, the pollutants you produce can cause acid rain that kills trees in a forest. That is a negative externality. Choice (C), marginal costs, are the costs of each step in a process. Thus, (C) is the correct answer.

3. **D** Clear-cutting is the removal of all trees from an area at the same time. This is typically done in areas that support fast-growing trees, like pines.

4. **A** Moderate irrigation with groundwater over a long period of time can cause serious problems, including a significant buildup of salts on the soil's surface, which makes the land unusable for crops. This condition is known as salinization.

5. **D** One of the major motivations for cutting down rainforests is to increase the amount of grazing land for cattle and other farm animals. All of the other options are problems associated with deforestation, but (D) benefits humans.

6. **A** Greenbelts are used in urban planning in order to increase green space and control the growth of cities. They are open or forested areas built at the outer edge of a city. Because no growth is permitted in them, they can increase the quality of life for people living near them. Sometimes satellite towns are built outside the greenbelt and connected to the city by highways and mass transportation methods.

7. **C** The Bureau of Land Management is responsible for the management of federal rangeland.

8. **C** The amount of copper ore in the Earth is limited. It is considered a nonrenewable resource. Choices (A), (B), and (D) are considered to be renewable because a renewable resource is one that will be available as long as humans use it in a sustainable manner.

9. **B** In the Tragedy of the Commons, a common resource is used by many people and then becomes depleted as these people do not regulate their consumption of the resource. Some sources say that 75 percent of the world's commercially usable fish are either overfished or are being fished at their maximum sustainable yield.

10. **A** Traditional industrialized agriculture consumes large amounts of energy and other resources. When all aspects are considered (growing crops, processing, and transportation), 17 percent of the United States' total commercial energy use goes into food production. Typically, for every unit of food energy eaten, it takes 10 units of fossil fuel energy to prepare and deliver the food.

11. **D** CITES is the international law that regulates the international trade in endangered species of both plants and animals. This is the correct answer. Choice (A) deals with species only in the United States. Choice (B) is another United States law that deals with marine mammals. Choice (C) is a law that sets how the United States will deal with environmental issues.

12. **B** Pesticide resistance is the only answer choice that represents a problem that results from the use of pesticides. Because pesticides have been used in large amounts, some species of pests have evolved traits that allow them to resist the action of pesticides.

13. **C** The uses of national parks are restricted to camping, fishing, and boating. Motor vehicles are permitted, but only in designated areas. Choice (D) describes the permitted uses of National Wildlife Refuge land, and (A) and (B) both describe acceptable uses of national forest lands.

14. **B** Sulfuric acid forms as water seeps through mines and carries off sulfur-containing compounds. The chemical conversion of sulfur-bearing minerals occurs through a combination of biological (bacterial) and inorganic chemical reactions.

15. **D** In 1995, most of the world's nations formed the WTO to establish ground rules for international commerce.

Free Response

1. Use the checklists below to determine if your responses are correct. We will use checklists like these when there are many different ways that you could have answered the question. Remember that your answers should be in paragraph form!

(a) Aspects

Positive Aspects	Negative Aspects
Low cost to implement this way of watering	Lots of water lost to evaporation
Little technology and training necessary	Not all areas well-adapted to this technique
Low cost to maintain this way of watering	Delivers more water than plants need
	Water distribution is uneven
	Promotes weed growth along with crops

(2 points maximum—1 point for a correct positive aspect and 1 point for a negative aspect)

(b) Alternatives

- Lining canals (Positive: reduces water lost to infiltration; Negative: expensive to do and uses resources)

- Leveling fields (Positive: water flows where needed; Negative: very expensive to level fields)

- Irrigating at night (Positive: avoids evaporation; Negative: requires careful planning and training)

- Irrigating only when necessary (Positive: less waste; Negative: difficult to time)

- Using drip irrigation (Positive: water drops right to roots; Negative: uses resources of plastic)

- Using center pivot (Positive: low waste as water is sprayed directly on to plants; Negative: equipment is very costly and runs on fossil fuels)
  (3 points maximum—1 point for identifying the process, 1 point for a correct positive impact, and 1 point for a correct negative impact)

(c) Possible negative impacts

- Saltwater intrusion: As the aquifer diminishes, nearby ocean water can migrate underground.

- Diminished water for domestic or industrial use

- Subsidence: As water is withdrawn, the soil settles and sinkholes can develop; these can damage buildings and destroy ecosystems.
  (2 points maximum—1 point for naming the impact and 1 point for the explanation)

(d) Possible positive and negative effects

Positive Effects	Negative Effects
Production of low-cost electricity	Costly to build
Reservoirs used for many recreational activities	Negative impact on local ecology
Provides flood control	Prevents silt recharging of floodplain
Irrigation water can be controlled	Decrease in fish migration and spawning
Durable	Great danger if breached

(3 points maximum out of 4 possible—1 point for each positive and negative impact)

CHAPTER 8 DRILL

Multiple Choice

1. **D** Net energy yield is a comparison between cost of extraction, processing, and transportation and the amount of useful energy derived from the fuel. For example, the net energy ration of natural gas used to heat homes is 4.9 and the net energy yield for electrical heating is 0.3. When the two are compared, the net yield from gas is much higher because the costs of getting the gas to your home is very small compared to the costs of running a nuclear power plant.

2. **D** This question asks you to calculate the amount of light produced by a bulb. You know the amount of energy going into the bulb (1.00 joules) and its efficiency (3 percent). So, 3 percent of 1.00 is 0.03. Choice (D) is the correct answer because the useful energy produced is light, not heat.

3. **C** Both methane and ethanol are created as bacteria break down biomass. An example of this is seen when farmers produce methane from decomposing manure to heat their barns. The manure is placed in underground pits, where bacteria break it down and release methane.

4. **D** The most common hybrid engines are gasoline-electric hybrids. The two engines work together to provide acceleration and power. When the car is driving slowly, less than 56 kph (35 mph), the electric engine powers the car.

5. **D** The only answer choice that describes a waste of energy is leaving room lights on. Most incandescent light bulbs have an energy efficiency rating of 3–5 percent. So, for every 100 units of energy, we only get 5 units of useful light.

6. **A** The correct formula for this problem is 4,500 tons × (2,000lb/1 ton) × (5,000 BTU/lb) = 4.5×10^{10}. It is much easier to use scientific notation here, so $(4.5 \times 10^3) \times (2 \times 10^3) \times (5 \times 10^3) = 45 \times 10^9$ or 4.5×10^{10}.

7. **C** CO_2 emissions are high in any process that releases energy by burning material. Therefore, burning coal, oil, diesel, and wood all generate CO_2. The processing and reprocessing of uranium into fuel rods generates a moderate amount of CO_2. The only non-combustion generation occurs by wind turbines.

8. **D** WATTS × TIME = kWh. 1,000 watts = 1 kWh, so $\frac{500}{1,000} = 0.5\,kWh$.

9. **B** Photovoltaic cells are constructed from silicon and boron. When sunlight strikes the cells, electrons are energized. These electrons can then flow freely, producing an electric current. By placing wires in the correct positions, this current can be used to power devices. Because the cells are expensive to make, the cost per kilowatt hour is high, but they are useful for applications in which there is no other source of electricity.

10. **C** Radioactive half-life is the amount of time it takes for half of a radioactive sample to degrade. In this question, there are 3 half-lives: the first is 2 curie → 1 curie, the second is 1 curie → 0.5 curie, and the third is 0.5 curie → 0.25 curie. According to the statement, each half-life lasts 20 years. So, $3 \times 20 = 60$ years.

11. **B** Appliances like microwaves do not have an off switch. They continuously draw power to remain in an "instant on" mode.

12. **D** Coal, oil, and natural gas are burned to generate heat and a nuclear reactor generates heat by radioactive decay. The heat turns water to steam and then steam spins a turbine that in turn spins a generator.

13. **D** The burning of coal causes air pollution in the form of sulfur and nitrogen oxides, which combine with atmospheric water to form acid rain. Acid rain is a leading cause of environmental harm to forests and lakes.

Free Response

1. **(a)** The core is the place in a nuclear reactor where nuclear reactions occur. It is made of very high-strength steel and other high-performance materials. Inside the core are the coolant, fuel rods, and moderators that control the reaction. The fuel rods are made of enriched uranium U-235. In the reactor, a chain reaction generates the heat, which is converted into steam to drive the electricity-producing turbines. The coolant prevents the melting of the uranium, or core, by removing the heat generated by the chain reaction. Some of the heat is used to turn water to steam that spins turbines. In most reactors, the heat produced in the core does not come in contact with the water that is vaporized; instead a device called a heat exchanger takes heat from the core (carried by water or molten sodium) and transfers it to water, which vaporizes.
(2 points maximum—1 point for each correct description)

(b) Currently, all highly radioactive waste is stored on the grounds of the power plant in deep pools of water. Some materials are stored in steel drums that are housed inside storage containers with walls made of lead and concrete. The United States is currently studying the possibility of opening a deep underground storage at Yucca Mountain, Nevada. Low-level radioactive materials are also stored in radioactive landfills.
(4 points maximum—2 points for a correct name and description of a storage technique)

(c) Because nuclear power plants do not use the combustion of fossil fuels to generate the heat to produce steam, there are no by-products of combustion. There is no particulate produced, no production of oxides of nitrogen or sulfur, and no release of heavy metals. Some CO_2 is generated, but this occurs primarily in the processing of the fuel rods.
(2 points maximum—1 point for describing the lowered emissions and 1 point for a correct example)

(d) Thermal pollution is generated from the turbines, which are powered by steam from the heat exchanger. These hot turbines must be cooled by water piped in from nearby oceans or rivers. This problem is mitigated by allowing the turbine coolant water to flow up a series of pipes built inside cooling towers. As cool air enters the bottom of the tower, it rises by convection and removes the heat from the water circulating in the pipes. The cooler water is then released.
(2 points maximum—1 point for describing how thermal pollution occurs and 1 point for a method to reduce it)

CHAPTER 9 DRILL

Multiple Choice

1. **A** Choices (B), (C), and (D) all contribute directly to the pollution of groundwater. Choice (A) can contribute indirectly if the exhaust is converted into a secondary pollutant (such as acid precipitation).

2. **B** Gray-air smog is a result of the burning of large volumes of coal to generate electricity and provide heat to homes. China uses a great deal of coal to meet its energy needs. The other cities listed have reasonably effective controls on their generation stations, and people in those countries do not use much coal to heat their homes.

3. **D** Sanitary landfills are designed to completely isolate garbage. These landfills contain clay and plastic liners that hold back and collect water as it passes through. The trash is compacted into the smallest area possible and covered with soil and more plastic and clay. As the trash decomposes, the methane generated is collected to be used for other purposes.

4. **D** This process is called eutrophication, and it occurs when organic material (or nutrients) is added to water. The waste supplies food to the bacteria, which thrive using the oxygen for metabolism. This lowers the level of oxygen in the water, which deprives fish and other aquatic animals.

5. **B** The presence of anaerobic organisms indicate that there is little oxygen in a body of water. The presence of trout, perch, and insect larvae would mean that the water was not very polluted. The absence of these indicator species would indicate that the water is polluted.

6. **C** Each person in the United States produces about 725 kgs (1,600 lbs) of waste each year. 54 percent of this waste is put in landfills, 34 percent is recycled, and most of the rest is burned (12 percent).

7. **D** The concentration of CO_2 in the atmosphere is increasing more rapidly than that of any other gas. Carbon dioxide is produced during the combustion of fossil fuels, and levels of CO_2 are expected to increase rapidly in the next 20 years. Water vapor also contributes to the greenhouse effect, but its levels have remained consistent for hundreds of years.

8. **B** In the correct order: solids are removed by a series of screens first. Next, the water is passed into tanks that contain bacteria. They remove 97 percent of the organic waste. Finally, chlorine is added to kill bacteria and denature viruses.

9. **D** PANs (peroxyacyl nitrates) are secondary pollutants. They are produced from the reaction of hydrocarbons, oxygen, and nitrogen dioxide. All the remaining options are primary pollutants, which are produced from burning fossil fuels.

10. **B** The repository was being built in Yucca Mountain, Nevada. President Obama delayed the storage of any radioactive material there; President Trump's fiscal 2019 budget included $120 million to restart the licensing process.

11. **A** When nitrogen oxides react with water vapor in the atmosphere, the result is nitric acid, HNO_3.

12. **C** Urban heat islands are created because of the presence of buildings, highways, factories, and automobiles, and the use of lights and machines that warm the surrounding air. This can cause the formation of clouds and can trap pollution near the Earth and prevent it from being diluted.

13. **D** Two of the most common AND deadly indoor air pollutants in developed countries are radon and tobacco smoke. Both are leading causes of lung cancer. While carbon monoxide, VOCs, and mold spores are also important indoor air pollutants in developed countries, soot from fuel is much more common in developing countries. CO_2 and methane are involved in atmospheric pollution, while nitric and sulfuric acids are components of acid rain.

14. **C** Because the ozone depletion that has occurred in recent times has, as a primary effect, the allowance of more UV radiation through to Earth's surface, damage caused by excess UV kills primary producers such as phytoplankton. In turn, this effect ripples up the food chain, decreasing numbers of crops and fish. The effects on human health include eye cataracts, skin cancers, and weakened immune systems.

15. **A** In a dose-response analysis, organisms are exposed to a toxin at different concentrations and the dosage that causes the death of the organism is recorded. The dosage of toxin it takes to kill 50 percent of the test animals is termed LD_{50}. If just the negative health effects are considered, ED_{50} is the point at which 50 percent of the test organisms show a negative effect from the toxin. The threshold dose is the dosage at which any negative effect occurs.

Free Response

Use the following lists to check your answers, but remember that you must write all of your answers in paragraph (essay) form.

1. **(a)** Primary pollutants include:

   Oxides of carbon—CO and CO_2

   Oxides of nitrogen—NO and NO_2

   Oxides of sulfur—SO_2

   Unburned hydrocarbons

   All of these primary pollutants are produced by the burning of fossil fuels.

   (3 points maximum—1 point for each correct pollutant and 1 point for the correct origin)

   (b) Secondary pollutants include:

   Sulfur trioxide—SO_3

   Nitric acid—HNO_3

   Sulfuric acid—H_2SO_4

   Hydrogen peroxide—H_2O_2

   Ozone—O_3

   Peroxyacyl nitrates or PANs—R-C(O)OONO$_2$

   Aldehydes—R-COH

   All of these secondary pollutants are formed in the atmosphere. When primary pollutants combine with water vapor and sunlight energy, the reactions that take place produce the products above.

   (3 points maximum—1 point for each correct pollutant and 1 point for a correct explanation of how they are produced)

(c) Photochemical smog is more likely to be found in industrialized nations because of their extensive use of fossil fuels (mostly coal and oil). Gray-air smog is found in areas where coal is the dominant fossil fuel.

(2 points maximum—1 point for photochemical smog explanation and 1 point for gray-air smog explanation)

(d) Ozone causes breathing problems, eye irritation, coughing, and reduced immune response. It also aggravates chronic diseases. CO can reduce the oxygen-carrying capacity of blood and cause dizziness and nausea. It can also retard fetal development. NO_2, SO_2, and PANs cause breathing problems, aggravate existing breathing problems, and increase susceptibility to disease.

(2 points maximum—1 point for the irritant and 1 point for the human effect)

CHAPTER 10 DRILL

Multiple Choice

1. **A** The Montreal Protocol was signed in 1987 by 36 nations concerned with the depletion of the ozone layer by chlorofluorocarbons (CFCs). The pact stated that CFCs had to be reduced by 35 percent between 1989 and 2000. There was a further refinement of the agreement in Copenhagen, Denmark, in 1992.

2. **D** The Marine Mammal Protection Act, established in 1972, is legislation that protects marine mammals in the world's oceans.

3. **D** Large federal projects that might have a large impact on the environment must produce an environmental impact statement. This is mandated by the 1969 National Environmental Policy Act.

4. **A** In 1962, Rachel Carson published the book *Silent Spring*. In this book, she describes the problems associated with the overuse of pesticides—mostly DDT. She explains how pesticides were affecting bird populations in the United States. The book, along with her advocacy, changed the way people thought about the impact of pesticides.

5. **D** The CAA enabled 110 of the most polluting power plants in 21 states to buy and sell pollution rights. This regulated how much pollution (SO_2) each plant could produce. If the plant produces an amount under the regulated level, it receives a "credit," which can be kept or sold to another plant that is producing pollution over the limit of permitted emissions.

6. **D** Environmental economics is the study of the impact of the goods and services economy, particularly market systems of allocation on environmental quality and ecological integrity. It also takes into consideration the economic basis for pollution problems and policy alternatives. Discussions of environmental issues always include monetary and economic considerations.

7. **B** The Endangered Species Act was passed by Congress in 1973 to provide protection for species that are threatened with extinction. This act authorizes federal agencies to undertake conservation programs to protect species listed as endangered or threatened and to purchase land to protect habitats.

8. **D** Holistic viewpoints center on the belief that we are one of many species on the planet and that we interact with all species. Holistic adherents also believe that if we harm the planet, we harm ourselves. The other options are all "planet management" viewpoints.

9. **C** Sustainability depends on the long-term utilization of resources; this is described in (C). The other options could cause resources to be used up more rapidly.

10. **D**   A cap-and-trade policy limits carbon dioxide in the atmosphere by implementing caps and offering incentives for reducing emissions.

11. **A**   WWF is an organization that is not associated with any branch of government.

Free Response

1. **(a)** The CAA sets two standards: National Ambient Air Quality Standards for six outdoor pollutants and the national emissions standards for over 100 toxic pollutants. The NAAQS sets limits on pollutants that people or the environment may be exposed to over a certain period of time. Carbon monoxide, ozone, sulfur dioxide, and nitrogen dioxide are compounds of concern to people near the plant. The second set of standards describes specific limits of hazardous air pollutants. Examples include lead, mercury, and radionuclides.

    (2 points maximum—1 to describe the goal of each standard [NAAQS and hazardous air pollutant] and 1 for a correct example of each)

    (b) Positive results include more employment, reliable electricity, lower cost electricity, less dependence on foreign energy sources, greater tax reviews, and a positive impact on the local economy.

    Negative impacts include more air pollution, more water pollution (thermal is an example), the issue of ash disposal, the need to build infrastructure to support the plant (railroads for coal, roads to and from the plant), and the loss of the natural beauty of the area.

    (4 points maximum—1 point for each positive and each negative impact)

    (c) ESA could be used if a threatened or endangered species is found in or near the area where the plant is to be built. The act allows for the conservation of ecosystems that support the endangered or threatened species. It is possible that the plant would not be built in order to conserve the ecosystem in which the species lives.

    (2 points maximum—1 point for defining the goal of the ESA and 1 point for stating that the plant could not be built)

    (d) PPA is designed to promote source reduction (stopping pollution from being produced). Sulfur reductions could occur by burning low-sulfur coal or by using fluidized-bed coal combustion. The sulfur and nitrogen oxides can be removed by mixing coal and limestone and burning it in a large volume of air. Another process is to wash the coal with water and remove pyritic sulfur particles.

    (2 points maximum—1 for correctly stating the PPA goal and 1 for a correct example of how to reduce pollutants before they are made. Note: No credit would be given for examples like scrubbers or catalysts because they clean up the pollutants after they are produced. PPA focuses on ways to lower the amounts produced, not how to clean them up after production.)

Part VI
Practice Tests

Practice Test 2

AP® Environmental Science Exam

SECTION I: Multiple-Choice Questions

DO NOT OPEN THIS BOOKLET UNTIL YOU ARE TOLD TO DO SO.

At a Glance

Total Time
1 hour and 30 minutes
Number of Questions
80
Percent of Total Grade
60%
Writing Instrument
Pencil required

Instructions

Section I of this examination contains 80 multiple-choice questions. Fill in only the ovals for numbers 1 through 80 on your answer sheet.

Indicate all of your answers to the multiple-choice questions on the answer sheet. No credit will be given for anything written in this exam booklet, but you may use the booklet for notes or scratch work. After you have decided which of the suggested answers is best, completely fill in the corresponding oval on the answer sheet. Give only one answer to each question. If you change an answer, be sure that the previous mark is erased completely. Here is a sample question and answer.

Sample Question   Sample Answer

Chicago is a

    (A) state
    (B) city
    (C) country
    (D) continent

Use your time effectively, working as quickly as you can without losing accuracy. Do not spend too much time on any one question. Go on to other questions and come back to the ones you have not answered if you have time. It is not expected that everyone will know the answers to all the multiple-choice questions.

About Guessing

Many candidates wonder whether or not to guess the answers to questions about which they are not certain. Multiple-choice scores are based on the number of questions answered correctly. Points are not deducted for incorrect answers, and no points are awarded for unanswered questions. Because points are not deducted for incorrect answers, you are encouraged to answer all multiple-choice questions. On any questions you do not know the answer to, you should eliminate as many choices as you can, and then select the best answer among the remaining choices.

GO ON TO THE NEXT PAGE.

ENVIRONMENTAL SCIENCE
Section I
Time—90 minutes
80 Questions

Directions: Each of the questions or incomplete statements below is followed by four suggested answers or completions. Select the one that is best in each case and then fill in the corresponding oval on the answer sheet.

Questions 1 and 2 refer to the following map.

The map below shows the major plate boundaries of the world and their boundaries and movement.

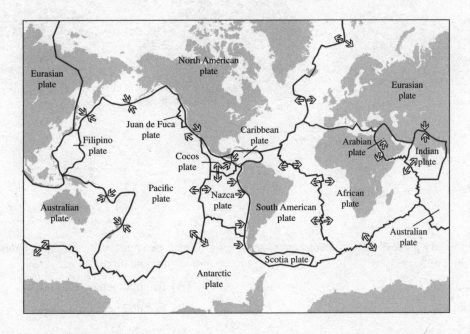

1. Which of the following plate boundaries most likely is characterized by seafloor spreading and rift valleys?

   (A) Nazca–South American boundary

   (B) Australian–Pacific boundary

   (C) Antarctic–Pacific boundary

   (D) Indian-Eurasian boundary

2. Which type of plate boundary typically results in mountain creation, island arcs, earthquakes, and/or volcanoes?

   (A) Convergent boundary

   (B) Divergent boundary

   (C) Transform boundary

   (D) Triple junction

GO ON TO THE NEXT PAGE.

Questions 3 and 4 refer to the following information and diagram.

A ring species is a connected series of neighboring populations, each of which can interbreed with the populations that neighbor it directly, but for which there are at least two end populations that are too distantly related to interbreed.

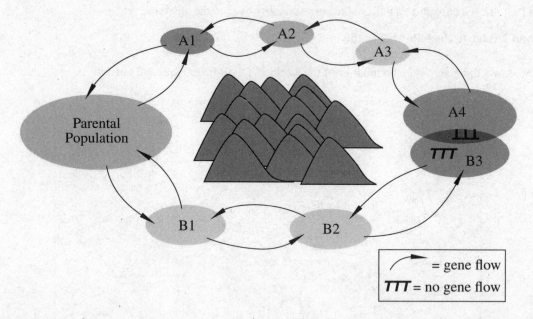

3. For which evolutionary process do ring species provide evidence most directly?

    (A) Speciation

    (B) Natural selection

    (C) Sexual selection

    (D) Extinction

4. Which two populations in the diagram above represent the end populations?

    (A) Parental population and A4

    (B) Parental population and B3

    (C) A1 and B1

    (D) A4 and B3

GO ON TO THE NEXT PAGE.

Questions 5 and 6 refer to the following chart.

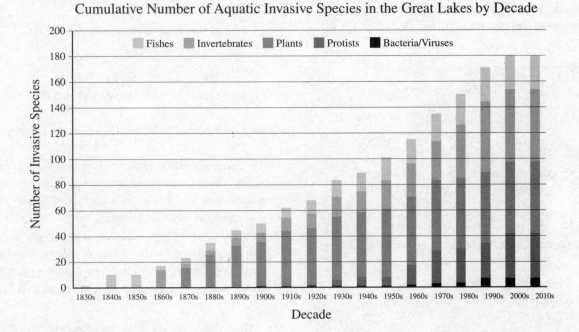

Cumulative Number of Aquatic Invasive Species in the Great Lakes by Decade

5. Based on the chart, which category of invasive species in the Great Lakes showed the greatest increase from the 1950s to the 1960s?

(A) Invertebrates

(B) Plants

(C) Protists

(D) Bacteria/Viruses

6. Which of the following represents the approximate percentage increase in the total number of invasive species in the Great Lakes between the 1980s and the 2000s?

(A) 13%

(B) 20%

(C) 30%

(D) 80%

GO ON TO THE NEXT PAGE.

Questions 7 and 8 refer to the following model.

The model below shows an example of primary ecological succession resulting in a deciduous forest.

Ecological Succession

Bare rock
↓
Lichen, Algae, Mosses, Bacteria
(Break down rock and leave organic debris
that together form soil)
↓
Grasses
(Add organic matter to soil and anchor it in place)
↓
Small herbaceous plants
(Continue to add organic matter to soil)
↓
Small bushes
(Add shelter and shade for other plants)
↓
Conifers
(Create additional habitats)
↓
Short-lived hardwoods such as
dogwood and red maple
(Can tolerate shade of conifers but
are short-lived and vulnerable to damage)
↓
Long-lived hardwoods
(More specialized, hardier hardwoods
such as oak and hickory)

7. What type of event might result in a bare area where this type of ecological succession could occur?

(A) Fire

(B) Tornado

(C) Clear-cutting

(D) Glacial retreat

8. Which of the following animal species is likely to exist in this area during the second stage (Lichen, Algae, Mosses, Bacteria)?

(A) mites, ants, spiders

(B) nematodes, flying insects

(C) lugs, snails, salamanders, frogs

(D) squirrels, foxes, mice, moles, birds

Questions 9 and 10 refer to the following information.

Soil is a complex, ancient material teeming with living organisms. Soil development is an intricate dance that involves four basic processes and six soil-forming factors. It takes hundreds to thousands to millions of years for a soil to develop its characteristic layers or profile. Any soil you see is a dynamic formation produced by the effects of climate and biological activity (organisms), as modified by topography (relief) and human influences, acting on parent materials over time.

9. All of the following could be components of the influence of human activity on soil formation EXCEPT

(A) Alteration of soil chemistry

(B) Compaction

(C) Decomposition

(D) Salinization

10. Which of the following is true about soils used by humans?

(A) Loamy soils composed of a balanced mixture of clay, silt, and sand are not ideal for plant growth.

(B) Soil is considered a nonrenewable resource.

(C) Monoculture practices in modern agriculture tend to enhance soil quality.

(D) The structure of soil (the extent to which it aggregates or clumps) is unimportant with respect to its arability.

GO ON TO THE NEXT PAGE.

Questions 11–13 refer to the following information.

One effect of global climate change is ocean acidification: the ongoing decrease in the pH of the oceans caused by the increase in atmospheric CO_2, which seawater takes up and dissolves. About 30–40% of the carbon dioxide that's released into the atmosphere due to human activities ends up in seawater and lakes and rivers. CO_2 reacts with water to form carbonic acid (H_2CO_3), and some of the carbonic acid molecules separate into bicarbonate ions (HCO_3^-) and hydrogen ions (H^+). The increase in hydrogen ions is interpreted as an increase in acidity. This increase can upset the balance of marine ecosystems by interfering with metabolism and immune response in some organisms, making it more difficult for organisms like coral and plankton to form calcium carbonate shells, and (along with ocean warming) contributing to coral bleaching and reef die-offs.

11. Which of the following is NOT a possible effect of ocean acidification that may impact human life and industry?

    (A) Coral bleaching

    (B) Disruption of marine food webs

    (C) Release of more CO_2 into the atmosphere

    (D) Putting endangered species more at risk due to habitat loss

12. Which is likely to be the best long-term solution to the problem of ocean acidification?

    (A) Reducing CO_2 emissions

    (B) Reducing overfishing and water pollution

    (C) Reforestation to add more oxygen to the atmosphere

    (D) Feeding iron to phytoplankton to speed up photosynthesis

13. The surface pH of the world's oceans has decreased since preindustrial times by about 0.11 pH units, which indicates an increase of almost 30% in the concentration of H^+ ions. If ocean surface pH is expected to drop by a further 0.3 to 0.5 pH units by 2100, approximately what range of possible increase does that represent in terms of H^+ ion concentration?

    (A) 82%–136%

    (B) 100%–216%

    (C) 173%–355%

    (D) 400%–600%

14. The goal of the second stage of a wastewater treatment plant is to

    (A) remove the large solid material

    (B) aerate the water

    (C) remove chemicals such as DDT or PCBs

    (D) lower the amount of organic material in the water

15. Which of the following organisms are the first to be adversely affected by thermal pollution in a stream?

    (A) Trees along the bank

    (B) Insect larvae in the water

    (C) Large fish migrating upstream

    (D) Bacteria in the water

GO ON TO THE NEXT PAGE.

Questions 16–19 refer to the following information and graph.

A scientist placed 100 fish eggs into each of seven solutions with different pH values. After 96 hours, the number of survivors was counted and converted into a percentage. The survival percentages are given in the graph below.

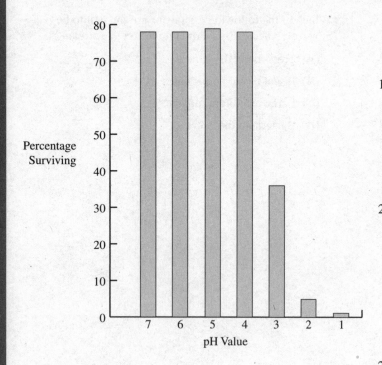

16. Which of the values below best represents the LD_{50} in this experiment?

    (A) 4.0

    (B) 4.5

    (C) 3.0

    (D) 2.5

17. Which of the following, if added to the solutions, would lead to an increase in pH?

    (A) Limestone

    (B) Carbon dioxide

    (C) Nitrogen dioxide

    (D) Sulfur dioxide

18. Which of the following best describes the goal of the experiment?

    (A) To observe how many fish would hatch at different pH values

    (B) To find out how many fish live in streams with different pH values

    (C) To understand how acid rain affects life in streams

    (D) To see what chemical is best at changing the pH of water

19. The pH value is a measure of the

    (A) amount of heavy metals in the water

    (B) biochemical oxygen demand (BOD) of the water

    (C) concentration of oxygen in the water

    (D) concentration of hydrogen ions in the water

20. Which of the following laws created the Superfund program?

    (A) Comprehensive Environmental Response, Compensation, and Liability Act (CERCLA)

    (B) Resource Conservation and Recovery Act

    (C) Clean Air Act

    (D) National Environmental Policy Act

21. Late fall frosts and the northward migration of some tree and plant species may indicate which of the following global changes?

    (A) Increased global temperatures

    (B) The effects of more ultraviolet light from the sun

    (C) A reduction in the volume of ice at the North and South Poles

    (D) Changes in global precipitation patterns

22. High infant mortality rates are likely to exist in countries that have

    (A) a strong and stable economy

    (B) high levels of education for adults

    (C) a stable food supply

    (D) high levels of infectious diseases

GO ON TO THE NEXT PAGE.

23. All of the following statements are true EXCEPT

    (A) energy can be converted from one form to another

    (B) energy input always equals energy output

    (C) energy and matter can generally be converted into each other

    (D) at each step of an energy transformation, some energy is lost to heat

24. Oxygen-depleted zones of the oceans, such as the one at the mouth of the Mississippi River, are most likely caused by

    (A) a reduction in the plant life in rivers that empty into the ocean near the dead zone

    (B) excessive fertilizers carried into the ocean, which cause algal blooms that lower the oxygen levels

    (C) thermal pollution in the ocean

    (D) acid precipitation falling onto the ocean

25. One potential benefit of using genetically modified foods is

    (A) the release of genes to other plants or animals

    (B) the resistance of crops against pesticides

    (C) their growth in monoculture leads to a reduction in biodiversity in the area

    (D) unknown effects on the ecosystem into which they are released

26. Which of the following compounds would probably supply the greatest amount of useful energy to humans?

    (A) The exhaust from a car

    (B) Unrefined aluminum ore

    (C) A glass bottle

    (D) A liter of gasoline

27. Which of the following choices gives the geologic eras in the correct sequence, from the oldest to the most recent?

    (A) Cenozoic—Mesozoic—Paleozoic—Precambrian

    (B) Precambrian—Paleozoic—Mesozoic—Cenozoic

    (C) Paleozoic—Precambrian—Cenozoic—Mesozoic

    (D) Paleozoic—Cenozoic—Precambrian—Mesozoic

28. Approximately what percentage of the world's solid waste does the United States produce?

    (A) 50 percent

    (B) 33 percent

    (C) 10 percent

    (D) 5 percent

29. Which of the following correctly describes conservation easement?

    (A) A process that conserves soil from erosion

    (B) A binding agreement that preserves land from further development in exchange for tax write-offs

    (C) An agreement that allows a developer to add new land to a housing project with little input from neighbors

    (D) A practice that prevents the breakdown of stream banks

30. The highest priority of the Clean Water Act is to provide

    (A) guidance in toxic chemical disposal

    (B) funds to reclaim old strip mines

    (C) policies to lessen the number of oil spills in the ocean

    (D) policies to attain fishable and swimmable waters in the United States

31. Which of the following best describes changes in the genetic composition of a population over many generations?

    (A) Evolution

    (B) Mutation

    (C) Natural selection

    (D) Biomagnification

32. Women have fewer and healthier children when all of the following are true EXCEPT

    (A) they have little education

    (B) they live where their rights are not suppressed

    (C) they have access to medicine and health care

    (D) the cost of a child's education is high

GO ON TO THE NEXT PAGE.

33. Which of the following is a true statement?

    (A) The population size of organisms following a logical population growth is not limited by a carrying capacity.

    (B) The population size of organisms following an exponential growth is limited by a carrying capacity.

    (C) Increasing the death rate of a population can lower the carrying capacity.

    (D) Increasing the rate of food production can increase the carrying capacity.

34. An increase in the amount of UV light striking the Earth as a result of ozone loss will cause which of the following?

    (A) Global climate change

    (B) Increased skin cancer rates in humans

    (C) Lowering of ocean water levels

    (D) An increase in CO_2 in the atmosphere

35. Ozone in the troposphere can result in all of the following EXCEPT

    (A) eye irritation

    (B) lung cancer

    (C) bronchitis

    (D) headache

36. Which of the following describes the amount of energy that plants pass on to herbivores?

    (A) The amount of solar energy in a biome

    (B) The First Law of Thermodynamics

    (C) The Net Primary Productivity (NPP) of an area

    (D) The Second Law of Thermodynamics

37. The Second Law of Thermodynamics relates to living organisms because it explains why

    (A) matter is never destroyed but it can change shape

    (B) plants need sunlight in order to survive

    (C) all living things must have a constant supply of energy in the form of food

    (D) the amount of energy flowing into an ecosystem is the same as the amount flowing out of that system

38. Acid deposition most severely affects amphibian species because amphibians

    (A) do not care for their young

    (B) are not mammals

    (C) need to live in both terrestrial and aquatic habitats

    (D) seldom reproduce

39. All of the following are internal costs of an automobile EXCEPT

    (A) car insurance

    (B) fuel

    (C) pollution and health care costs

    (D) raw materials and labor

40. Scrubbers are devices installed in smokestacks to

    (A) reduce the amount of materials such as SO_2 in the smoke they discharge

    (B) clean out the stack so smoke can move rapidly upward

    (C) reduce the amount of sulfur in coal before it is burned

    (D) reduce the amount of toxic ash produced

41. After ore is mined, the unusable part that remains is placed in piles called

    (A) overburden

    (B) seam waste

    (C) leachate

    (D) tailings

42. All of the following are examples of externalities EXCEPT

    (A) a construction worker purchasing an automobile that reduces his commute time to work.

    (B) a factory producing air pollution that leads to acid rain in the neighboring forest area.

    (C) the construction of a new football stadium leading to increased income for local businesses.

    (D) driving electric and hybrid vehicles reducing the amount of greenhouse gas emissions.

GO ON TO THE NEXT PAGE.

43. Which fuel contains the greatest amount of sulfur?

    (A) Wood

    (B) Natural gas

    (C) Nuclear reactor fuel rods

    (D) Coal

44. Biological reserves are areas that allow countries to

    (A) concentrate agricultural production in one area

    (B) set aside critical habitats to ensure the survival of species

    (C) control the flow of rivers and storm waters

    (D) provide grazing land in order to ensure economic growth

45. Which of the following countries has the shortest population doubling time?

    (A) Denmark

    (B) Australia

    (C) United States

    (D) Kenya

46. Which of the following processes leads to an increase in atmospheric water content?

    (A) Precipitation

    (B) Transpiration

    (C) Infiltration

    (D) Condensation

47. Full cost pricing of a refrigerator would include

    (A) adding the cost of employee salaries to the total cost

    (B) the refrigerator's total impact on the environment

    (C) the cost of transporting the refrigerator to the retail store

    (D) the value of the refrigerator if it was donated to a nonprofit group

48. Since 2000, the atmospheric concentration of which of the following has decreased as a result of anthropogenic activity?

    (A) Methane

    (B) Carbon dioxide

    (C) Chlorofluorocarbons

    (D) Nitrous oxide

49. During a society's postindustrial state, the population will exhibit

    (A) rapid growth with a low birth rate and high death rate

    (B) slow growth with a slowing birth rate and a low death rate

    (C) rapid growth with a high birth rate and low death rate

    (D) zero growth with a low birth rate and low death rate

50. Which of the following is NOT a disadvantage of old-style landfills?

    (A) They generate gases that can be recovered and used as fuel.

    (B) Bad odors come from these landfills.

    (C) Toxic wastes leach into ground water.

    (D) Subsidence of the land after the landfill is filled.

51. The international treaty concerning endangered species (CITES) has tried to protect endangered species by taking which of the following steps?

    (A) Making more countries keep these species in zoos

    (B) Paying the debts of member countries in order to relieve the pressure to sell endangered species

    (C) Developing a list of endangered species and prohibiting trade in those species

    (D) Restoring endangered habitats

52. In 2000, the population of Country A was 50,000. If Country A has a constant birth rate of 33 per 1,000 and a constant death rate of 13 per 1,000, when will the population of Country A equal 200,000?

    (A) 2035

    (B) 2050

    (C) 2070

    (D) 2105

GO ON TO THE NEXT PAGE.

53. Salt intrusion into freshwater aquifers, beach erosion, and the disruption of costal fisheries are all possible results of which of the following?

    (A) Rising ocean levels as a result of global warming

    (B) More solar ultraviolet radiation on the Earth

    (C) More chlorofluorocarbons in the atmosphere

    (D) Reduced rates of photosynthesis

54. The chemical actions that produce compost would best be described as

    (A) photosynthesis

    (B) augmentation

    (C) respiration

    (D) decomposition

55. Which of the sources below would produce non-point source pollution?

    (A) The smokestack of a factory

    (B) A volcano

    (C) A pipe leading into a river from a sewage treatment plant

    (D) A large area of farmland near a river

56. A nation's gross domestic product represents

    (A) the ability to provide health care

    (B) the amount of goods it imports

    (C) the amount of its economic output

    (D) the quality of its environment

57. Which of the following mining operations requires people and machinery to operate underground?

    (A) Mountain top removal

    (B) Contour stripping

    (C) Dredging

    (D) Shaft sinking

58. A country's total fertility rate (TFR) indicates which of the following?

    (A) The life expectancy of women in the country

    (B) The average number of babies born to women between the ages of 14 and 45

    (C) The number of babies under one year of age who die per 1,000

    (D) The total use of contraceptives in the country

59. The wastes stored in Love Canal contaminated the surrounding area by all of the following methods EXCEPT

    (A) leaching into the ground water

    (B) fumes from burning the wastes

    (C) runoff into a nearby stream

    (D) spilled drums of waste

60. All of the following are currently used by humans to produce energy EXCEPT

    (A) nuclear fusion

    (B) nuclear fission

    (C) harnessing solar energy

    (D) harnessing the Earth's internal heat

61. In sea water, carbon is mostly found in the form of

    (A) phosphoric acid

    (B) carbon disulfide

    (C) bicarbonate ions

    (D) methane gas

62. Acid rain and snow harm some areas more than others because certain areas

    (A) have more bacteria in the soil than others

    (B) have a lesser ability to neutralize the acids

    (C) are at a higher elevation than the unaffected areas

    (D) are closer to lakes than are the unaffected areas

63. Which one of the following does NOT store a great deal of phosphorus?

    (A) Rocks

    (B) Water

    (C) Atmosphere

    (D) Living organisms

64. The addition of oxygen to the early Earth's atmosphere most likely occurred through the process of

    (A) volcanic outgassing

    (B) photosynthesis

    (C) meteorite impact

    (D) respiration by animals

GO ON TO THE NEXT PAGE.

65. Scientists use which of the following to estimate environmental risks to humans?

    I.   Animal studies
    II. Epidemiological studies
    III. Statistical probabilities

(A) II only

(B) I and II only

(C) I and III only

(D) I, II, and III

Questions 66–69 refer to the following graph.

A group of students did a Biological Oxygen Demand (BOD) study along a 30-mile section of a stream. The data they obtained are given in the graph below.

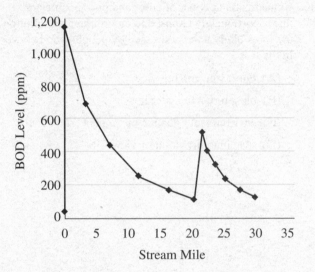

66. Which of the following best describes the type of pollution at mile 0?

(A) Point source

(B) Acid deposition

(C) Secondary pollutant

(D) Deep well

67. The BOD at mile 12 is approximately

(A) 700 ppm

(B) 220 ppm

(C) 200 ppm

(D) 175 ppm

68. The BOD test is designed to directly measure

(A) how much light can pass to the bottom of the stream

(B) the amount of nitrates in the water

(C) the amounts of coliform bacteria

(D) the rate at which oxygen is being consumed by microorganisms

69. Anaerobic bacteria, sludge worms, and fungi are most likely to be found in which part of this stream?

(A) 0 to 5 miles

(B) 10 to 15 miles

(C) 15 to 20 miles

(D) 25 to 30 miles

70. Riparian zones are important parts of lands because they are

(A) the area where most cattle feed when they graze

(B) an area of diverse habitats along the banks of rivers

(C) important buffers against wind

(D) areas where varying amounts of light cause different layers of plant growth

71. Which of the following is a disadvantage of fish farming?

(A) It can allow for genetic engineering, which leads to bigger yields.

(B) It is very profitable.

(C) It can lead to large die-offs due to disease.

(D) It can reduce the pressure to harvest wild species.

72. The form of nitrogen that plants can use directly is

(A) nitrates

(B) nitrites

(C) N_2 gas

(D) methane

GO ON TO THE NEXT PAGE.

73. Which of the following best describes the effects of a thermal inversion?

    (A) Cold ocean water moves to the surface and warm water sinks.

    (B) Warm, polluted air rises and mixes with cool upper air, and pollutants escape.

    (C) Warm river water cools when it enters the ocean.

    (D) Polluted air at the surface cannot rise because it is blocked by warm air above it.

74. Shifting taxes to tax pollution and waste rather than taxing the cost of products will allow people to

    (A) maximize profit

    (B) increase the tax base in a city

    (C) use tax money for local schools

    (D) shift to a pattern of more sustainable development

75. Which of the following molecules is most damaging to stratospheric ozone?

    (A) H_2O

    (B) CO_2

    (C) Chlorofluorocarbons

    (D) N_2O

76. Which of the following ocean zones has the highest levels of nutrients?

    (A) Coastal zone

    (B) Euphotic zone

    (C) Bathyal zone

    (D) Abyssal zone

77. Samples of atmospheric gases from past eras can most easily be obtained from which of the following sources?

    (A) Methane gas trapped in oil reserves

    (B) Different types of sedimentary rock

    (C) Gases trapped in polar ice caps

    (D) Mud samples from eutrophic lakes

78. Acid deposition on soil kills beneficial decomposers. Which of the following cycles would be most affected by the loss of decomposers?

    (A) Sulfur cycling

    (B) Phosphorus cycling

    (C) Hydrologic cycling

    (D) Nitrogen cycling

79. Which of the following is a trace element necessary for plant growth?

    (A) Carbon

    (B) Nitrogen

    (C) Phosphorous

    (D) Magnesium

80. Concerns that people of color and poor people are disproportionately exposed to environmental pollution are most likely to be addressed by people who believe in the

    (A) Earth stewardship view

    (B) planetary manager view

    (C) environmental justice movement

    (D) sustainability point of view

END OF SECTION I

ENVIRONMENTAL SCIENCE
SECTION II
Time—70 minutes
3 Questions

Directions: Answer all three questions, which are weighted equally; the suggested time is about 23 minutes for answering each question. Where calculations are required, clearly show how you arrived at your answer. Where explanation or discussion is required, support your answers with relevant information and/or specific examples.

1. The following editorial is excerpted from a recent edition of the Hilltop Express:

Hilltop Express

New Pests Invade Farm

A new species of corn-infesting insect has recently been discovered in a local farmer's field. Bill Jones stated: "Last week a section of my corn field was covered in small black beetles. They can fly from plant to plant, and they eat large holes in the leaves. I called the county extension agent Sarah Smith and she came out and identified them. I'm going to start spraying tomorrow morning." In a telephone interview with Sarah, she stated that this species was new to the county and has the potential for causing real damage to the corn crop. She stated that the adults do most of the damage to growing leaves.

The grubs live near the base of the plant and feed on bacteria and other organisms living in the soil. She added that the beetle was resistant to the most common pesticide, NOGrub. NOGrub, she commented, had been tried in another county and was not found to be effective. The editors of the Hilltop Express realize the potential dangers to the county's most important cash crop. We urge the county agents to recommend a series of new pesticide treatments to control this new menace to our livelihood.

(a) **Explain** how the beetle might have become resistant to NOGrub. Assume that NOGrub had been applied to a population of beetles in another county.

(b) If the county agents do not have information on which pesticide is most effective against the beetle, the county plans to investigate by trying several pesticides on controlled sections of beetle-infested crop.
   i.   **Identify** the independent and dependent variables in such a trial.
   ii. **Describe** an effective control group and the experimental groups for this experiment.
   iii. **Identify** TWO environmental factors that must be controlled to keep the experiment from producing skewed results.

(c) **Identify** TWO negative impacts of using chemical pesticides on the surrounding ecosystem.

(d) One strategy for combating pests is Integrated Pest Management (IPM).
   i.   **Describe** IPM.
   ii. **Identify** one benefit and one difficulty the county would likely encounter in using IPM to control this outbreak.

GO ON TO THE NEXT PAGE.

2. The map below shows two cities: City *X* and City *Z*, separated by several kilometers.

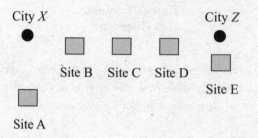

City *X*

City *Z*

Site B   Site C   Site D

Site E

Site A

Students from a high school in between the two cities studied soil pH values at the sites labeled A through E on the map. The results of the pH study are given in the following table:

Site	pH value
A	6.2
B	5.6
C	5.0
D	4.5
E	4.3

(a) Refer to the table and diagram above to answer the following questions.
   i.   **Describe** one point source for the pollution that caused the change in the soil's pH as shown.
   ii. In the description, **identify** a fuel that could create the pollution.

(b) Assume that the fuel identified in (a)(i) is the source of the pollution.
   i.   **Identify** one primary and one secondary pollutant that can cause the change in the soil's pH.
   ii. **Describe** the process that causes the change in the pH.

(c) **Propose** one possible method to reduce the air pollutants that are causing the pH change.

(d) **Identify** and **describe** one provision of the Clean Air Act of 1990 that could be used to control and reduce the emissions.

GO ON TO THE NEXT PAGE.

3. According to the United States Energy Information Administration, the consumption of natural gas by the United States increases at 8 percent per year. The U.S. receives its supplies from a variety of international and domestic locations. Natural gas is used in the home, for industry, and for power generation.

   (a) **Calculate** the approximate number of years it would take to double the consumption of natural gas in the United States. **Show** all your work.

   (b) **Identify** one method by which natural gas is recovered and transported.

   (c) **Describe** two benefits to the environment that would occur if the United States switched from coal to natural gas-fired electric power generation.

   (d) Some people advocate increasing the use of coal versus natural gas for the production of electricity. Give one argument that the proponents of coal might use to **justify** their position.

   (e) Others advocate for non-hydrocarbon fuel alternatives.
      i. **Identify** one non-hydrocarbon fuel alternative.
      ii. **Describe** ONE drawback of the alternative identified in (e)(i).

STOP

END OF EXAM

GO ON TO THE NEXT PAGE.

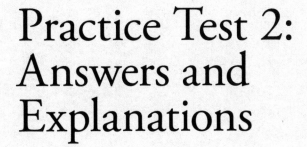

Practice Test 2:
Answers and
Explanations

PRACTICE TEST 2 ANSWER KEY

1.	C	21.	A	41.	D	61.	C
2.	A	22.	D	42.	A	62.	B
3.	A	23.	C	43.	D	63.	C
4.	D	24.	B	44.	B	64.	B
5.	C	25.	B	45.	D	65.	D
6.	B	26.	D	46.	B	66.	A
7.	D	27.	B	47.	B	67.	B
8.	A	28.	B	48.	C	68.	D
9.	C	29.	B	49.	D	69.	A
10.	B	30.	D	50.	A	70.	B
11.	C	31.	A	51.	C	71.	C
12.	A	32.	A	52.	C	72.	A
13.	B	33.	D	53.	A	73.	D
14.	D	34.	B	54.	D	74.	D
15.	B	35.	B	55.	D	75.	C
16.	C	36.	C	56.	C	76.	D
17.	A	37.	C	57.	D	77.	C
18.	A	38.	C	58.	B	78.	D
19.	D	39.	C	59.	B	79.	D
20.	A	40.	A	60.	A	80.	C

PRACTICE TEST 2 EXPLANATIONS

Section I—Multiple-Choice Questions

1. **C** The map shows whether plates at each boundary are moving apart (divergent) or toward each other (convergent). A boundary characterized by seafloor spreading and rift valleys should be a divergent boundary. Since (A), (B), and (D) are convergent boundaries, only (C) can be correct.

2. **A** Mountain creation is a result of orogenic belts (also called collision boundaries), which are convergent boundaries between two continental plates. Subduction zones, or convergent boundaries between oceanic and continental plates, usually cause volcanoes to form, along with deep trenches. Convergent boundaries between two oceanic plates can form island chains. All of these types can result in earthquakes due to the compression and tension between colliding plates.

3. **A** Speciation is the process by which populations evolve to become distinct species. A ring species is an example of a species that is "caught in the act" of speciation: the end populations. If looked at in the absence of the other populations, they would be considered distinct species because of their inability to interbreed. However, the connectedness of the ring shows that the parent population was originally one species.

4. **D** The passage states that the end populations are those "that are too distantly related to interbreed." Looking at the key, you can see that there is gene flow between each set of neighboring populations except for A4 and B3. There's no evidence in the diagram as to whether the parental population can interbreed with other populations besides those it neighbors.

5. **C** Looking at the chart, you can see that for most types of organisms there was no significant increase between the bars for the 1950s and the 1960s. However, the protists group just about doubled. Therefore, protists showed the greatest increase between those decades.

6. **B** To find a percentage change, divide the difference by the original amount, and then multiply by a factor of 100 to get the value into percentage form. The total number of invasive species in the Great Lakes in the 1980s was about 150, while in the 2000s it was about 180. The difference is $180 - 150 = 30$. So, divide: $30 \div 150 = 0.2$. This is equivalent to 20%.

7. **D** Primary succession begins in virtually lifeless areas, such as those that result from lava flows or glacial retreat. Secondary succession takes place where an existing community has been cleared by a disturbance event (such as a fire, tornado, or human impact) but the soil has been left intact. Since the description tells you this example is of primary ecological succession (and the model begins with bare rock, not soil), only (D) can be correct.

8. **A** While ecological succession theory was developed mostly by botanists, the stages include animal species as well as plants. The most likely species to survive in an early stage such as that dominated by lichens, algae, mosses, and bacteria are the smallest: tiny insects that can feed on these primary producers and live in the cracks between them. Therefore, (A) is the best choice.

9. **C**   The effects of human activity must increasingly be acknowledged as a factor in soil development. Human use of fertilizers and pesticides, and pollution and its side effects such as acid rain, can alter soil chemistry. Roads with vehicle traffic and heavy machinery for agriculture and construction can compact soil, and irrigation and depletion of groundwater can cause salinization. Decomposition, on the other hand, is the province of other organisms such as bacteria and fungi that live within the soil, so (C) does NOT represent a human influence.

10. **B**   The question asks which statement is true: only (B) is correct. Soil is considered nonrenewable because it requires a great length of time to form. Loamy soils are considered ideal for plant growth; monocultural agriculture tends to decrease soil quality by decreasing biodiversity and leaching specific nutrients, and soil needs good structure to be arable.

11. **C**   Coral bleaching, (A), affects whole ecosystems built around coral reefs, which can impact human fishing (as well as leisure activities such as snorkeling). That's one example of the larger category of problems that (B) represents: any marine ecosystem that is disrupted by acidification can become imbalanced and this affects its food webs, which directly impacts human fishing. Choice (D) affects humans less directly, but habitat loss causing further endangerment to already-threatened species may cause extinctions that could have big repercussions to ecosystems and food chains. However, release of more CO_2 into the atmosphere is not a potential problematic effect of ocean acidification. CO_2 emissions cause ocean acidification, not the other way around, and acidification actually removes CO_2 from the atmosphere ((C) is correct).

12. **A**   The only effective solution currently considered for the problem of ocean acidification is the reduction of CO_2 emissions on a global scale, (A). This makes sense, since the problem is a direct result of these emissions. Reducing overfishing and water pollution, (B), will certainly help with other problems and help increase resilience of marine organisms to acidification, but doesn't actually address the problem itself. Reforestation, (C), will help with air quality but won't directly impact CO_2 emissions or ocean acidification. Finally, iron fertilization, (D), has been considered as a possible mitigation approach to counter the effects of acidification, but hasn't been shown to be effective or free from unwanted side effects. If you don't know anything about iron fertilization, you can still tell that (A) is the best choice because you know it directly addresses the problem at hand.

13. **B**   Keep in mind that pH is a logarithmic scale. If a decrease of 0.11 pH units corresponds to an increase of almost 30% in H^+ ions, that means that $\log_{10} x - \log_{10} y \approx 0.11$. Remember that $\log_b x - \log_b y = \log_b \frac{x}{y}$, so $\log_{10} \frac{x}{y} \approx 0.11$; in other words, $10^{0.11} \approx \frac{x}{y}$. So $\frac{x}{y} \approx 1.28824....$ and you can see that that is equivalent to almost a 30% increase. So use the same process to find equivalent percentage increases for the range of pH decreases given: if the pH drops by 0.3 units, then $10^{0.3} \approx \frac{x}{y}$, so $\frac{x}{y} \approx 1.9953$, which is approximately a 100% increase. And if the pH drops by 0.5 units, then $10^{0.5} \approx \frac{x}{y}$, so $\frac{x}{y} \approx 3.1623$, which is approximately a 216% increase. Choice (B) is correct.

14. **D**   Wastewater treatment plants use a three-step process in removing waste from water. The main goals of sewage treatment are to remove the solid waste and reduce the biological oxygen demand (BOD). This BOD is a measure of how much organic material is in the water and how much bacteria can live in the

water. The first stage of wastewater treatment is the mechanical removal of solid objects, and screens of various sizes are used to perform this task. In the second stage, the water is sprayed on a bed of rocks that harbor billions of bacteria. The bacteria break down the organic molecules and lower the BOD. After the remaining liquid is treated, chlorine is put into the water to control bacteria populations.

15. **B**   Insect larvae are the organisms most vulnerable to changes in abiotic factors like thermal pollution. As is true with the young of most species, the larval forms constitute the most vulnerable stage of an insect's life cycle. Species whose populations rise and fall with changes in abiotic factors are called *indicator species*. The presence of an indicator species in a certain ecosystem tells ecologists that the stream water is ideal to live in—if a fragile indicator species inhabits the region, then it must be safe for all species. If the indicator species is not there, then the water is not optimally clean.

16. **C**   The LD_{50} is the value at which 50% of a population dies. If you read the graph, you can see that that value falls closest to the solution with the pH of 3.0. LD_{50} values are important because they help us understand the health risks of certain materials. As you might imagine, if a chemical is very toxic, it has a very low LD_{50}. For example, the nerve agent VX has an LD_{50} of 0.14 mg/kg body weight! It is extremely harmful to humans: this LD_{50} value tells you that it only takes 10.2 mg of nerve agent VX to kill half of the test population of 155-pound males! For the test, just remember that the smaller the LD_{50} number, the more toxic a chemical is.

17. **A**   Choices (B), (C), and (D) are all pollutants involved in the production of acid precipitation that would lead to a decrease in the pH, not an increase. Therefore, these choices are incorrect. Limestone, (A), is a base and would increase the pH of the solution. Choice (A) is correct.

18. **A**   The purpose of this test is to observe how many fish hatch at different pH values, plain and simple. If you chose any other answer, then you might have been reading too much into the experiment. Remember not to make any inferences—you must answer the question only on the basis of the information you're given. In this experiment, the survival percentage is the dependent variable, and the pH is the independent variable. The data collected relate to hatching survival rates. pH is one of the most important abiotic factors for aquatic organisms. If the pH value shifts out of the ideal range, then the essential proteins in the cells that constitute fish eggs can denature. If that happens, the young will not survive.

19. **D**   Only (D) correctly defines pH. pH is the measure of the number of hydrogen ions in a solution. Chemically speaking, it is the negative logarithm (base 10) of the hydrogen ion concentration. So, if a solution has 10^{-7} hydrogen ions in it, then $-\log 10^{-7} = -(-7) = 7$. The pH of the solution is 7; the solution is neutral. The pH scale runs from 0–14; acidic solutions have a pH less than 7, while basic solutions have a pH greater than 7. Logarithms are used to keep the numbers simple because the range of hydrogen ion concentrations can be as high as 100 trillion!

20. **A**   The Comprehensive Environmental Response, Compensation, and Liability Act (CERCLA) is a law that was passed to establish the "Superfund." This law created a tax on the chemical and petroleum industries and provides broad federal authority to respond directly to releases or threatened releases of hazardous substances that may endanger public health or the environment. Choice (B) is incorrect:

RCRA established a system for managing non-hazardous and hazardous solid wastes in an environmentally sound manner. Specifically, it provides for the management of hazardous wastes from their point of origin to their point of final disposal ("cradle to grave"). RCRA also promotes resource recovery and waste minimization. Choice (C) is responsible for air pollution control. Finally, (D) mandates that environmental impact studies be done before major construction projects are started.

21. **A** Increased global temperatures are responsible for the phenomena in most of the other answer choices! Global warming is more formally referred to as tropospheric warming. You learned that the atmosphere contains much H_2O vapor as well as trace amounts of CO_2, CH_4, and NO_2. These gases reflect infrared radiation that is, itself, reflected off Earth's surface as it is warmed by the sun. This process of tropospheric warming is called the "greenhouse effect," and it is a natural phenomenon. However, as humans burn huge amounts of fossil fuels, the excess CO_2 and NO_2 enter the atmosphere. These gases increase the heat-storing capacity of the atmosphere, which causes average global temperatures to increase. This, in turn, can cause late fall frosts and the northward migration of tree and plant species.

22. **D** This is an easy one. All of the answer choices are "positive" answer choices—which you would not think would cause high infant mortality—except (D). In fact, (D) is one strong indicator of countries that exhibit high infant mortality. Generally, high infant mortality rates occur in nations where the Gross Domestic Income (GDI) is low (less than $4,000). These nations cannot offer infant support services such as medicine, clean drinking water, sewage removal, and food—and this leads to high rates of infectious disease. Choices (A), (B), and (C) all occur in countries that have low infant mortality rates.

23. **C** Choice (C) is the only false statement among the answer choices. This question requires that you be familiar with the first two laws of thermodynamics. The first law states that energy cannot be created or destroyed. That is, the total energy of a system will always remain the same. The second law states that at each energy transformation, some energy is lost to the surroundings in an unusable form (usually as heat). For example, as a flashlight runs using battery power, some of the energy from the batteries goes toward warming the light bulb, which is not useful work.

24. **B** The area of the Gulf of Mexico that extends from just south of the Mississippi River mouth to Texas receives nutrient-rich water from both the Mississippi and Atchafalaya Rivers. The water from these rivers carries runoff fertilizers from farms that lie in or near the watersheds for the rivers. These nutrients promote a rapid rise in the algae population in the Gulf waters, which in turn causes eutrophication. When the algal cells die, bacteria living in the water decompose them, or break them down. These decomposers use oxygen for metabolism, and eventually the water becomes extremely oxygen poor. Low levels of oxygen often make these bodies of water uninhabitable for other aquatic organisms.

25. **B** Choice (B) is a positive effect, while all of the other answer choices cite negative effects of genetically modified foods. Genetically modified foods (GMFs) are crop plants to which genes from other species have been added. For example, the "Flavr Savr" tomato, sold from 1994 to 1997, contained select genes from fish! These genes were inserted into the tomato's genes so that the tomatoes would not freeze in a frost. Additional positive aspects for GMFs include increased yields, increased resistance to pests, resistance against pesticides, and the potential to grow in habitats that previously did not support the plants.

26. **D** "Energy usefulness" is defined as how helpful an energy source is to humans. A source is more useful if it costs little to harvest, transport, and use. Useful energy is most commonly in a concentrated form. For example, coal is highly useful compared to wind energy; (A), (B), and (C) are all examples of low-quality energy and matter.

27. **B** Review Chapter 4 if you have trouble remembering the basics of the geologic time scale. You will not be expected to memorize this chart, but you should be familiar with the most recent era and the most talked-about ones. Choice (B) correctly starts with the earliest era, the Precambrian (600 million years ago), then the Paleozoic (500 to 250 million years ago), then the Mesozoic (250 to 65 million years ago), and finally, the Cenozoic (65 million years ago to today)!

28. **B** Although the United States holds only 4.6 percent of the world's population, it produces 33 percent of the world's solid waste.

29. **B** A conservation easement is a voluntary agreement that allows a landowner to limit the type or amount of development on his or her property, while retaining private ownership of the land. The easement is signed by the landowner, who is the easement donor, and the conservancy, who is the party receiving the easement. The conservancy accepts the easement with understanding that it must enforce the terms of the easement in perpetuity. After the easement is signed, it is recorded with the County Register of Deeds and applies to all future owners of the land. In conservation easements, the owner usually receives a tax break for the promise that he or she will not build on that land.

30. **D** Choice (D) is the defining policy statement of the CWA. This law governs the discharge of pollutants into the waters of the United States. It gives the Environmental Protection Agency the authority to implement pollution control programs and sets wastewater standards for industry. The Clean Water Act also continues to set water quality standards for all contaminants in surface waters. The act makes it unlawful for any person to discharge any pollutant from a point source into navigable waters unless a permit was obtained under its provisions. It also funds the construction of sewage treatment plants under the construction grants program, and recognizes the need to address the critical problems posed by non-point source pollution.

31. **A** Evolution is the change in the genetic makeup of a population over time. For this exam, the key is to remember that individuals do not evolve and that only populations can evolve. Choice (B) is not a good answer because a mutation is a change in DNA. Choice (C) is also incorrect—evolution can occur in ways other than natural selection. Finally, we can eliminate (D) because it refers to an increase, or magnification, of toxins in a food chain.

32. **A** This question requires you to understand the factors that determine the number of offspring that women are expected to have from a statistical standpoint. Generally speaking, prosperity, high income, and a high level of education are all factors that lower a woman's total fertility rate (the number of offspring she will have). Statistically, the less wealthy and less educated a woman is, the higher her expected total fertility rate will be. Choices (B) through (D) all list factors that lower the total fertility rate, so the correct answer is (A).

33. **D**   The carrying capacity is the maximum number of organisms that can live sustainably in a habitat. Increasing the rate of food production can increase the carrying capacity, so (D) is the correct answer. Organisms with a carrying capacity have a logistical population growth, so (A) and (B) are incorrect. Choice (C) is also incorrect because carrying capacity is not affected by the death rate.

34. **B**   As chlorine atoms (from CFCs) catalyze the breakdown of ozone molecules in the stratosphere, more ultraviolet light is expected to reach Earth's surface. These UV rays will break down DNA molecules in human skin cells, which can lead to cancers, cataracts, and other illnesses. Of course, other animals and plants can also be affected. While you may have been tempted to choose (A), (B) is a better option because it is more specific—(A) is much too general to be correct. Choice (C) is an opposite and (D) is a result of global warming.

35. **B**   Ozone is generated as a secondary pollutant in photochemical smog. Now, this is an EXCEPT/NOT/LEAST question, so remember that you're looking for the incorrect answer! The exposure to tropospheric ozone can cause all of the symptoms given in (A), (C), and (D). However, (B), lung cancer, is primarily caused by smoking cigarettes.

36. **C**   The Net Primary Productivity (NPP) is the rate at which all the plants in an ecosystem produce net useful chemical energy. It is equal to the difference between the rate at which the plants in an ecosystem produce useful chemical energy and the rate at which they use some of that energy through cellular respiration. It is calculated by taking the Gross Primary Productivity (the amount of sugar that the plants produce in photosynthesis) and subtracting from that the amount of energy the plants need for growth, maintenance, repair, and reproduction. NPP is measured in kilocalories per square meter per year ($kcal/m^2/y$). Choice (A) is incorrect because it denotes the amount of energy available for photosynthesis. Choice (B) is no good; it merely states that energy cannot be created or destroyed. Choice (D) can also be eliminated; it states that at each energy transfer, some energy is lost to the environment.

37. **C**   Choice (C) is the only option that relates to the Second Law of Thermodynamics. Recall that the second law states that at each step of energy transformation, a large amount of energy is lost as heat to the environment. When we eat food, our bodies convert the energy in the food's chemical bonds to chemical bonds in our molecules. During this conversion, most energy is lost as heat (which is why our bodies are warm!).

38. **C**   Acid deposition affects species by damaging habitats and by reducing or contaminating food sources through uptake of toxic levels of metals. Species such as amphibians, which require both aquatic and terrestrial environments, are most at risk. For example, in the acid-sensitive areas of eastern Canada, 16 of the 17 amphibian species have more than 50 percent of their ranges affected by acidic deposition. Monitoring amphibian populations may provide a biological indication of changes in acid deposition.

39. **C**   Internal costs are all the costs that are incurred by the buyer and seller of the item in question. For example, the cost of gasoline and the purchase price of the car are both buyer costs. The costs of metal, plastics, employee salaries, and plant maintenance are internal costs of the seller. Choice (C) is the correct answer because it is not an internal cost, but an external one; it is a cost paid by society. Health care costs that arise as a result of air pollution and the cost of building roads are both external costs.

40. **A** Scrubbers are devices that are installed in smokestacks in order to reduce the amount of pollutants that rise up the stack. They extract the pollutants in different ways: by trapping dust in filter material, using cyclonic air motion to concentrate dust in one area, or spraying water in the stack to trap both dust and gases that are water-soluble. Choice (B) is simply not the best answer; (C) occurs before smoke is produced; and (D) is simply untrue—in fact, the amount of ash is greater when scrubbers are in place because scrubbers trap materials.

41. **D** After ore is mined, the unusable parts that are placed into piles that are called tailings. This is a question that you'd either know the answer to or you wouldn't. Remember that if you come to a question like this and have absolutely no idea which choice is correct, pick your favorite letter and move on. You can also circle it and come back to it in the next pass—maybe you'll remember the answer if you give your subconscious mind a chance to do its thing. Let's look at the other answer choices: (A), overburden, is the rock that's removed from the ground in the surface mining of coal, so it's wrong. We made up (B), we hope you didn't choose it! Choice (C), leachate, is the liquid that percolates through a mine or landfill.

42. **A** An externality is an unanticipated consequence experienced by unrelated third parties. Choices (B), (C), and (D) are all examples of externalities. Choice (A) is not an externality because the effect is experienced by a related party (the construction worker who purchased the automobile).

43. **D** The fossil fuel coal is sedimentary rock that's derived from decayed and fossilized plant and animal materials. During the fossilization process, sulfur is incorporated into the rocks; this sulfur comes from bacteria that perform the decomposition of the material, as well as from the decaying material itself. None of the other fuels contains as much of the element sulfur (S).

44. **B** Choice (B) is the best definition of a biological reserve. These reserves are areas of habitat that are crucial to the survival of species. Strictly speaking, a biological reserve is an area of land and/or water designated as having protected status for purposes of preserving certain biological features. Reserves are managed primarily to safeguard these features and provide opportunities for research into the problems underlying the management of natural sites and of vegetation and animal populations. Regulations are usually imposed that control public access and prevent people from disturbing these areas. All other answer choices would lower the biodiversity of an area.

45. **D** The doubling time is the amount of time it takes for the population of a country to double. The country with the shortest doubling time is the country with the most rapid population growth. Among the answer choices, Kenya is the only one with a rapid population growth, so (D) is the correct answer. Denmark is experiencing zero growth, so (A) is wrong. The United States and Australia are experiencing slow growth, so (B) and (C) are also incorrect.

46. **B** Transpiration is the evaporation of water from plants into the air. This process increases the atmospheric water content. Choice (B) is the correct answer.

47. **B** Choice (B) is the definition of full cost pricing: Full cost includes the environmental effects. All the other answer choices are part of the external costs paid by the buyer or seller.

48. **C** In 1987, the Montreal protocol was signed. The protocol called for the worldwide end of chlorofluoro-carbon (CFC) production. As the result of the protocol, the atmospheric concentration of CFCs has decreased, allowing for the ozone hole to recover. Choice (C) is correct. Choices (A), (B), and (D) are all greenhouse gases that have been continually *increasing* in atmospheric concentration, not decreasing.

49. **D** In the postindustrial state, the population has reached its carrying capacity and will exhibit a zero growth rate, or even drop below a zero growth rate.

50. **A** Methane is a useful biofuel that is produced from the decomposition of organic molecules in a landfill. This gas can be trapped, purified, and used to power vehicles. All of the other effects that are listed are undesirable results of the existence of old landfills.

51. **C** The CITES treaty (the Convention on International Trade in Endangered Species of Wild Fauna and Flora) entered into force in 1975. It was first discussed in the late 1960s when people recognized that there was no way to internationally protect endangered or threatened species. CITES is international in scope and is designed to halt the killing of endangered or threatened species for food, collection, or medicinal purposes by penalizing those who collect, trade, or buy those species.

52. **C** The growth rate of Country A is equal to its birth rate subtracted by its death rate: 33 births per 1,000 – 13 deaths per 1,000 = an increase of 20 persons per 1,000, or a growth rate of 2%. The doubling time can be calculated using the rule of 70:70 divided by 2 is equal to 35. This means that it takes 35 years for the population of Country A to double. In order for the population of Country A to reach 200,000, the population needs to double twice. This would take 70 years, so the answer is 2070, which is (C).

53. **A** Increased ocean levels is the most logical cause of the phenomena listed in the question. The increase is due to two main factors: The first factor is that as water warms, it expands. This is known as thermal expansion. Secondly, as the average global temperature increases, glaciers and other ice formations will melt; this will increase the volume of the world's oceans. All of the answer choices represent events that could be caused by an increase in the volume of the world's oceans.

54. **D** Compost is formed from decaying plants and other organic material. In the composting process, both bacteria and fungi decompose large organic molecules into smaller molecules. Compost can then be used to fortify and condition soil. While it might have been tempting, (C) is too general an answer, because all living organisms undergo respiration.

55. **D** Non-point pollution sources are defined as pollution that comes from a broad, ill-defined area, such as a large area of farmland. Since most farmers in the area presumably fertilize their crops, you cannot point to one specific area as the only source. Choices (A) through (C) are all point sources of pollution.

56. **C** Choice (C) represents the standard definition of GDP. It is calculated by totaling the value of all the goods and services in the country for a specific period and is often considered a measure of the standard of living in a country.

57. **D** Choice (D) is the correct answer; all other answer choices are open-surface mining practices.

58. **B**   The total fertility rate in a country represents the average number of children a woman will have during her reproductive lifetime (between ages 14 to 45). The total fertility rate is affected by many factors, including level of education, culture, and the country's standard of living.

59. **B**   Choice (B), or the fumes produced from the burning of wastes, is the only process that does not directly contribute to the contamination of the surrounding soil. When the EPA started to clean up the wastes from Love Canal, they had to deal with wastes that contaminated the surrounding land by all the other methods listed in the answer choices.

60. **A**   Humans have not yet managed to make nuclear fusion viable. Choice (A) is correct. The other choices are all different methods currently used by humans to produce energy.

61. **C**   The bicarbonate ion (HCO_3^-) is produced when atmospheric carbon dioxide reacts with ocean water. Bicarbonate ions act as a buffer in seawater and allow the pH to be relatively stable, ranging between 7.5 and 8.5. Choice (A), phosphoric acid, contains no carbon atoms. Neither methane nor carbon disulfide exist in large quantities in, so (B) and (D) are wrong as well.

62. **B**   Choice (B) is correct because areas rich in limestone or other basic minerals can neutralize the acid deposition products. Limestone is composed mostly of calcium carbonate ($CaCO_3$). This dissolves in water to form bicarbonate ions (HCO_3^-), which act as the buffering agent.

63. **C**   Atmosphere (C) is the only answer choice that does not store significant amounts of phosphorus. Phosphorus is rarely present in gases. Additionally, it is relatively insoluble in water. It is found most frequently in rocks and minerals.

64. **B**   The current hypothesis for the formation of large volumes of oxygen in the atmosphere is called the autotrophic hypothesis. According to the hypothesis, the early atmosphere had almost no O_2. It was not until the evolution of photosynthesis that the levels of O_2 increased. As for the other answer choices, volcanic outgas mostly consists of oxides of carbon, nitrogen, and hydrogen sulfide; meteorites could not carry gases effectively; and animals produce CO_2 not O_2.

65. **D**   "Animal studies" refers to tests conducted on animals in laboratories, under controlled conditions. For example, this type of study is used to determine the dosage of a chemical that will kill 50 percent of a test population (LD_{50}). Epidemiological studies are done by examining human populations exposed to certain risks (for example, smoking). Statistical probabilities quantify the likelihood that a given event will occur.

66. **A**   This is a single dose of the pollutant from a point source. You can determine this because the BOD drops off, meaning that the pollutant is being diluted by unpolluted water. Choice (B) is incorrect because it is a secondary air pollutant. Choice (C) is a class of pollutants that result from the mixing of combustion products in the atmosphere. Choice (D), deep well, is a location where toxic materials are stored deep underground. It is unlikely that it would pollute the stream.

67. **B**   This question tests your ability to estimate based on the graph. Look carefully at the axes and scale. It is helpful to use your pencil to draw a line from the 12-mile mark in the x-axis up to the line, and then

another line from that point to the *y*-axis. You should be able to estimate 220 ppm (parts per million). Choice (A) is too high an estimate. Choices (C) and (D) are estimates that are too low.

68. **D** The BOD, or Biological Oxygen Demand, test determines the rate at which microorganisms take oxygen out of the water. This test is an estimate of the amount of biodegradable organic matter in the water and is an indirect indicator of water quality. Choice (A) is the turbidity test—the number of particles that scatter light in the water. Choice (B), nitrates, are important plant nutrients and are measured by tests that can quantitatively determine the concentration of the chemical ions. Choice (C), coliform bacteria, live in the intestinal tracts of animals and are an indication that fecal matter is in the water. This is not a BOD test because the coliform test measures the number of living cells, not their biological activity.

69. **A** You should recognize that the three organisms live in conditions with very low oxygen content. Looking carefully at the graph, you will see that the BOD is highest from 0 to 5 miles. Because the BOD is highest here, the amount of oxygen in the water is the lowest. Remember, high demand means low amounts of oxygen in the water! Choices (B), (C), and (D) are areas where the BOD is low—so there is more oxygen in the water. The three listed organisms do not live in areas of high oxygen content.

70. **B** Choice (B) is a correct definition of a riparian zone. It is an ecotone between a river or stream and land. Riparian zones act as buffer areas, absorbing excess water and pollutants that travel from the land to the water. Choice (A) defines rangeland; (C) is the role of a tree line; (D) is the understory of a forest.

71. **C** This is the only answer that is a disadvantage. Choices (A), (B), and (D) are all advantages of fish farming. When a fish (or any animal) population is dense, parasites and diseases can flourish.

72. **A** There is a little saying that might help you remember which form of nitrogen is utilized by plants. A mnemonic is: "The plants ate the nitrate." Notice that nitrate has the same last three letters as *ate*. Choices (B) and (C) are parts of the nitrogen cycle, but are not absorbed by plants. Choice (D) is not a nitrogen-containing compound.

73. **D** Thermal inversions occur when a warm layer of air prevents polluted air from rising up over a city. Normally, cool upper air allows the warm air to rise, and the pollutants are dispersed along with the warm air. When an inversion occurs, the upper air is warmer and the polluted air cannot rise. This traps the pollution close to Earth's surface. Choices (A) and (C) both deal with water; thermal inversions are atmospheric phenomena. Choice (B) is the pattern in a normal atmosphere.

74. **D** Choice (D) is the correct answer, based on the principles of an environmentally sustainable economy. This type of economy is defined as a low-waste society that uses conservation, recycling, and reduction to keep its use of nonrenewable resources at sustainable levels. In a sustainable society the focus is on preserving that society for future generations.

75. **C** Chlorofluorocarbons (CFCs) contain the element chlorine, which catalyzes the breakdown of ozone in the stratosphere. The energy for this reaction comes from the sun's ultraviolet radiation. It is estimated that one atom of chlorine will break down 100,000 molecules of ozone during the time it remains in the atmosphere. Choice (A) is a normal component of the atmosphere; (B) is a greenhouse gas; (D) causes acid deposition.

76. **D**   The abyssal zone is the deepest region of the ocean and the zone with the highest levels of nutrients due to decaying plant and animal matter that sinks down from the zones above. Choice (D) is correct.

77. **C**   The ice cap that lies over much of Greenland is more than two miles thick and has trapped gases from the atmosphere from hundreds of thousands of years ago. Other samples—(B) and (D)—do not list an actual sample; methane is produced in the oil-making process, among other sources, so (A) is wrong as well.

78. **D**   Bacteria are vital for the nitrogen cycle. They convert nitrogen from one form to another, and eventually into nitrates, which plants can use. Choices (A), (B), and (C) are not dependent on bacteria or fungi in the soil.

79. **D**   By definition, trace elements are needed in small amounts. Choices (B) and (C) are typical components of fertilizers and are needed in large quantities; (A) is supplied by the atmosphere in large quantities.

80. **C**   Choice (C) states a fundamental tenet of the environmental justice movement. Choice (A), the Earth stewardship view, is a system of beliefs that includes the belief that people can learn to live in natural harmony with the planet. Those who have the viewpoint in (B) believe that humans are the best planet managers and that our technology and understanding of systems will help us make the correct decision; and sustainability, (D), is the point of view that all people must live in ways that allow Earth to be sustainable.

Section II—Free-Response Questions

Question 1
Question 1 refers to the editorial on page 351.

(a) **Explain** how the beetle might have become resistant to NOGrub. Assume that NOGrub had been applied to a population of beetles in another county.

In a sexually reproducing population, there is natural variation in the genetic makeup of the population. There are large numbers of beetles, so some of them have a genetic makeup that allows for their survival in an environment with NOGrub in it. When the pesticide is applied, the susceptible beetles die and the resistant ones survive. The survivors reproduce in a habitat of lessened competition; therefore, their genes are passed along.
(1 point maximum—for the idea of natural selection/selective survival, and growth of resistant populations)

(b) If the county agents do not have information on which pesticide is most effective against the beetle, the county plans to investigate by trying several pesticides on controlled sections of beetle-infested crop.

i. **Identify** the independent and dependent variables in such a trial.

The independent variable in the trial is which pesticide is used. The dependent variable is its effect on the beetles: for example, what percentage of beetles die.

ii. **Describe** an effective control group and the experimental groups for this experiment.

A control group for the experiment would be a section of beetle-infested crop on which no pesticide was used. The experimental groups would be separate sections of beetle-infested crop, each treated with one of the pesticides under consideration.

iii. **Identify** TWO environmental factors that must be controlled to keep the experiment from producing skewed results.

Environmental factors:

- Density and health of beetles' food source (the crop): the control and experimental groups should be similar in terms of what the beetles have to eat.

- Presence of predators: the control and experimental groups should not differ in terms of likelihood that beetles will be preyed upon.

- Disturbance: crop sections should not be disturbed, or steps should be taken to ensure the amount of disturbance is equal between groups.

- Irrigation: all crop sections should receive the same amount of water.

- Environmental conditions: the sections should be similar in terms of shade, shelter, and landscape features.

(4 points maximum—1 for correct identification of the independent and dependent variables, 1 for correct description of the control and experimental groups, and 1 for each factor to control for—maximum of 2)

(c) **Identify** TWO negative impacts of using chemical pesticides on the surrounding ecosystem.

Negative impacts of using chemical pesticides on the surrounding ecosystem include the following:

- Pesticides can migrate to other habitats, thus harming other organisms.

- Pesticides can kill beneficial insects, which act as predators of destructive insects.

- Pesticides can combine with other chemicals and create other toxic chemicals.

- Pesticides can be accidentally ingested by humans and cause illness.
  (2 points maximum—1 for each correct impact)

(d) One strategy for combating pests is Integrated Pest Management (IPM).

    i. **Describe** IPM.

- IPM employs a combination of biological, chemical, and physical (any two of the three) methods for pest control.

- Goal of IPM is to curtail (or eliminate) pesticide use.

- IPM generally does not eradicate the pest population, but brings it to a tolerable level.

    (1 point for showing understanding of one of the listed aspects of IPM)

    ii. **Identify** one benefit and one difficulty in using IPM to control this outbreak.

Benefits of Using Integrated Pest Management	Difficulties in Using Integrated Pest Management
Lowers costs for pesticides	Needs expert knowledge of insect life cycle
Reduces amounts of pesticides in environment	Takes time to implement management practices
Several weaknesses of insects can be exploited	Initial costs might be higher
Reduces use of fertilizers	Governments subsidize pesticide use
Improves crop yields	Training takes a long time
Lowers genetic resistance issues	Methods for one area might not apply in other areas

    (2 points maximum—1 for a correct benefit, and 1 for a correct difficulty)

Question 2

The map shows two cities: City *X* and City *Z*, separated by several kilometers. Students from a high school in between the two cities studied soil pH values at the sites labeled A through E on the map. The results of the pH study are given in the following table:

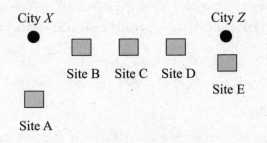

Site	pH value
A	6.2
B	5.6
C	5.0
D	4.5
E	4.3

(a) Refer to the table and diagram above to answer the following questions.

 i. **Describe** one point source for the pollution that caused the change in the soil's pH as shown.

 ii. In the description, **identify** a fuel that could create the pollution.

Point sources could include the following:

• Electricity generation plants

• Industry sites

• Cars or trucks

• Homes

The fuels involved would be coal or oil. (Natural gas and nuclear would not be accepted as correct.)

(2 points maximum—1 for the point source and 1 for the fuel)

(b) Assume that the fuel identified in (a)(i) is the source of the pollution.

 i. **Identify** one primary and one secondary pollutant that can cause the change in the soil's pH.

 ii. **Describe** the process that causes the change in the pH.

Primary pollutants are either SO_x or NO_x (saying just "sulfur" or "nitrogen" is not correct, as they do not directly produce the acid reaction). Also, suspended particulate matter can carry acidifying chemicals.

As the fuel is burned, SO_x or NO_x are produced; these enter the atmosphere and react with water vapor to create sulfuric acid or nitric acid. Farther downwind, these chemicals fall as wet or dry deposition and acidify the soil.

(4 points maximum—1 for identifying the primary pollutant and 3 for describing the process of acid formation and deposition)

(c) **Propose** one possible method to reduce the air pollutants that are causing the pH change.

Methods of reducing the air pollutants that are causing the pH change include:

- Reducing factory emissions of SO_x or NO_x by using scrubbers to remove chemicals from smoke

- Using cleaner-burning fuels (natural gas, low-sulfur coal, etc.)

- Reducing demand for electricity, which lowers production

- Adding catalytic converters to lower pollution from cars and trucks

   (2 points maximum—1 for the method and 1 for the correct description)

(d) **Identify** and **describe** one provision of the Clean Air Act of 1990 that could be used to control and reduce the emissions.

Provisions of the Clean Air Act of 1990 that could be used to control and reduce the emissions include:

- National Ambient Air Quality standards—a set of maximum permissible levels for pollutants

- Emissions trading policy—each year, certain factories are given "rights" or "credits" to release set amounts of pollutants; these credits can be bought, sold, or traded to reduce a company's liability under the CAA

- National emission standards for hundreds of toxic pollutants (e.g., mercury)

   (2 points maximum—1 for the policy and 1 for the correct description of the policy)

Question 3

According to the United States Energy Information Administration, the consumption of natural gas by the United States increases at 8 percent per year. The U.S. receives its supplies from a variety of international and domestic locations. Natural gas is used in the home, for industry, and for power generation.

(a) **Calculate** the approximate number of years it would take to double the consumption of natural gas in the United States. **Show** all your work.

Apply the rule of 70 to calculate the doubling time. The formula is $\frac{70}{8}$ = 8.75. In other words, it would take about nine years to double the consumption. (Either an exact or rounded figure would be considered correct.)

(2 points maximum—1 point for the correct setup, 1 point for the answer)

(b) **Identify** one method by which natural gas is recovered and transported.

Natural gas is recovered through a series of pipes and valves fitted over a drilled hole. These are often associated with oil wells, coal beds, or natural gas pockets. The gas can then be transmitted through piping or cooled and compressed into Liquefied Natural Gas (LNG) and transported by specially fitted trucks, trains, or ships. Some natural gas is collected as a product of the anaerobic decay of organic material in landfills.

(2 points maximum—1 for a correct description of a collection method, 1 point for a correct transport method)

(c) **Describe** two benefits to the environment that would occur if the United States switched from coal to natural gas-fired electric power generation.

The environment would benefit from the following:

- Less habitat destruction from the mining of coal

- Less acid mine drainage due to fewer mines being dug

- Less aesthetic damage done to the environment because fewer mines are dug

- Reduced SO_2 emissions, which will reduce the levels of acid deposition

- Reduced CO_2 emissions, which will reduce the levels of climate-changing gases

- Reduced soot emissions, which will reduce the formation of industrial (gray) smog

- Less fly ash produced

- Less mercury or radioactive materials released by the burning of coal

- Longer boiler life because gas produces fewer products that harm the boilers

- Less cost in building new generation plants, because gas combustion produces fewer dangerous by-products

  (2 points maximum—1 point for benefit, 1 point for description. Remember, you receive credit only for your first two answers.)

(d) Some people advocate increasing the use of coal versus natural gas for the production of electricity. Give one argument that the proponents of coal might use to **justify** their position.

Coal-use proponents could make the following arguments:

- There is a larger supply of domestic coal than natural gas, so the United States will be less dependent on foreign energy sources.

- Coal is safer to use. In an accident, natural gas can explode, harming people or facilities.

- The infrastructure for transporting coal already is in place. We would have to do little to increase the production and transportation of coal.

- It might be cheaper to refit existing coal plants with scrubbers to clean up toxic emissions than to build a new, cleaner-burning gas plant.

(2 points maximum—1 point for argument, 1 point for description)

(e) Others advocate for non-hydrocarbon fuel alternatives.

- i. Identify one non-hydrocarbon fuel alternative.

- ii. Describe ONE drawback of the alternative identified in (e)(i).

Possible answers:

Energy Source	Drawbacks
Nuclear	• Safety concerns (such as potential for radiation leaks) • Waste disposal concerns (such as radioactive or toxic waste disposal) • Raw materials storage (safe storage of radioactive materials) • Thermal pollution (very hot reactor cooling fluids)
Hydroelectric	• Water loss via evaporation at dam site • Dam may hinder fish and other aquatic migration • Silting • Habitat destruction/alteration (via flooding upstream, or alteration of water quality downstream) • Potential for displacing populations or wildlife living in dam's flood zone • Potential for seismic activity below reservoir • Potential for plants killed by flooding to produce methane gas (a greenhouse gas)
Solar	• Initial costs are high • Material for solar panels uses hydrocarbon fuels • Solar panels lack aesthetic appeal • Disposal of toxic chemicals (from rechargeable batteries)
Wind	• Initial costs are high • Turbines lack aesthetic appeal • Sound and vibration produced are annoying • Turbines may affect bird migration patterns
Geothermal	• Geographical limitations (There are only certain places in the world where geothermal energy can be harvested.) • Thermal pollution • Toxic chemicals (potential for releasing H_2S into environment)

(2 points maximum—1 point for listing energy source, 1 point for listing related drawback)

HOW TO SCORE PRACTICE TEST 2

Section I: Multiple-Choice

_____ × 1.125 = _____
Number Correct   Weighted
(out of 80)   Section I Score
                              (Do not round)

Section II: Free-Response

Question 1   _____   × 2 =   _____
                (out of 10)   (Do not round)

Question 2   _____   × 2 =   _____
                (out of 10)   (Do not round)

Question 3   _____   × 2 =   _____
                (out of 10)   (Do not round)

AP Score Conversion Chart
Environmental Science

Composite Score Range	AP Score
107-150	5
90-106	4
73-89	3
56-72	2
0-55	1

Sum = _____
          Weighted
       Section II Score
       (Do not round)

Composite Score

_____ + _____ = _____
   Weighted   Weighted   Composite Score
Section I Score   Section II Score   (Round to nearest
                                        whole number)

Note: This score sheet is to help you estimate your approximate score for the official exam, not your actual score.

Practice Test 3

AP® Environmental Science Exam

SECTION I: Multiple-Choice Questions

DO NOT OPEN THIS BOOKLET UNTIL YOU ARE TOLD TO DO SO.

Instructions

Section I of this examination contains 80 multiple-choice questions. Fill in only the ovals for numbers 1 through 80 on your answer sheet.

Indicate all of your answers to the multiple-choice questions on the answer sheet. No credit will be given for anything written in this exam booklet, but you may use the booklet for notes or scratch work. After you have decided which of the suggested answers is best, completely fill in the corresponding oval on the answer sheet. Give only one answer to each question. If you change an answer, be sure that the previous mark is erased completely. Here is a sample question and answer.

Sample Question   Sample Answer

Chicago is a   (A) ● (C) (D)

    (A) state
    (B) city
    (C) country
    (D) continent

Use your time effectively, working as quickly as you can without losing accuracy. Do not spend too much time on any one question. Go on to other questions and come back to the ones you have not answered if you have time. It is not expected that everyone will know the answers to all the multiple-choice questions.

About Guessing

Many candidates wonder whether or not to guess the answers to questions about which they are not certain. Multiple-choice scores are based on the number of questions answered correctly. Points are not deducted for incorrect answers, and no points are awarded for unanswered questions. Because points are not deducted for incorrect answers, you are encouraged to answer all multiple-choice questions. On any questions you do not know the answer to, you should eliminate as many choices as you can, and then select the best answer among the remaining choices.

At a Glance

Total Time
1 hour and 30 minutes
Number of Questions
80
Percent of Total Grade
60%
Writing Instrument
Pencil required

GO ON TO THE NEXT PAGE.

ENVIRONMENTAL SCIENCE
Section I
Time—90 minutes
80 Questions

Directions: Each of the questions or incomplete statements below is followed by four suggested answers or completions. Select the one that is best in each case and then fill in the corresponding oval on the answer sheet.

Questions 1 and 2 refer to the following graph.

A survivorship curve graph showing populations of humans, chickens, and a tree species is shown below.

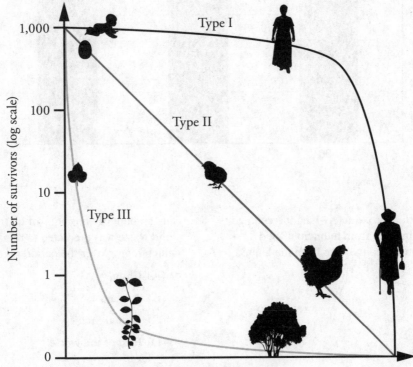

Survivorship indicates the probability that a given organism will live to a certain age.

1 The survivorship curve for the tree species above is an example of a Type III curve. Which of the following species would show a similar curve?

   (A) Moose

   (B) Songbirds

   (C) Sea turtles

   (D) Oysters

2. Which of the following is an accurate description of a Type II curve?

   (A) High survival probability in early and middle life, followed by a rapid decline in later life

   (B) Roughly constant survival probability regardless of age

   (C) High mortality in early life, followed by relatively high survival probability for those surviving the bottleneck

   (D) Relatively high survival probability in early and late life, with a bottleneck of high mortality in the middle of the lifespan

GO ON TO THE NEXT PAGE.

Questions 3 and 4 refer to the following chart.

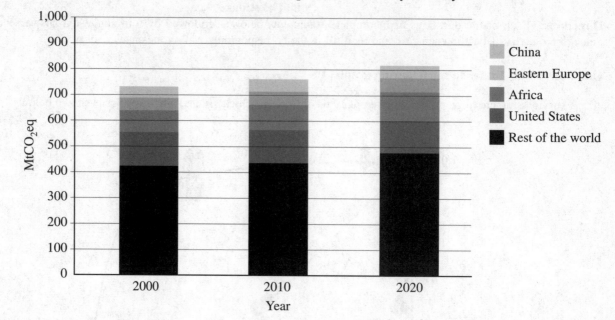

Methane Emission Predictions from Municipal Solid Waste by Country: 2000–2020

3. According to the chart, approximately what percentage of global methane emissions from municipal solid waste in 2010 were predicted to come from the United States?

   (A) 7%

   (B) 17%

   (C) 57%

   (D) 73%

4. The amount of emissions of methane from municipal solid waste was predicted to remain the most stable in which region over the period given?

   (A) China

   (B) Africa

   (C) United States

   (D) Rest of the world

GO ON TO THE NEXT PAGE.

Questions 5 and 6 refer to the following map.

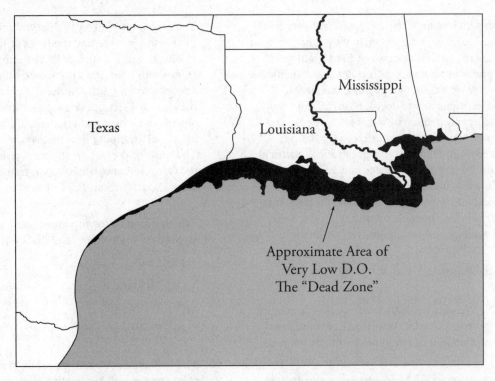

Approximate Area of
Very Low D.O.
The "Dead Zone"

5. The map shows an area of eutrophication ("dead zone") in the Gulf of Mexico. Which of the following accurately describes the chain of events leading to its formation?

   (A) Mississippi River carries fertilizer and sewage into the Gulf—phytoplankton and zooplankton experience major population decrease—detritus from dead phytoplankton and zooplankton feeds bacteria at sea floor—bacteria metabolize available oxygen while decomposing this food source

   (B) Mississippi River carries fertilizer and sewage into the Gulf—phytoplankton and zooplankton experience population explosion—less detritus from dead phytoplankton and zooplankton to feed bacteria at sea floor—bacteria produce less oxygen while decomposing this food source

   (C) Mississippi River carries fertilizer and sewage into the Gulf—phytoplankton and zooplankton experience major population decrease—less detritus from dead phytoplankton and zooplankton to feed bacteria at sea floor—bacteria produce less oxygen while decomposing this food source

   (D) Mississippi River carries fertilizer and sewage into the Gulf—phytoplankton and zooplankton experience population explosion—detritus from dead phytoplankton and zooplankton feeds bacteria at sea floor—bacteria metabolize available oxygen while decomposing this food source

6. Which of the following is NOT an effect of the presence of this "dead zone"?

   (A) Collapse of the shrimp and shellfish industries in the area

   (B) Nothing that depends on oxygen can grow in the zone from May to September

   (C) Reproductive problems in fish involving decreased size of reproductive organs, low egg counts, and lack of spawning

   (D) Mass migration of fish to other areas of the Gulf

GO ON TO THE NEXT PAGE.

Questions 7 and 8 refer to the following information.

Ecosystem services are benefits that humans receive from the ecosystems in nature when they function properly. There are four categories: **provisioning services**: providing humans with water, food, medicinal resources, raw materials, energy, and ornaments; **regulating services**: waste decomposition and detoxification, purification of water and air, pest and disease control and regulation of prey populations through predation, and carbon sequestration; **cultural services**: use of nature for science and education, therapeutic and recreational uses, and spiritual and cultural uses; and **supporting services** (the ones that make other services possible): primary production, nutrient recycling, soil formation, and pollination.

7. Which of the following is an example of a regulating service?

   (A) In New York City, authorities worked to restore the polluted Catskill Watershed, restoring soil absorption and filtration of chemicals via soil and its microbiota, improving water quality.

   (B) Restoration of wild bee populations to an area relieves large-scale farms from having to import non-native honeybees to pollinate crops.

   (C) In the Roman Empire, water-powered mills produced flour from grain, and were also utilized for power to saw timber and cut stone.

   (D) In modern ecotherapy, patients are encouraged to spend time in nature to enhance their physical, mental, and emotional well-being.

8. All of the following are negative consequences to humans of the disruption of ecosystem services EXCEPT

   (A) lower crop yields due to depletion of soil nutrients and microbiota

   (B) outbreaks of diseases carried by insect vectors due to habitat loss in predator species

   (C) higher costs of mining due to depletion of mineral resources

   (D) greater availability of habitats in different stages of ecological succession for scientific study due to human development

Questions 9–11 refer to the following information.

The greenhouse effect is a natural process that warms the Earth's surface and the layer of the atmosphere closest to Earth's surface by reducing radiative loss from Earth's surface to space. In other words, the sun's energy warms Earth's surface, and some is radiated back toward space; certain gases in the troposphere absorb some heat and reradiate it downward again, effectively trapping heat that would otherwise be lost. The most important greenhouse gases are carbon dioxide, methane, water vapor, nitrous oxide, and chlorofluorocarbons (CFCs).

9. Which of the following greenhouse gases is NOT a significant contributor to global climate change?

   (A) CFCs

   (B) Methane

   (C) Water vapor

   (D) Carbon dioxide

10. The enhanced, or anthropogenic, greenhouse effect refers to a strengthening of the greenhouse effect due to human activities that are contributing to global climate change. Which of the following is NOT true of this anthropogenic effect?

    (A) Depletion of the stratospheric ozone layer is a contributing factor.

    (B) It is mainly due to increases in emissions of greenhouse gases and pollutants, as well as to changes in land use.

    (C) Some of its effects include sea level rise, changes in ocean properties, and an increase in extreme weather events.

    (D) Possible effects on the biosphere include changes in biomes, mass migrations and extinctions, spread of disease vectors, and changes in population dynamics.

11. Global warming potential (GWP) is a measure of how much heat a greenhouse gas can trap in the atmosphere, relative to the reference point of carbon dioxide. Which of the following greenhouse gases has the highest GWP?

    (A) Chlorofluorocarbons

    (B) Nitrous oxide

    (C) Carbon dioxide

    (D) Methane

GO ON TO THE NEXT PAGE.

12. Exposure to which of the following noises would cause the most damage to a person's hearing?

    (A) A vacuum cleaner

    (B) A chain saw

    (C) A factory

    (D) The firing of a rifle

13. The phrase that best defines population density is

    (A) the number of individuals in a certain geographic area

    (B) the rate at which a population increases

    (C) the maximum number of individuals that a habitat can sustain

    (D) the time it takes for a population to increase to carrying capacity

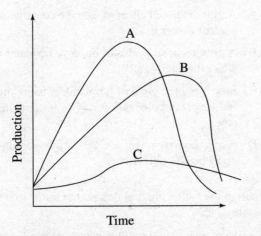

14. The graph above represents possible depletion curves of a nonrenewable resource. Curve C best describes the resource as it

    (A) has just been discovered and no technology exists to use the resource

    (B) is quickly used up and is not recycled

    (C) is newly discovered and in high demand

    (D) has expanding reserves and consumption is reduced

15. The shrinking of the Aral Sea and the ecological disaster that followed was mainly caused by

    (A) the diversion of the sea's two feeder rivers for agricultural use

    (B) withdrawing groundwater from the area

    (C) a major earthquake that hit the region

    (D) the massive use of pesticides

16. Which of the following is a negative result of overfishing a particular species of edible fish?

    I.   Loss of so many fish that there is no longer a breeding stock
    II. The removal of non-target species
    III. The reduction of other species that rely on the edible species as food

    (A) II only

    (B) I and II only

    (C) II and III only

    (D) I, II, and III

17. Which type of irrigation results in the greatest amount of water lost to evaporation and runoff?

    (A) Flood irrigation

    (B) Drip irrigation

    (C) Furrow irrigation

    (D) Spray irrigation

GO ON TO THE NEXT PAGE.

Questions 18–21 refer to the following graph.

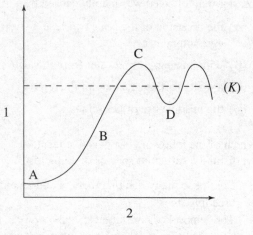

18. The population is growing at its highest rate at which letter?

    (A) A

    (B) D

    (C) B

    (D) C

19. Which of the following phrases best describes line *K* in the graph above?

    (A) The rate of population growth

    (B) The carrying capacity of the environment

    (C) The time it takes for the population to double in size

    (D) The birth rate of the population

20. The best label for the axis labeled 2 in the graph above is

    (A) time

    (B) population number

    (C) logistic growth rate

    (D) environmental resistance

21. Which of the following statements is true concerning the events occurring at point D in the graph above?

    (A) Environmental resistance is high.

    (B) Environmental resistance is low.

    (C) The population will continue to fall.

    (D) The population is above its carrying capacity.

22. The long-term storage of phosphorus and sulfur is in which of the following forms?

    (A) Rocks

    (B) Water

    (C) Plants

    (D) Atmosphere

23. The movement of sections of the Earth's lithosphere is known as

    (A) mass depletion

    (B) plate tectonics

    (C) background extinction

    (D) migration

24. Which of the following best defines the Green Revolution?

    (A) An international effort to stop the construction of nuclear power plants

    (B) A group whose goal is to improve how nations affect the environment

    (C) Increasing the yield of farmland by using more fertilizer, better irrigation, and faster growing crops

    (D) A method that makes a viable soil conditioner by using household waste

25. Which of the following best defines the term "infant mortality rate"?

    (A) How many children live in each square hectare

    (B) The number of births in a population

    (C) The number of infant deaths per 1,000 people aged zero to one

    (D) The difference between the birth rate and the death rate in a population

26. The release of chemicals from underground storage tanks is most likely to pollute which of the following?

    (A) A landfill

    (B) The atmosphere

    (C) The ecotone

    (D) Aquifers

GO ON TO THE NEXT PAGE.

27. The United States Congress failed to ratify which of the following international agreements that is designed to control the release of carbon dioxide?

    (A) CITES agreement

    (B) Kyoto Protocol

    (C) Montreal Protocol

    (D) Clean Air Act

28. Which of the following is the most sustainable way to ensure sufficient energy for the future?

    (A) Find more fossil fuels

    (B) Develop more effective solar power generators

    (C) Build more nuclear reactors

    (D) Reduce waste and inefficiency in electricity use and transmission

29. Which of the following best describes the goals of the CAFE standards?

    (A) Reduce pollution by coal-fired power plants

    (B) Improve the quality of air around cities

    (C) Protect certain endangered species

    (D) Improve the fuel efficiency of automobiles in the United States

30. Which of the following best describes the use of DDT?

    (A) It supplies needed nitrogen to plants.

    (B) It kills weeds and unwanted plants.

    (C) It decreases the amount of pollution from car exhaust.

    (D) It is an insecticide.

31. Which of the following best explains why pesticides become ineffective over time?

    (A) Pesticide manufacturers learn how to cut corners and sell an inferior product.

    (B) Pests acquire resistance to the pesticide during exposure to the pesticide.

    (C) Pests resistant to the pesticide survived and passed on their resistance to the next generation.

    (D) Pests acquired resistance to the pesticide and passed on their resistance to the next generation.

32. For every ton of plastic recycled, 0.3 fewer tons of plastic is required to be manufactured. If 150 million tons of plastic are produced and 50 million tons of plastic are recycled each year in North America, how many tons of plastic need to be newly manufactured each year?

    (A) 1.5 million tons

    (B) 15 million tons

    (C) 115 million tons

    (D) 135 million tons

33. All of the following statements about the degradation of soil by climate change are true EXCEPT

    (A) increased flooding in coastal areas leads to soil desalinization

    (B) increased global temperature produces more deserts

    (C) increased evaporation of water leads to soil salinization

    (D) increased global temperature leads to increased loss of organic material in soil

Freshwater Usage in the United States, 2000

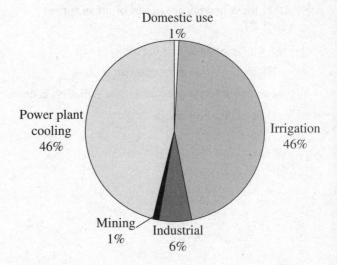

34. According to the diagram above, cooking, showering, and using toilets accounted for approximately what percentage of total water use?

    (A) 92 percent

    (B) 46 percent

    (C) 6 percent

    (D) 1 percent

GO ON TO THE NEXT PAGE.

35. One result of increased troposphere temperatures that is observed today is

    (A) an increase in skin cancers in people

    (B) an increase of 10 to 20 cm in the average global sea level

    (C) more radon seepage into people's homes

    (D) a deeper permafrost in Arctic regions

36. All of the following are major components of air pollution EXCEPT

    (A) sulfur dioxide

    (B) lead

    (C) arsenic

    (D) ozone

37. Which of the following is the root cause of habitat loss, especially in less developed nations?

    (A) Road building

    (B) Poverty

    (C) Conversion of forest to farmland

    (D) Capturing exotic animals for resale

38. All of the following are causes of urban sprawl EXCEPT

    (A) increased number of roads

    (B) higher crime rates in urban areas

    (C) higher levels of air and noise pollution in cities

    (D) increased oil prices

39. The motion of tectonic plates accounts for most of Earth's

    (A) CO_2 emissions

    (B) river formation

    (C) changes of season

    (D) volcanic activity

40. "K" and "r" are used to describe which of the following aspects of populations?

    (A) The place in a habitat where these organisms live

    (B) The number of males and females in the population

    (C) The reproductive tactics used by the populations

    (D) The time it takes a population to double

41. Convectional heating and cooling of the atmosphere transfers which of the following to other parts of the Earth?

    I. Heat
    II. Moisture
    III. Nutrients

    (A) I only

    (B) II only

    (C) I and II only

    (D) I, II, and III

GO ON TO THE NEXT PAGE.

Questions 42 and 43 refer to the following diagram.

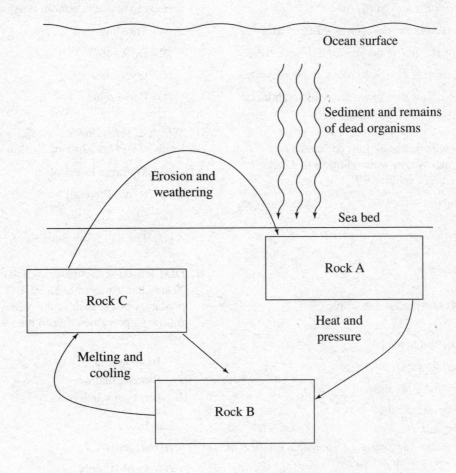

42. Which of the following gives the correct sequence in the rock cycle shown above?

    (A) Rock A—Metamorphic/Rock B—Igneous/ Rock C—Sedimentary

    (B) Rock A—Sedimentary/Rock B—Metamorphic/ Rock C—Igneous

    (C) Rock A—Igneous/Rock B—Sedimentary/ Rock C—Metamorphic

    (D) Rock A—Sedimentary/Rock B—Igneous/ Rock C—Metamorphic

43. Which of the following rock types would contain the greatest number of fossils?

    (A) Igneous rock only

    (B) Metamorphic rock only

    (C) Sedimentary rock only

    (D) Sedimentary and igneous rock

GO ON TO THE NEXT PAGE.

44. The North Atlantic Current provides which of the following for Europe and North America?

    (A) Fish to feed predators such as killer whales

    (B) Warm water that moderates land temperatures

    (C) Large amounts of CO_2 to promote photosynthesis

    (D) Mineral-rich waters to reduce depleted mineral reserves

45. Which of the following indoor air pollutants is composed of microscopic mineral fibers that can produce lung cancer in humans?

    (A) Nitrogen oxides

    (B) Asbestos

    (C) Radon

    (D) Formaldehyde

46. Which of the following is an example of commensalism?

    (A) Cheetahs and antelope

    (B) Bees and flowers

    (C) Humans and tapeworms

    (D) Barnacles and whales

47. Which two countries together are responsible for 40% of global greenhouse gas emissions?

    (A) United States and Great Britain

    (B) United States and China

    (C) China and India

    (D) Great Britain and Russia

48. Nuclear reactors use which of the following to absorb neutrons in the reactor core?

    (A) A steam condenser

    (B) Control rods

    (C) A heat exchanger

    (D) Fuel rods

49. Riparian areas are vital to the preservation of high-quality

    (A) mountain slopes

    (B) grazing land

    (C) rivers and streams

    (D) ocean beaches

50. Which of the following fishing techniques is most damaging to ocean bottom ecosystems?

    (A) Trawling

    (B) Drift nets

    (C) Long lines

    (D) Purse seine

51. Which of the following treaties is responsible for lower levels of CFC production worldwide?

    (A) Montreal Protocol

    (B) Kyoto Protocol

    (C) Clean Air Act

    (D) The Rio Earth Summit of 1972

52. DDT is an insecticide sprayed to control insects. Years after it was introduced, DDT was found in large predatory birds, such as the osprey. Which of the following processes caused the DDT to be found in the osprey?

    I.   Biomagnification
    II. Bioremediation
    III. Bioaccumulation

    (A) I only

    (B) III only

    (C) I and III only

    (D) I, II, and III

53. Doing which of the following could most cost effectively reduce acid rain and acid deposition?

    (A) Reducing the use and waste of electricity

    (B) Making taller smokestacks

    (C) Adding lime to acidified lakes

    (D) Moving power plants to desert areas

54. Which of the following forms of radiation is most harmful to humans?

    (A) Alpha

    (B) Gamma

    (C) Beta

    (D) Infrared

GO ON TO THE NEXT PAGE.

55. Which of the following ecosystems does NOT use solar energy as its ultimate energy source?

   (A) Pond

   (B) Deep-sea hydrothermal vent

   (C) Rain forest

   (D) Tundra

56. All of the following are true about CO_2 sequestering EXCEPT

   (A) it can be accomplished by pumping CO_2 into carbonated beverages

   (B) it can be accomplished by pumping CO_2 into crop lands

   (C) it can be accomplished by pumping CO_2 deep under the ocean floor

   (D) it can be accomplished by pumping CO_2 deep underground into dried-up oil wells

57. Which of the following is NOT true concerning invasive species?

   (A) They can outcompete native species in a habitat.

   (B) They are highly specialized and have narrow niches.

   (C) They alter the biodiversity of the area they are invading.

   (D) They are introduced into a habitat and are not native.

58. The Second Law of Thermodynamics is best exemplified by which of the following?

   (A) The amount of solar radiation going into an ecosystem is equal to the total amount of energy going out of that system.

   (B) The amount of carbon in the atmosphere has increased due to the combustion of fossil fuels.

   (C) As electricity is transmitted through wires, some of the power is lost to the environment as heat.

   (D) Wind-generated electricity has more power than electricity generated at a hydropower plant.

59. Salinization of soil can be caused by all of the following EXCEPT

   (A) flooding in coastal areas

   (B) rising temperatures

   (C) excessive irrigation

   (D) drip irrigation

60. Which of the following is true about early loss populations, such as fish?

   (A) The chances of an adult dying are about the same as a child dying.

   (B) The maturation process is slow.

   (C) The populations are close to the carrying capacity.

   (D) Many individuals die at an early age.

GO ON TO THE NEXT PAGE.

Questions 61 and 62 refer to the illustration of succession below.

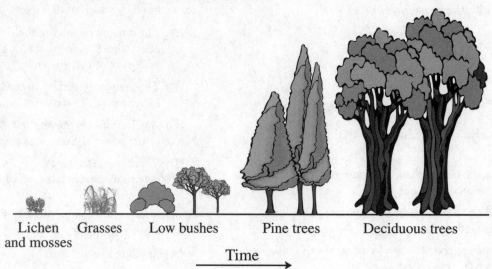

Lichen and mosses Grasses Low bushes Pine trees Deciduous trees

Time

61. A farmer stops farming a certain tract of land, and small bushes soon grow there. The land then progresses to the deciduous tree stage. This process is known as

    (A) pioneer succession

    (B) wetland succession

    (C) secondary succession

    (D) primary succession

62. According to the diagram, low species diversity and small-sized plants are characteristics of which stage of succession?

  I.   Late-stage succession

  II. Midstage succession

  III. Early-stage succession

    (A) I only

    (B) II only

    (C) III only

    (D) I and III only

GO ON TO THE NEXT PAGE.

63. Ozone depletion is occurring most rapidly in the Earth's polar regions because

    (A) the atmosphere is thicker at the poles, so ozone destruction is easier to observe

    (B) large amounts of chlorofluorocarbons (CFCs) can accumulate on ice crystals formed in the cold atmosphere

    (C) the upper atmosphere winds form a pattern of high- and low-pressure systems that can cause the destruction of ozone

    (D) the solar UV radiation is stronger at the poles, promoting the breakdown of ozone

64. Smaller forest fires are beneficial to forests for all of the following reasons EXCEPT

    (A) combustion of dried leaves or needles, which reduces the threat of large fires

    (B) burning the crowns of trees

    (C) germinating seeds of certain plant species

    (D) making burned matter available as a nutrient

65. What are the negative impacts of dams on ecosystems?

    I.   Loss of silt in the river downstream from the dam
    II. Generation of low pollution electricity
    III. Loss of terrestrial biodiversity in areas surrounding the dam

    (A) I only

    (B) III only

    (C) I and III only

    (D) I, II, and III

66. Which of the following best illustrates the process of evolution?

    (A) A parasite population becomes resistant to a drug

    (B) Rabbits can have brown fur in summer and white fur in winter

    (C) Frogs burrow deep into the mud during winter

    (D) A baby is born and has a different color hair than his or her parents

67. Which of the following is a renewable energy source?

    (A) Crude oil

    (B) Coal

    (C) Natural gas

    (D) Hydrogen cells

68. The energy necessary to produce stratospheric ozone comes from which of the following?

    (A) Sunlight

    (B) Radioactive decay

    (C) Magma

    (D) Wind

69. Which of the following chemicals can cause lung irritation in the troposphere but is very helpful to humans in the stratosphere?

    (A) O_2

    (B) O_3

    (C) Chlorofluorocarbons

    (D) H_2SO_4

70. Which of the following is true of primary and secondary pollutants?

    (A) Primary pollutants rise up the smokestack before secondary pollutants are formed.

    (B) Primary pollutants are formed from secondary pollutants interacting in the water.

    (C) Secondary pollutants are formed from primary pollutants interacting in the atmosphere.

    (D) Secondary pollutants are directly created by the burning of coal and primary pollutants from the burning of oil.

71. A sample of radioactive iodine-131 is found to have an activity level of 4×10^{-6} curies and a half-life of 8 days. How much time must pass before the activity level of the radioactive waste drops to 2.5×10^{-5} curies?

    (A) 4 days

    (B) 8 days

    (C) 16 days

    (D) 32 days

GO ON TO THE NEXT PAGE.

72. Coal, oil, and natural gas were all formed as a result of

    (A) the decay of organic matter

    (B) the movement of magma in volcanoes

    (C) sedimentary rock turning into metamorphic rock

    (D) the radioactive decay occurring inside Earth

73. All of the following are negative impacts of food production EXCEPT

    (A) increased erosion

    (B) air pollution from fossil fuels

    (C) bioaccumulation of pesticides

    (D) lower death rates

74. Soils found in mid-latitude grasslands would be most accurately described as having

    (A) a high acid content with little organic matter

    (B) a deep layer of humus and decayed plant material

    (C) a layer of permafrost right below the O-horizon

    (D) a high content of iron oxides and very little moisture

75. All of the following are useful methods for reducing domestic water use EXCEPT

    (A) using low-flow shower heads

    (B) using low-flush-volume toilets

    (C) fixing leaks as soon as they start

    (D) lowering the temperature of the water heater

76. Biodiversity is a direct result of which of the following?

    (A) Deforestation

    (B) Respiration

    (C) Erosion

    (D) Evolution

77. Students studying a river found high levels of fecal coliform bacteria. They concluded that

    (A) this water is fit to swim in

    (B) a nearby treatment plant added chlorine to the waste water

    (C) they can safely drink the water

    (D) untreated animal waste was put into the water

78. During an El Niño-Southern Oscillation, weather events change in which of the following areas?

    (A) The Pacific and Indian Oceans

    (B) The Atlantic and Indian Oceans

    (C) The Arctic Sea

    (D) The Indian and Antarctic Oceans

79. Which of the following pairs correctly matches the source of gray water with its most frequent use in the home?

    (A) Dishwasher and sink water used to flush toilets

    (B) Flushed toilet water used to irrigate garden plants

    (C) Dishwasher and sink water used to irrigate garden plants

    (D) Water collected from rainfall used to flush toilets

80. The Clean Water Act did all of the following EXCEPT

    (A) set water quality standards for all contaminants in surface waters

    (B) make it unlawful for any person to discharge any pollutant from a point source into navigable waters

    (C) demand that an environmental impact statement be prepared for any major development

    (D) fund the construction of sewage treatment plants

END OF SECTION I

ENVIRONMENTAL SCIENCE
Section II
Time—70 minutes
3 Questions

Directions: Answer all three questions, which are weighted equally; the suggested time is about 23 minutes for answering each question. Where calculations are required, clearly show how you arrived at your answer. Where explanation or discussion is required, support your answers with relevant information and/or specific examples.

1.   A class wished to determine the LD_{50} of a particular herbicide, Chemical X, on seedlings of the most common type of pest plant at a local tree farm. Using standard laboratory apparatus and glassware, they accurately made the following dilutions: 1.0M, 10^{-1}M, 10^{-2}M, 10^{-3}M, 10^{-4}M, and 10^{-5}M. They grew the seedlings under standard conditions, varying only the concentrations of Chemical X. Finally, they determined the percentage of seedlings that germinated at each concentration.

   (a) Use the description of the experiment above to answer the following questions.
   i.   **Identify** a reasonable hypothesis that this experiment could test.
   ii. **Describe** the experimental control group and offer ONE method for performing repeated trials.

   (b) **Identify** a set of hypothetical results.
   i.   Using the axes below, graph your results. Properly **identify** and label the axes and provide a title for the graph.
   ii. **Calculate** and identify the LD_{50} concentration on your graph. **Show** your work.

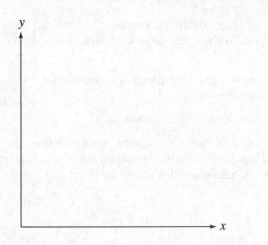

   (c) **Describe** ONE positive outcome and TWO negative outcomes of using herbicides in the environment.

   (d) Given the possible negative effects of herbicide use, **propose** ONE alternative strategy the tree farm could employ to control this pest plant.

GO ON TO THE NEXT PAGE.

2. The diagram below illustrates the demographic transition model of the relationship between economic status and population.

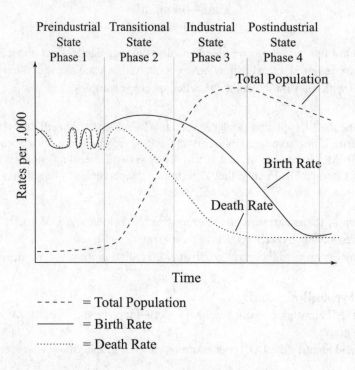

(a) In Phases 2 and 3, there is a large difference between the birth rate and the death rate. **Describe** the effects on the overall population as a result of this difference. **Explain** why the population doubling time during these phases is short.

(b) Choose ONE of the four phases and **identify** an economic factor that would account for the differences between birth rate and death rate.

(c) **Describe** ONE biological method of birth control.

(d) Population experts have reported that in some developing countries, the population is experiencing a reverse transition from Phase 2 to Phase 1. **Describe** what would happen to a country's population during such a reverse transition and **make a claim** about one event that could cause this reverse transition.

GO ON TO THE NEXT PAGE.

3. Under certain conditions, an internal combustion car engine produces approximately 3 grams of NO_x per kilometer driven. In Country C, there are 300 million cars, and each car is driven only 20,000 km per year.

   (a) **Calculate** the number of metric tons of NO_x produced by the cars in Country C under these conditions in one year. 1 metric ton = 1,000,000 g.

   (b) **Identify** a secondary pollutant that is derived from the NO_x produced by Country C, **explain** how it is produced, and **explain** how that pollutant travels to adjacent countries.

   (c) **Describe** ONE abiotic and ONE biotic impact that the NO_x pollution will have on any countries adjacent to Country C.

   (d) **Propose** and **describe** one method that Country C could employ to reduce the amount of emitted NO_x.

STOP

END OF EXAM
___

Practice Test 3:
Answers and
Explanations

PRACTICE TEST 3 ANSWER KEY

1.	D	21.	B	41.	C	61.	C
2.	B	22.	A	42.	B	62.	C
3.	B	23.	B	43.	C	63.	B
4.	A	24.	C	44.	B	64.	B
5.	D	25.	C	45.	B	65.	C
6.	D	26.	D	46.	D	66.	A
7.	A	27.	B	47.	B	67.	D
8.	D	28.	D	48.	B	68.	A
9.	C	29.	D	49.	C	69.	B
10.	A	30.	D	50.	A	70.	C
11.	A	31.	C	51.	A	71.	D
12.	D	32.	D	52.	C	72.	A
13.	A	33.	A	53.	A	73.	D
14.	D	34.	D	54.	B	74.	B
15.	A	35.	B	55.	B	75.	D
16.	D	36.	C	56.	A	76.	D
17.	C	37.	B	57.	B	77.	D
18.	C	38.	D	58.	C	78.	A
19.	B	39.	D	59.	D	79.	C
20.	A	40.	C	60.	D	80.	C

PRACTICE TEST 3 EXPLANATIONS

Section I—Multiple-Choice Questions

1. **D** Type III curves tend to describe *r*-selected species, which have high growth rates and produce many offspring, relatively few of which survive to adulthood. Oysters are a good example of an *r*-selected species, while moose are *K*-selected. Songbirds and sea turtles are likely to have Type II survivorship curves, showing them to be somewhere on the continuum between *r*- and *K*-selection.

2. **B** Type II curves are more or less diagonal lines, like that of the chickens in the graph. This shape represents a constant slope, or roughly equal mortality rates along the curve. Choice (A) describes Type I curves, (C) describes Type III curves, and (D) is not a pattern that's commonly seen among most species.

3. **B** On the bar chart, the middle column represents 2010. The portion of the bar that represents the United States' contribution is the second portion from the bottom. Use your answer sheet as a ruler to get approximate values for the top and bottom of this box on the *y*-axis: about 425 and about 550. Do the same for the column total: it's about 750. The units are not important since this is a percentage question. Subtract the top and bottom values for the United States portion to find the U.S. contribution: 550 − 425 = 125. To find the percentage of the total, divide this value by the total and multiply the result by 100: 125 ÷ 750 ≈ 0.167, or about 16.7%. Since the question asks for an approximation and the answer choices are whole numbers, round up to the nearest whole number. Choice (B), 17%, is correct.

4. **A** The sections of each column representing Africa, the United States, and the rest of the world each increase between 2000 and the 2020 predicted value. However, the section representing China remains about the same size.

5. **D** The dead zone was created because of the fertilizer and sewage that the Mississippi River carries into the Gulf. Since the water containing these pollutants is very nutrient-rich and warm, it creates a food source for phytoplankton and zooplankton in the surface waters there, creating a population explosion among these organisms. Since there is then more detritus from dead phytoplankton and zooplankton feeds, the bacteria at the sea floor have abundant food for decomposition, but they metabolize and use up the available oxygen while doing so, creating the hypoxic ("dead") zone. Choice (D) correctly captures this chain of events.

6. **D** The hypoxic zone stays in place from May to September, when cooler and wetter weather helps break it up: during this time, so little oxygen is available that most organisms can't grow there, (B). Another effect is reproductive problems for fish who develop in low-oxygen areas near the dead zone, (C). Both of these contribute to the decline of the fishing industries that depend on ocean life there, (A). While it is easy to imagine that fish would simply migrate, many are simply rendered unconscious and die quickly, and habitat loss kills others, (D).

7. **A** A regulating service is one that provides benefits from the regulation of some ecosystem process. Water purification, (A), is a good example. Pollination, (B), is a supporting service; energy generation, (C), is a provisioning service; ecotherapy, (D), is a therapeutic, and therefore cultural, service.

8. **D** All the choices are consequences of disruption to ecosystem services, but (D) is a positive, not a negative consequence: habitat disruption in this case is providing a cultural service (that of science and education). Lower crop yields and higher costs of mining are examples of negative consequences to provisioning services, while outbreaks of disease are an example of a negative consequence to regulating services.

9. **C** While water vapor, (C), is a greenhouse gas, it doesn't contribute significantly to global climate change because it has a short residence time in the atmosphere. All the other choices are important contributors to global climate change.

10. **A** The depletion of stratospheric ozone, while an important environmental issue that also has anthropogenic causes, is not directly linked to the greenhouse effect. All the other choices are true about human involvement in the greenhouse effect.

11. **A** Chlorofluorocarbons have the highest GWP, followed by nitrous oxide, and then methane. Carbon dioxide has a GWP value of 1, since it is used as a baseline.

12. **D** Noise is measured in decibels (dB). A vacuum cleaner has a loudness of about 70 dB, a chain saw has a loudness of about 100 dB, the average factory has a loudness of about 80 dB, and a rifle has a loudness of about 160 dB. The rifle would most likely cause the greatest damage to a person's hearing because it is the loudest.

13. **A** Population density is a measure of individuals per unit area. On land, it is measured as the number of individuals per square area (hectare, meter, etc.), while in aquatic environments, the measure is based on the number of individuals per unit of volume (milliliters, liters, etc.).

14. **D** This graph illustrates what can happen to supplies of a nonrenewable resource, such as oil. Curve A shows a resource which is being rapidly consumed, which is not being recycled, and for which no new reserves are found. Curve B shows a resource which can be partially recycled, for which there are more reserves, and which is being conserved. Curve C is seen when the use of the resource is reduced and there are relatively ample reserves.

15. **A** The Aral Sea, located in Uzbekistan and Kazakhstan (both countries were part of the former Soviet Union), is a saline lake. It is in the center of a large, flat desert basin. In the past few decades, the Aral Sea's volume has decreased by 75 percent because of both natural and human causes. Choice (B) would be a result, not a cause; (C) is not correct because earthquakes do not cause water loss; (D) is not correct because pesticides would cause water pollution, not water loss.

16. **D** All three of these events can occur when fish populations are overharvested. The U.S. National Fish and Wildlife Foundation found that 14 major commercial fish species are almost depleted. Also, one-fourth of the annual fish catch is by-catch or nontarget organisms.

17. **C** In flood irrigation, water flows via gravity to all parts of the field. In this process, about 20% of the water is lost to evaporation and runoff. Drip irrigation uses above-ground or below-ground pipes to deliver water directly to the plant's roots; this method is highly effective. Furrow irrigation involves cutting furrows between crop rows and filling them with water: about 1/3, or 33%, of the water is lost to evaporation and runoff. Finally, spray irrigation is more efficient than furrow irrigation but still loses about 25% of water.

18. **C** Watch your letters! Option B is (C). Choice (C) represents the exponential growth phase of this population. In exponential growth, birth rates are high and environmental resistance is low, so the population rapidly increases over a short period. Also note that the line is at its steepest slope.

19. **B** In population graphs such as this logistical growth curve, the label (K) represents the carrying capacity. The carrying capacity represents the maximum number of individuals that the habitat can sustain for a long time. The biotic potential (or how rapidly the organisms can reproduce) is balanced by the environmental resistance (factors that lower a population). Note how the population size modulates above and below the carrying capacity.

20. **A** Population growth is studied over time, so the independent axis (x) is time. Choice (B) is the dependent variable, population number, which is on the vertical axis (y). Choice (C) is the name of the growth pattern; it is not a variable. Choice (D) refers to environmental effects, factors that would slow down population growth.

21. **B** "Environmental resistance" refers to those factors that slow the growth of a population. Unfavorable abiotic factors (e.g., the climate is too hot or too dry) or biotic factors (e.g., predators or disease) will limit the number of survivors. At point D, the population is about to increase, so environmental resistance is not high—it's low.

22. **A** Phosphates are most commonly found as phosphate salts containing phosphate ions (PO_4^{-3}). Phosphorous does not dissolve easily in water and doesn't form a gas at normal temperatures and pressures. Most of the Earth's sulfur is stored in rocks as sulfate salts (SO_4^{-2}). Both phosphorus and sulfur are only held in living organisms for short periods.

23. **B** The movement of tectonic plates in the Earth's lithosphere occurs because these plates are floating on the semi-liquid magma underneath them. Choices (A) and (C) both deal with the loss of species, and (D) deals with the movement of populations.

24. **C** The Green Revolution started in the 1950s; farmers used new methods of farming, such as breeding high-yield crops, better irrigation processes, and large amounts of fertilizer and pesticides.

25. **C** Mortality figures are calculated as the number of deaths per 1,000 people aged zero to one. For example, in 1987, 9,889 babies were born in State X. In that same year, 116 infants (aged zero to one) died in the state. The infant mortality rate would be calculated as

$(116 \div 9{,}889 = .0117) \times 1{,}000 = 11.7$

So, the mortality rate is 11.7.

26. **D**   Underground storage tanks can leak chemicals into the surrounding ground; these chemicals might then move into the water that lies underground, also known as the aquifer. Choice (A) may pollute aquifers as well, so it is not a good answer. Choice (B) is not associated with underground events. Choice (C) is the boundary between two habitats; it is too broad an answer.

27. **B**   Choice (A): CITES is an agreement that concerns the international trade of endangered species. Choice (B): The Kyoto Protocol is the correct answer. Choice (C): The Montreal Protocol concerns the emissions of CFCs. Choice (D): The Clean Air Act deals with air pollution in the United States; it is not an international treaty.

28. **D**   Choice (D) is the best answer. Choices (A) and (C) both require the use of more mineral (and therefore nonrenewable) resources. Choice (B) consumes mineral resources and is also costly. Sustainability is the ability of people to utilize a resource over the long term (for many generations). This can be achieved by reducing demand and finding ways to reduce waste as much as possible. Replacing incandescent light bulbs with florescent light bulbs is one example of a way to decrease waste.

29. **D**   The "Energy Policy Conservation Act," enacted into law by Congress in 1975, added Title V, "Improving Automotive Efficiency," to the Motor Vehicle Information and Cost Savings Act and established CAFE standards for passenger cars and light trucks. The act was passed in response to the 1973–1974 Arab Oil Embargo. The near-term goal was to double new car fuel economy by model year 1985. Choice (D) clearly states the goal of the Corporate Average Fuel Economy policy.

30. **D**   DDT is an organic insecticide first produced in 1873. It is a neurotoxin and causes nerve cells to fire continuously, leading to spasms and death. It is dangerous because it can accumulate in the fatty tissues of all animals, and humans that are exposed to it can become sick. Birds exposed to DDT lay eggs with shells that are too thin for the embryo to survive incubation.

31. **C**   Resistance cannot be acquired or developed during exposure to the pesticide, so eliminate (B) and (D). In order for pests to survive the application of the pesticide, there must have already existed pests with resistance to the pesticide. The pests with resistance to the pesticide survive and reproduce so that the next generation of pests is pesticide resistant. This explains why pesticides become ineffective over time. Therefore, (A) can be eliminated, and (C) is the correct answer.

32. **D**   The amount of plastic that is conserved each year by recycling is equal to $\dfrac{0.3 \text{ tons of plastic}}{1 \text{ ton recycled plastic}} \times$ 50 million tons of recycled plastic = 15 tons of plastic from recycling. As 150 million tons of plastic are produced each year, the amount of plastic that needs to be newly manufactured is equal to 150 million tons – 15 tons of plastic from recycling = 135 million tons of newly manufactured plastic each year. Choice (D) is correct.

33. **A**   Although climate change does increase the amount of flooding in coastal areas, flooding leads to soil salinization, not desalinization. Choice (A) is correct. Choices (B), (C), and (D) are all true statements.

34. **D** Domestic water use includes all uses of water around the home, and the chart indicates that it is responsible for 1 percent of all water use.

35. **B** Sea levels are rising because water is entering the oceans from melting ice caps and glacial melting. Another contributor to rising sea levels is the thermal expansion of water. Choice (A) is a result of decreased ozone in the stratosphere. Choice (C) is not correct: radon seepage does not correlate with atmospheric temperatures. Choice (D) would occur if there were lower temperatures, not higher temperatures.

36. **C** Arsenic is a groundwater pollutant released by mining activities and is not an air pollutant. The major components of air pollution are sulfur dioxide, particulates, lead, ozone, nitrogen dioxide, and carbon monoxide.

37. **B** Poverty is a force that drives many people to exploit the land they live on. Farming, trade in animals, and building roads are all methods that people use to make money.

38. **D** Increased oil prices would make it less affordable for people to fuel their cars to commute to and from their jobs. This would lead to less urban sprawl, so (D) is correct. The other choices are all reasons why people migrate out of the city and into the suburbs.

39. **D** When two of these massive rock plates collide, often one slips under the other; this process is called subduction. The collision of two plates causes the rock layer of the lithosphere to crack. At these sites, magma may rise from the molten exterior of the Earth, and a volcano is formed. Choice (A) is a human activity; (B) can be caused by many geological processes; and (C) is an effect of the relationship between Earth and the sun.

40. **C** "r" and "K" describe the different ways that populations reproduce. r-strategists reproduce at their biological potential, the maximum reproductive rate. These populations produce as many offspring as possible, and many of the offspring will not live to reproductive age. These populations are characterized by rapid increases and rapid declines, as seen, for example, in insect populations and weeds. K-strategists produce few offspring and nurture them carefully; humans are examples of K-strategists.

41. **C** Convectional cooling is an important process that moves heat and moisture around the planet. As warm, moist air rises, it cools. This cooling causes the water vapor to condense and fall as rain. The now cool, dry air becomes denser as it sinks to Earth's surface, where more moisture and heat are picked up and convection starts again.

42. **B** Sedimentary rock is formed by the compression of eroded rock, silt, and the remains of dead organisms. Under heat, pressure, and stress within the mantle, sedimentary rock can form metamorphic rock. Deeper in the mantle, the metamorphic rock melts and cools, forming igneous rock.

43. **C** Since sedimentary rock is made of the remains of dead organisms, it contains the greatest number of fossils. Choice (D) is not correct because the melting and cooling process that forms igneous rock destroys fossils, which are usually quite fragile.

44. **B** The North Atlantic Current circulates water that was warmed by the sun (at the equator) into the northern latitudes, thus warming the land masses. When it cools, the water becomes more dense and it picks up large amounts of CO_2 and salt. This cold, salty water circulates into the equatorial areas, cooling them while releasing CO_2 and salt.

45. **B** Choice (A) is a gas formed by combustion. Choice (C) is a gas produced by the radioactive decay of uranium. Choice (D) is a gas that comes from furniture stuffing and foam insulation. Asbestos is a mineral used in insulating pipes and as a fire retardant. If not completely confined it can crumble, and the fibers can be inhaled into the lungs, which can cause certain types of cancer.

46. **D** In commensalism, one organism benefits while the other is neither helped nor hurt. Barnacles grow on whales, where they can "hitch a ride" to nutrient-rich areas. The whale is unaffected by the barnacles, so this is an example of commensalism. Choice (D) is correct. Choice (B) is an example of mutualism, (C) is an example of parasitism, and (A) is an example of predator and prey.

47. **B** The United States and China together emit 40% of all global greenhouse gases. In 2016, both countries ratified the Paris Agreement to lower greenhouse gas emissions. Choice (B) is correct.

48. **B** Choice (A) is used to turn steam into water so that it can be turned back into steam and spin the turbines. Choice (C) is the site where the heat from the core heats water, turning it into steam that spins the turbines. Choice (D) contains uranium, which is the fuel. Control rods contain compounds, such as cadmium, that absorb neutrons; this reduces their ability to perpetuate a chain reaction. Choice (B) is correct.

49. **C** Riparian zones are the areas of vegetation that abut a river or stream, so choice (C) is correct. They form a corridor of vegetation along the banks of the river or stream. They increase habitat diversity for organisms living in or near the water and can absorb excessive nitrogen, phosphorous, and pesticides, preventing them from entering the water.

50. **A** Trawling involves dragging a net across the ocean floor. This disrupts the habitat and catches a wide variety of bottom-dwelling species. Drift nets float in the water; long lines include baited hooks, and they are towed behind boats; purse seine involves catching surface fish as a net is drawn up from below.

51. **A** In 1987, 36 nations met in Montreal and signed the Montreal Protocol, which cut the emissions of CFCs by about 35 percent between 1989 and 2000. In 1992, the protocol was updated in a meeting in Denmark to accelerate the phasing out of ozone-depleting chemicals. Choice (B): Kyoto Protocol dealt with CO_2 levels. Choice (C): CAA affected the United States only. Choice (D): The Rio Earth summit dealt with many global environmental problems, but Montreal was specific to ozone depletion.

52. **C** Bioaccumulation is the process by which chemicals remain and accumulate in the bodies of animals. These chemicals come from food and cannot be metabolically removed from the tissues. Generally, they are fat-soluble. Biomagnification occurs when a predator eats several organisms and each individual prey has a bit of the chemical in its tissues. Bioaccumulation and biomagnification work together to increase the levels of toxic materials in the bodies of animals. Bioremediation is the removal of toxic compounds by living organisms.

53. **A** A reduction in the creation of pollution is always the least expensive method. Choice (D) is impractical because air pollution can still spread to other areas. Choice (C) does not help acid deposition on land or in the atmosphere. Choice (B), taller stacks, would just spread pollution to other locations.

54. **B** Gamma radiation can penetrate most materials and, of the listed forms of radiation, it is the most harmful to humans. When gamma radiation gets into a cell and damages the DNA, cancer can result. Choice (A): Alpha rays can be stopped by a piece of paper. Choice (C): Beta rays can be stopped by wood or clothing. Choice (D): Infrared radiation is heat, such as that which comes from a stove.

55. **B** The producers in the deep sea hydrothermal vent ecosystem do not capture sunlight and use it to perform photosynthesis. Instead, they use chemical energy from the hot, mineral-rich water that comes out of the vents. The producers in these vents are bacteria that fall into the domain Archaea. The other animals that live near the vents—tube worms, crabs, and many others—all depend on the bacteria for food. Choices (A), (C), and (D) all refer to ecosystems in which producers use radiant energy to carry out photosynthesis.

56. **A** All of the answer choices represent correct processes except (A): adding it to beverages. In this process, CO_2 would reenter the atmosphere, which is the opposite of the long-term storage goal of sequestering.

57. **B** Choice (B) is not necessarily true. Invasive species are *r*-selected organisms; they are usually small adults, have many offspring, sexually mature very quickly, and are generalists. Choices (A), (C), and (D) all correctly describe invasive species.

58. **C** Choice (A) is an example of the First Law of Thermodynamics. Choice (B) exemplifies the law of mass conservation. Choice (D): The usefulness of electricity is the same regardless of how it is produced. The Second Law of Thermodynamics states that there is a loss of energy at each energy transformation.

59. **D** Drip irrigation is a form of irrigation designed to prevent salinization of soil from excessive irrigation. Choice (D) is correct. The other choices are all causes of salinization of soil.

60. **D** Many of the young die in species that are considered early loss; one example of this is seen in fish. In some fish species, most of the individuals die in the first weeks after hatching. Choice (A) describes species such as birds, and (B) and (C) both describe late-loss species, like humans and elephants.

61. **C** Secondary succession is defined as succession that begins on land that was disturbed by an (often human) activity. Primary succession begins on land that was exposed to abiotic factors; e.g., when lava cools or when a glacier leaves an area.

62. **C** By definition, early-stage species are small and exhibit little diversity. At the early stages of succession, an area is populated mostly by *r*-selected organisms. Middle stage and late-stage communities generally have larger adult species and many more *K*-selected organisms.

63. **B** Steady winds create polar vortices, which can trap large amounts of CFCs from other parts of the world. In winter, ice crystals form in the upper atmosphere. The CFCs accumulate on the crystals, and when the polar spring returns, the crystals melt and release the CFCs. Then, the free Cl breaks down the ozone.

64. **B** The crown is the top, or most rapidly growing part, of the tree. In very large forest fires, the crowns of the trees burn, weakening or killing the trees. Leaves or needles burning in smaller fires take away fuel for future fires and break down matter to ash, which provides nutrients for the trees. Fires are also responsible for allowing the seeds of many plant species to germinate.

65. **C** Dams can have detrimental impacts on local environments and populations. These detrimental effects include high construction costs, high CO_2 emissions from decaying biomass, flooding of natural areas, conversion of terrestrial ecosystems to aquatic ecosystems, danger of collapse, and blocking migratory fish. Choice II, the generation of low pollution electricity, is a positive effect, so eliminate any answer choices that include II.

66. **A** Evolution is the change in the genetic makeup of a population over time. The key term is *genetic makeup*—also referred to as the gene pool. In this type of question, look for a phrase or word that implies the passage of time; in this question, it's *becomes*. That should give you a hint that, at one time, the population was not resistant and now it is resistant. All the options except (A) are temporary changes. Choice (D) is also incorrect because the change occurs in one individual, not a whole population.

67. **D** Hydrogen cells, (D), are a renewable energy source. The fuel, hydrogen, can be produced by electrolysis of water. The waste product of hydrogen cells is water, which can be reused to produce more hydrogen fuel. Choices (A), (B), and (C) are all nonrenewable energy sources and are, therefore, incorrect.

68. **A** Ozone formation occurs when sunlight (UV) binds an atom of oxygen to a diatomic oxygen molecule (O_2). Neither radioactive decay nor magma's heat influences events in the stratosphere. Wind energy does not cause the chemical reaction between O_2 and oxygen.

69. **B** Ozone is a major component of air pollution, especially pollution that results from the combustion of fossil fuels. In the stratosphere, however, ozone blocks large amounts of UV light from the sun. Choice (A): All animals breathe O_2, so it is vital to survival. Choice (C): Chlorofluorocarbons destroy O_3, so it is harmful in the stratosphere. Choice (D): H_2SO_4, or sulfuric acid, is harmful anywhere it's found in the atmosphere.

70. **C** Primary pollutants are created via combustion (coal burning, wood burning, etc.) or from volcanic activity. Examples of pollutants produced this way include oxides of sulfur and nitrogen. Secondary pollutants are created when primary pollutants combine in the atmosphere. These reactions are very complex and involve energy from the sun, water vapor, and the mixing effects of air currents.

71. **D** In order for the activity level of the radioactive iodine-131 sample to drop to 2.5×10^{-5} curies, four half-lives must pass. As the half-life of iodine-131 is 8 days, a total of 32 days must pass.

Days Elapsed	Activity Level (curies)
0	4×10^{-6}
8	2×10^{-6}
16	1×10^{-6}
24	$0.5 \times 10^{-6} = 5 \times 10^{-5}$
32	2.5×10^{-5}

72. **A** Choices (B) and (C) are both geological processes. For (D), radioactive materials do not form fossil fuels. Coal and other fossil fuels all formed from plants and other organic material that lived some 300 to 400 million years ago. These plants and animals decayed, and the remains were exposed to tremendous pressures and temperatures.

73. **D** Countries that have more food will have healthier children and adult populations, and thus they will have lower death rates. Increased food production results in increased erosion because after crops are harvested, the bare ground is more readily eroded. Approximately 90 percent of applied pesticides do not reach their targets and can end up in water and soil. The heavy use of large machinery means that more plants can be harvested, but these machines are powered by fossil fuels.

74. **B** Mid-latitude grasslands alternate between rapid growth and drought. Drought and cold winter temperatures permit only a small amount of decomposition each year, which leads to the accumulation of organic matter. In North America, this layer can be more than 30 m thick. Choice (A) describes coniferous forest soils, (C) is arctic tundra, and (D) describes desert soil.

75. **D** Choice (D) might reduce the cost of heating the water, but it will not slow down water consumption. All the other answer choices describe ways to use less water.

76. **D** Evolution results in the development of populations that are fit to live in certain habitats. Sexual reproduction allows for genetic diversity in populations. Organisms with slightly different adaptations can survive in different niches within a habitat. So, as different types of species evolve in different niches, the biodiversity in a habitat increases.

77. **D** Because fecal coliform bacteria come from animal (and human) waste, the answer is (D). Choices (A) and (C) are not correct, as fecal bacteria can cause a number of diseases. Chlorine is used to kill fecal bacteria, so (B) is also incorrect.

78. **A** In an El Niño, the winds that normally blow from the eastern Pacific to the western Pacific weaken or stop. This shifts the precipitation from the western Pacific to the central Pacific and can alter weather events over as much as two-thirds of the globe.

79. **C** Gray water is water from sources such as washing machines, showers, dishwashers—almost any water-using appliances, except toilets. In areas of limited rainfall, gray water is stored in underground tanks and used as irrigation water for gardens, lawns, etc. It may constitute 50 percent to 80 percent of domestic wastewater. Generally, it is not used to irrigate crop plants.

80. **C** Choices (A), (B), and (D) are all provisions of the CWA. Choice (C) is a provision of the National Environmental Policy Act, so it is the exception and the correct answer.

Section II—Free-Response Questions

Question 1

A class wished to determine the LD_{50} of a particular herbicide, Chemical X, on seedlings of the most common type of pest plant at a local tree farm.. Using standard laboratory apparatus and glassware, they accurately made the following dilutions: 1.0M, 10^{-1}M, 10^{-2}M, 10^{-3}M, 10^{-4}M, and 10^{-5}M. They grew the seedlings under standard conditions, varying only the concentrations of Chemical X. Finally, they determined the percentage of seedlings that germinated at each concentration.

(a) Use the description of the experiment above to answer the following questions.

   i. **Identify** a reasonable hypothesis that this experiment could test.

   You should have provided a clearly written statement that describes how the independent variable (Chemical X concentration) will affect the dependent variable (germination of seeds). Some examples might include: "As the concentration of Chemical X increases, the amount (or percentage) of seed germination decreases"; "As the amount of Chemical X decreases, the amount of seed germination increases"; or "The concentration of Chemical X does not affect the rate of seed germination."

   ii. **Describe** the experimental control group and give ONE method for performing repeated trials.

   The control group would include seeds grown under exactly the same conditions as the experimental group, except that they would have no Chemical X added to their environment.

   You should indicate that the seeds were germinated under the various conditions. Trials might include many seeds in one container or many containers each with a single seed, all watered with the same amount of Chemical X.

   (3 points maximum—1 point each for the hypothesis, control, and repeated trials)

(b) **Identify** a set of hypothetical results.

   i. Using the axes below, graph your results. Properly **identify** and label the axes and provide a title for the graph.

   ii. **Calculate** and identify the LD_{50} concentration on your graph. **Show** your work.

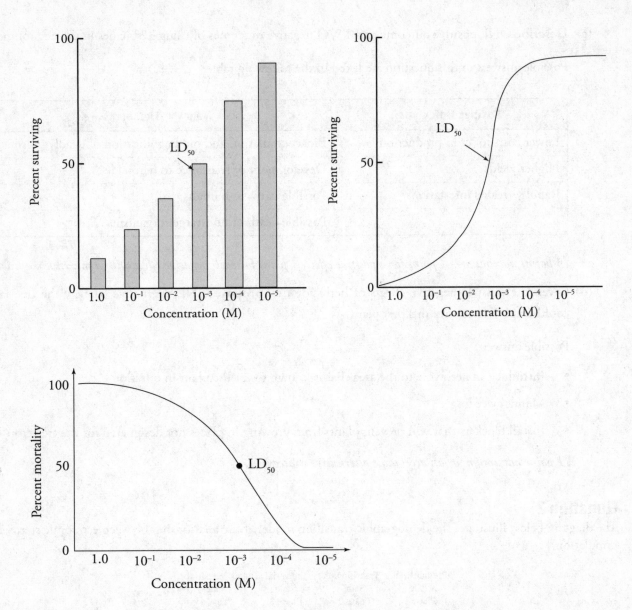

The independent variable (x-axis) is concentration (1.0M, 10^{-1}M, 10^{-2}M, 10^{-3}M, 10^{-4}M, and 10^{-5}M). The dependent variable (y-axis) is percentage surviving or mortality percentage. The LD_{50} value is indicated by the point at which less than 50 percent of the seeds germinated. For the example data here, since the percentage surviving/mortality percentage 50% corresponds to a concentration of 10^{-3}M, the LD_{50} point on the graph is labeled there. Students can use either a histogram (left) or line graph (right) for a correct graph type.

(3 points maximum—1 point for a correct x-axis and y-axis, 1 point for reasonable line or histogram graph data, and 1 point for illustrating LD_{50})

(c) **Describe** ONE positive outcome and TWO negative outcomes of using herbicides in the environment.

Possible answers to this question are listed in the following table.

Positive Outcomes	Negative Outcomes
Lower costs for crop production	Bioaccumulation and biomagnification of deadly poisons
Higher yields	Development of resistance to herbicide
Rapidly treated infestations	Possible harm to humans
	Possible death of non-target organisms

(3 points maximum—1 point for a positive effect, 1 point for each negative effect with a maximum of two)

(d) Given the possible negative effects of herbicide use, **propose** ONE alternative strategy the tree farm could employ to control this pest plant.

Possible answers:

* Introduce an herbivore to the area that is known to eat the plant in question

* Manual weeding

* Install blockers that will prevent plants from growing in places not designated for the tree crop

(1 point maximum for an acceptable alternative strategy)

Question 2

The diagram below illustrates the demographic transition model of the relationship between economic status and population.

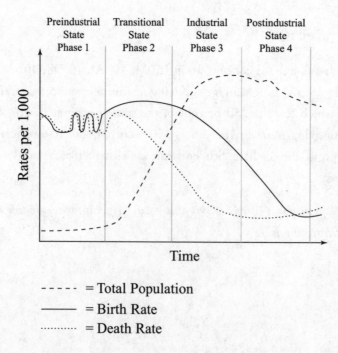

(a) In Phases 2 and 3, there is a large difference between the birth rate and the death rate. **Describe** the effects on the overall population as a result of this difference. **Explain** why the population doubling time during these phases is short.

With a large number of children being born and fewer dying, the population would grow very rapidly. Doubling times would be very short because the percent growth would be very large. The Rule of 70 states that when the percent change is large, the doubling time is rapid (e.g., 70/4 percent = 17.5 years doubling time, while 70/1 percent = 70 years doubling time). When the large number of children mature and procreate, there will be even more children born. This is known as population inertia. *(2 points maximum—1 for effects and 1 for doubling time)*

(b) Choose ONE of the four phases and **identify** an economic factor that would account for the difference between birth rate and death rate. **Explain** how it accounts for the difference.

- Phase 1—Birth rates (BR) and death rates (DR) are approximately equal. While there are a large number of births, they are offset by high death rates due to starvation, disease, unsanitary water, poverty, or any factor that would increase the death rate.

- Phase 2—The death rate falls due to better economic conditions (more food, better health care, more jobs means there is more money to spend). The birth rate is still high.

- Phase 3—The birth rate begins to drop due to more education opportunities for women, an increase in the cost of a child's education, an increase in the cost of raising children, more available and effective forms of birth control, fewer children needed for child labor, more children surviving past infancy, and any other factor that would lower birth rate.

- Phase 4—The birth rate and death rate are now about equal. Any factor that lowers the birth rate and maintains a low death rate is acceptable.
(2 points maximum—1 point for the factor and 1 point for the explanation)

(c) **Describe** ONE biological method of birth control.

There are three basic methods.

- Surgical—In females, examples include tubal ligation or cutting of oviducts (Fallopian tubes), hysterectomy (removal of the uterus), and removal of the ovaries. In males, examples include removal of the testicles and vasectomy (cutting of the vas deferens).

- Hormonal—Birth control hormones for women include "the pill" and variations, such as hormonal implants. In men, hormonal methods and chemical control are still under development.

- Obstructive/Non-hormonal Chemical—Methods include cervical cap, condoms, abstinence, vaginal ring, intrauterine device, vaginal pouch, diaphragm, spermicide, rhythm method, withdrawal, and douche.
(2 points maximum—1 for the method and 1 for the explanation)

(d) Population experts have reported that in some developing countries the population is experiencing a reverse transition from Phase 2 to Phase 1. **Describe** what would happen to a country's population during such a reverse transition and **make a claim** about one event that could cause this reverse transition.

A backward transition would mean that the death rate is increasing while a high birth rate is maintained. This means that the overall population would grow very slowly, or even decrease, depending on how close the birth rate and death rate are to each other. Events that might cause a rapid rise in death rate are war, civil unrest, abject poverty, a rapid rise in infectious diseases (such as AIDS), or natural disaster.

(4 points maximum—2 for the description of the effect and 2 for naming the event and explanation)

Question 3

Under certain conditions, an internal combustion car engine produces approximately 3 grams of NO_x per kilometer driven. In Country C, there are 300 million cars and each car is driven only 20,000 km per year.

(a) **Calculate** the number of metric tons of NO_x produced by the cars in Country C under these conditions in one year. 1 metric ton = 1,000,000 g.

3×10^8 cars $\times 2 \times 10^4$ km per year $\times 3$ g NO_x per km = 18×10^{12} or 1.8×10^{13} g NO_x. 1 metric ton is 1×10^6 g, so the final answer is 1.8×10^7 tons of NO_x.

(2 points maximum—1 point for the setup and 1 point for the correct answer)

(b) **Identify** a secondary pollutant that is derived from the NO_x produced by Country C, **explain** how it is produced, and **explain** how that pollutant travels to adjacent countries.

Nitric acid is a secondary pollutant produced by the addition of NO_x to atmospheric water. This acidified water is carried by wind and rain to the countries downwind and released as acid deposition (in the form of acid rain, snow, fog, sleet, or hail). PANS (peroxyacetyl nitrates) result from NO reacting with hydrocarbons.

(2 points maximum—1 point for the explanation of nitric acid production and 1 point for the explanation of acid deposition)

(c) **Describe** ONE abiotic and ONE biotic impact that the NO_x pollution will have on any countries adjacent to Country C.

See the table below for possible answers to this question.

Abiotic Effects	Biotic Effects
Water acidification	Death of young fish; loss of primary productivity in bodies of water
Soil acidification	Death of decomposers in soil (bacteria, fungi)
Rain acidification: Acid may burn plant leaves or bark, making plants more susceptible to disease.	Death of mycorrhizal fungi living with plant roots and loss of nutrients to plants
Increased transpiration might lead to lower soil water levels.	Loss of cutin layer on leaves means more transpiration out of plants

(4 points maximum—1 point each for the abiotic factor and description and for the biotic factor and description)

(d) **Propose** and **describe** ONE method that Country C could employ to reduce the amount of emitted NO_x.

Some methods Country C could use include the following:

- Using cleaner-burning fuels

- Adding or increasing efficiency of catalytic converters

- Increasing public transportation in order to lower the number of cars on roadways

- Taxing or employing other incentives to get people to use alternate-fuel vehicles
  (2 points maximum—1 point for the method and 1 point for the description)

HOW TO SCORE PRACTICE TEST 3

Section I: Multiple-Choice

_____ × 1.125 = _____
Number Correct   Weighted
(out of 80)   Section I Score
                                    (Do not round)

Section II: Free-Response

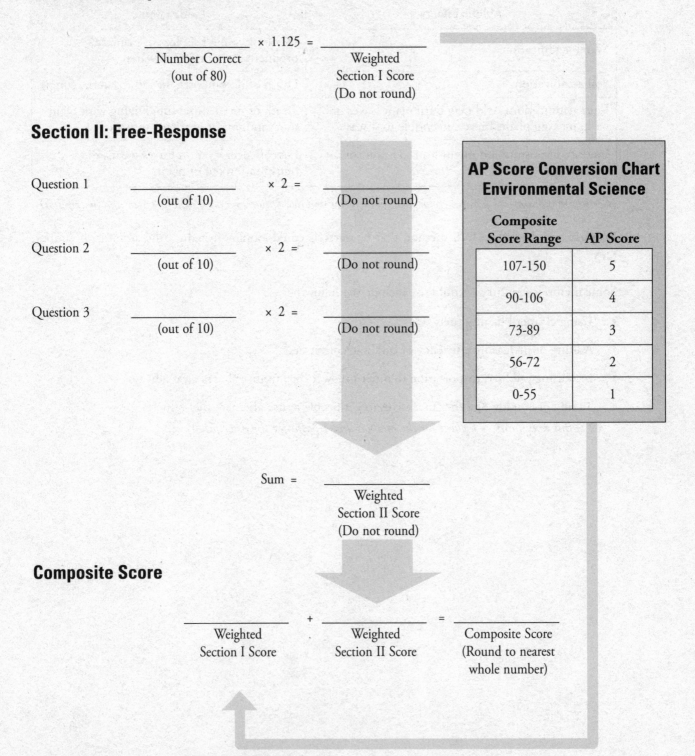

Question 1   _____   × 2 =   _____
              (out of 10)   (Do not round)

Question 2   _____   × 2 =   _____
              (out of 10)   (Do not round)

Question 3   _____   × 2 =   _____
              (out of 10)   (Do not round)

AP Score Conversion Chart Environmental Science

Composite Score Range	AP Score
107-150	5
90-106	4
73-89	3
56-72	2
0-55	1

Sum =   _____
         Weighted
         Section II Score
         (Do not round)

Composite Score

_____   +   _____   =   _____
Weighted   Weighted   Composite Score
Section I Score   Section II Score   (Round to nearest
                                             whole number)

Note: This score sheet is to help you estimate your approximate score for the official exam, not your actual score.

The Princeton Review

OUR NAME: _____

(nt) Last   First   M.I.

NATURE: _____ **DATE:** ___/___/___

ME ADDRESS: _____

(at)   Number and Street

City   State   Zip Code

NE NO. : _____

(nt)

IORTANT: Please fill in these boxes exactly as shown on the back cover of your test book.

5. YOUR NAME

First 4 letters of last name				FIRST INIT	MID INIT
A	A	A	A	A	A
B	B	B	B	B	B
C	C	C	C	C	C
D	D	D	D	D	D
E	E	E	E	E	E
F	F	F	F	F	F
G	G	G	G	G	G
H	H	H	H	H	H
I	I	I	I	I	I
J	J	J	J	J	J
K	K	K	K	K	K
L	L	L	L	L	L
M	M	M	M	M	M
N	N	N	N	N	N
O	O	O	O	O	O
P	P	P	P	P	P
Q	Q	Q	Q	Q	Q
R	R	R	R	R	R
S	S	S	S	S	S
T	T	T	T	T	T
U	U	U	U	U	U
V	V	V	V	V	V
W	W	W	W	W	W
X	X	X	X	X	X
Y	Y	Y	Y	Y	Y
Z	Z	Z	Z	Z	Z

. TEST FORM _____

3. TEST CODE   4. REGISTRATION NUMBER

0	A	0	0	0	0	0	0	0	0	0	0
1	B	1	1	1	1	1	1	1	1	1	1
2	C	2	2	2	2	2	2	2	2	2	2
3	D	3	3	3	3	3	3	3	3	3	3
4	E	4	4	4	4	4	4	4	4	4	4
5	F	5	5	5	5	5	5	5	5	5	5
6	G	6	6	6	6	6	6	6	6	6	6
7		7	7	7	7	7	7	7	7	7	7
8		8	8	8	8	8	8	8	8	8	8
9		9	9	9	9	9	9	9	9	9	9

6. DATE OF BIRTH

Month	Day		Year	
JAN				
FEB				
MAR	0	0	0	0
APR	1	1	1	1
MAY	2	2	2	2
JUN	3	3	3	3
JUL		4	4	4
AUG		5	5	5
SEP		6	6	6
OCT		7	7	7
NOV		8	8	8
DEC		9	9	9

7. SEX

MALE

FEMALE

The Princeton Review®

---

ction I Start with number 1 for each new section.
If a section has fewer questions than answer spaces, leave the extra answer spaces blank.

A B C D	21. A B C D	41. A B C D	61. A B C D
A B C D	22. A B C D	42. A B C D	62. A B C D
A B C D	23. A B C D	43. A B C D	63. A B C D
A B C D	24. A B C D	44. A B C D	64. A B C D
A B C D	25. A B C D	45. A B C D	65. A B C D
A B C D	26. A B C D	46. A B C D	66. A B C D
A B C D	27. A B C D	47. A B C D	67. A B C D
A B C D	28. A B C D	48. A B C D	68. A B C D
A B C D	29. A B C D	49. A B C D	69. A B C D
A B C D	30. A B C D	50. A B C D	70. A B C D
A B C D	31. A B C D	51. A B C D	71. A B C D
A B C D	32. A B C D	52. A B C D	72. A B C D
A B C D	33. A B C D	53. A B C D	73. A B C D
A B C D	34. A B C D	54. A B C D	74. A B C D
A B C D	35. A B C D	55. A B C D	75. A B C D
A B C D	36. A B C D	56. A B C D	76. A B C D
A B C D	37. A B C D	57. A B C D	77. A B C D
A B C D	38. A B C D	58. A B C D	78. A B C D
A B C D	39. A B C D	59. A B C D	79. A B C D
A B C D	40. A B C D	60. A B C D	80. A B C D

YOUR NAME: _____

(Print)   Last   First   M.I.

SIGNATURE: _____   **DATE:** ___ / ___ / ___

HOME ADDRESS: _____

(Print)   Number and Street

    City   State   Zip Code

PHONE NO. : _____

(Print)

IMPORTANT: Please fill in these boxes exactly as shown on the back cover of your test book.

2. TEST FORM

3. TEST CODE

4. REGISTRATION NUMBER

5. YOUR NAME

| First 4 letters of last name | | | | | FIRST INIT | MID INIT |

6. DATE OF BIRTH

Month	Day	Year
JAN		
FEB		
MAR		
APR		
MAY		
JUN		
JUL		
AUG		
SEP		
OCT		
NOV		
DEC		

7. SEX

- MALE
- FEMALE

The Princeton Review®

Section I

Start with number 1 for each new section.
If a section has fewer questions than answer spaces, leave the extra answer spaces blank.

OUR NAME: _____
(nt)   Last   First   M.I.

NATURE: _____   **DATE:** ___/___/___

ME ADDRESS: _____
(it)   Number and Street

City   State   Zip Code

NE NO. : _____
(nt)

ORTANT: Please fill in these boxes exactly as shown on the back cover of your test book.

TEST FORM

6. DATE OF BIRTH

Month	Day	Year
JAN		
FEB		
MAR	(0)(0)	(0)(0)
APR	(1)(1)	(1)(1)
MAY	(2)(2)	(2)(2)
JUN	(3)(3)	(3)(3)
JUL	(4)(4)	(4)(4)
AUG	(5)(5)	(5)(5)
SEP	(6)(6)	(6)(6)
OCT	(7)(7)	(7)(7)
NOV	(8)(8)	(8)(8)
DEC	(9)(9)	(9)(9)

3. TEST CODE   4. REGISTRATION NUMBER

(0)(A)(0)(0)(0)(0)(0)(0)(0)(0)(0)(0)
(1)(B)(1)(1)(1)(1)(1)(1)(1)(1)(1)(1)
(2)(C)(2)(2)(2)(2)(2)(2)(2)(2)(2)(2)
(3)(D)(3)(3)(3)(3)(3)(3)(3)(3)(3)(3)
(4)(E)(4)(4)(4)(4)(4)(4)(4)(4)(4)(4)
(5)(F)(5)(5)(5)(5)(5)(5)(5)(5)(5)(5)
(6)(G)(6)(6)(6)(6)(6)(6)(6)(6)(6)(6)
(7)(7)(7)(7)(7)(7)(7)(7)(7)(7)
(8)(8)(8)(8)(8)(8)(8)(8)(8)(8)
(9)(9)(9)(9)(9)(9)(9)(9)(9)(9)

7. SEX
( ) MALE
( ) FEMALE

The **Princeton Review®**

5. YOUR NAME

First 4 letters of last name				FIRST INIT	MID INIT

(A)(A)(A)(A)   (A)(A)
(B)(B)(B)(B)   (B)(B)
(C)(C)(C)(C)   (C)(C)
(D)(D)(D)(D)   (D)(D)
(E)(E)(E)(E)   (E)(E)
(F)(F)(F)(F)   (F)(F)
(G)(G)(G)(G)   (G)(G)
(H)(H)(H)(H)   (H)(H)
(I)(I)(I)(I)   (I)(I)
(J)(J)(J)(J)   (J)(J)
(K)(K)(K)(K)   (K)(K)
(L)(L)(L)(L)   (L)(L)
(M)(M)(M)(M)   (M)(M)
(N)(N)(N)(N)   (N)(N)
(O)(O)(O)(O)   (O)(O)
(P)(P)(P)(P)   (P)(P)
(Q)(Q)(Q)(Q)   (Q)(Q)
(R)(R)(R)(R)   (R)(R)
(S)(S)(S)(S)   (S)(S)
(T)(T)(T)(T)   (T)(T)
(U)(U)(U)(U)   (U)(U)
(V)(V)(V)(V)   (V)(V)
(W)(W)(W)(W)   (W)(W)
(X)(X)(X)(X)   (X)(X)
(Y)(Y)(Y)(Y)   (Y)(Y)
(Z)(Z)(Z)(Z)   (Z)(Z)

Section I

Start with number 1 for each new section.
If a section has fewer questions than answer spaces, leave the extra answer spaces blank.

(A)(B)(C)(D)   21. (A)(B)(C)(D)   41. (A)(B)(C)(D)   61. (A)(B)(C)(D)
(A)(B)(C)(D)   22. (A)(B)(C)(D)   42. (A)(B)(C)(D)   62. (A)(B)(C)(D)
(A)(B)(C)(D)   23. (A)(B)(C)(D)   43. (A)(B)(C)(D)   63. (A)(B)(C)(D)
(A)(B)(C)(D)   24. (A)(B)(C)(D)   44. (A)(B)(C)(D)   64. (A)(B)(C)(D)
(A)(B)(C)(D)   25. (A)(B)(C)(D)   45. (A)(B)(C)(D)   65. (A)(B)(C)(D)
(A)(B)(C)(D)   26. (A)(B)(C)(D)   46. (A)(B)(C)(D)   66. (A)(B)(C)(D)
(A)(B)(C)(D)   27. (A)(B)(C)(D)   47. (A)(B)(C)(D)   67. (A)(B)(C)(D)
(A)(B)(C)(D)   28. (A)(B)(C)(D)   48. (A)(B)(C)(D)   68. (A)(B)(C)(D)
(A)(B)(C)(D)   29. (A)(B)(C)(D)   49. (A)(B)(C)(D)   69. (A)(B)(C)(D)
(A)(B)(C)(D)   30. (A)(B)(C)(D)   50. (A)(B)(C)(D)   70. (A)(B)(C)(D)
(A)(B)(C)(D)   31. (A)(B)(C)(D)   51. (A)(B)(C)(D)   71. (A)(B)(C)(D)
(A)(B)(C)(D)   32. (A)(B)(C)(D)   52. (A)(B)(C)(D)   72. (A)(B)(C)(D)
(A)(B)(C)(D)   33. (A)(B)(C)(D)   53. (A)(B)(C)(D)   73. (A)(B)(C)(D)
(A)(B)(C)(D)   34. (A)(B)(C)(D)   54. (A)(B)(C)(D)   74. (A)(B)(C)(D)
(A)(B)(C)(D)   35. (A)(B)(C)(D)   55. (A)(B)(C)(D)   75. (A)(B)(C)(D)
(A)(B)(C)(D)   36. (A)(B)(C)(D)   56. (A)(B)(C)(D)   76. (A)(B)(C)(D)
(A)(B)(C)(D)   37. (A)(B)(C)(D)   57. (A)(B)(C)(D)   77. (A)(B)(C)(D)
(A)(B)(C)(D)   38. (A)(B)(C)(D)   58. (A)(B)(C)(D)   78. (A)(B)(C)(D)
(A)(B)(C)(D)   39. (A)(B)(C)(D)   59. (A)(B)(C)(D)   79. (A)(B)(C)(D)
(A)(B)(C)(D)   40. (A)(B)(C)(D)   60. (A)(B)(C)(D)   80. (A)(B)(C)(D)

NOTES

NOTES

NOTES

NOTES